Green Energy and Technology

More information about this series at http://www.springer.com/series/8059

Lorenzo Battisti

Wind Turbines in Cold Climates

Icing Impacts and Mitigation Systems

Springer

Lorenzo Battisti
DICAM
University of Trento
Trento
Italy

ISSN 1865-3529 ISSN 1865-3537 (electronic)
Green Energy and Technology
ISBN 978-3-319-05190-1 ISBN 978-3-319-05191-8 (eBook)
DOI 10.1007/978-3-319-05191-8

Library of Congress Control Number: 2015930332

Springer Cham Heidelberg New York Dordrecht London

Printed on acid-free paper

Springer International Publishing AG Switzerland is part of Springer Science+Business Media (www.springer.com)

Foreword

This book covers most of the engineering aspects concerning wind turbines operating in cold climates, with a specific part devoted to the analysis of icing and methods for its mitigation. The interest in wind turbines icing as well as in predicting its effect on load, control system, power production, etc., has increased significantly in the recent years. The research and development area in ice accretions analysis and anti-icing design is rather multidisciplinary: it encompasses several research fields and competences as meteorology, aerodynamics, heat transfer and ice physics, together with the knowledge of wind turbine operation, economics, manufacturing details and not last, norms and regulations.

The hostile environment of cold climates can lead to a potential reduction in the turbine technical availability. Extreme events cause additional loads, damages and sudden failure. The air density alteration (low temperatures, high elevations) changes the energy harvest and has a major impact on the control strategy. Low temperatures affect physical properties of materials and normal operation on electronic devices. Finally, icing determines higher loads and fatigue, vibrations and energy losses. To mitigate these effects additional ancillary equipments (anti-icing and/or de-icing systems) are required.

Despite the enormous work carried out in the aeronautical field on the topic of icing hazards (in the general references list at the end of the book some relevant contributions used in this work have been listed), very little has appeared on wind turbine icing. Apart from a few exceptions, researchers have to handle the problem with tools originally developed for aircraft and helicopter applications. The few infield observations and open reports on iced wind turbines are of little help in customising and tuning these models. Wind turbines are often not surveilled and it is complex to link occasional losses of power or general malfunctioning to icing events. Decay of power curves is occasionally reported in icing conditions but data on meteorological conditions associated to the event are rarely available from plant owners or manufacturers. Even rarer is it to get information about the ice roughness on the blades that caused the drop in power.

As shown for instance in Chap. 3, aircraft experiences on typical airfoils in use, also for wind turbine blades, indicate that a few millimetres of ice on the leading edge (2–5 mm) of an about 1 m chord wing, causes a drop of maximum lift coefficient between 20 and 50 % and a decay of aerodynamic efficiency till 80 %. This situation would certainly lead to a very large drop for a wind turbine.

Additional problems would be caused by the associated anticipation of the stall, which will bad cope with the original schedule of the control systems, and the mass unbalance caused by ice that unevenly grows on blades. Nevertheless, from the economic point of view, it is not always recommended to equip wind turbines with anti-icing systems to prevent ice formation. Light ice contamination can be occasionally tolerated on rotor, or the turbine can be safely shutdown during heavy icing. These strategies raise the further problem of reliable and efficient ice detection, a new and challenging area of technology development.

To encompass this huge horizon of aspects, I made the decision to face the topic from the point of view of a systemic assessment of the matter, providing the necessary engineering tools to approach the matter. The theory presented in the various chapters has been simplified to an easily and ready implementable model and to give direct access to the order of magnitude of phenomena, rather than pointing to a detailed numerical solution of the single issue.

Nevertheless, several items have been treated, when necessary, with finite elements and finite difference models, which can be a basis for more sophisticated analysis. Despite these numerical treatments, attention has been paid to get relevant engineering conclusions helpful for the practical comprehension of the matter.

This book has been built on more than 10 years of activity in the field, and collects some relevant results presented at conference presentations, articles, Ph.D. work and courses. In particular, it extends the matter developed in the courses held by the author during the period 2004–2009 in his Master's course on Wind energy of the DTU, at Lyngby (DK).

Trento, August 2014 Lorenzo Battisti

Preface

To accomplish the general task of the book, the matter has been divided into five chapters, each of them developing progressively the basic knowledge for the comprehension of the subject. Due to the number of topics analysed, the calculations and examples developed, both on fluid dynamics and thermodynamics, the work at the end took the form of a handbook where the scientific analysis has resulted in technical methods suited to face general design tasks.

Chapter 1, after introducing characters and typology of cold climates, reviews the special equipments needed to safely exploit wind energy systems in such sites, and gives an updated picture of the current installations in the world. Then the analysis points the effect of site elevation for inland plants where some aspects are often neglected or even ignored either in turbine design or wind park development. As instance, deviation from standard density has a series of consequences both on power curve and loads, and if counteracting means are not adopted, the turbines will suffer from poor performance and ruptures. An overview of offshore icing is also given. A brief overview of issues concerning operations in icing conditions is given, and example of reduction in annual energy harvest brings some indication of the relevance of the problem.

Chapter 2 addresses specifically the topic of the effect of icing on wind turbines. General icing characteristics are discussed, together with simple ice growth models on wind turbines. The problem of ice detection and the behaviour of iced sensors are then discussed. The results of an experimental campaign, undertaken with a dedicated wind measurement station at 2000 m a.s.l. with heated and not heated anemometers to enlighten the phenomenon of direct icing and icing persistence, are presented. A simple procedure to deduce the number of icing days in the year is thus given. Short-term forecast approaches are reviewed and discussed. A probabilistic-based methodology has been further proposed to evaluate the icing period on a site with little information. Ice growth on blades involves also safety of people and goods because of ice throw. Icing risk has been approached with a dedicated model based on a Monte Carlo method. Finally, the economic risks of adopting or not adopting ice prevention system is discussed with the help of a dedicated break-even model. The analysis aims to asses the minimum number of icing days that makes the investment viable.

Chapter 3 is dedicated to the study of the aerodynamic performances of ice contaminated airfoils. Generalities on aerodynamics of contaminated profile are explained, with the help of an exhaustive review and comment of the results currently available. Successively, the type of ice contamination is analysed and classified according to the major effect on boundary layer and aerodynamics. To close the substantial gap in systematic studies in the field, which lacks a general assessment of the problem, a more consistent classification of ice shapes is proposed. These effects were quantitatively used to evaluate the drop in power curve of a wind turbine. A model based on WT-perf BEM code has been developed to predict the contaminated power curve of wind turbine. With regard to loads, an analysis of the effect on aeroelastic loads have been accomplished by modification of the Flex-5 code. In order to solve the basic lack of realistic ice shapes and masses on the blades usually employed, the concepts of arbitrary contamination levels and ice frequency levels have been introduced to create realistic damage scenarios of the turbine in actual icing environments.

Chapter 4 models the physics of the water impingement and ice formation mechanism. Body discretization, external flow and temperature field, body wetness, have been analysed. The content gives the fundamentals for the design of anti-icing or de-icing systems. The icing process will be described from the thermo-fluid dynamic point of view. The aim is not to detail the ice growing process, but rather to give methods to determine the water mass flow captured by the aerodynamic profile, the impingement limits and the heat flows involved in the process on the surface, as ice prevention systems are designed to keep the surface reasonably clean of ice. To this aim the general theory of droplet trajectory includes the fixed cylinder case, the collision efficiency calculation for profiles at zero and other than zero AoAs. Calculation of the difference between translating and rotating blade on impinging water is presented and discussed. A numerical example for the NACA 44XX profile of the Tjærborg wind turbine rotor is given. Finally, some relevant conclusions applied to wind turbines are drawn. The chapter analyses the water mass balance at the surface, and the thermo-fluid dynamic processes at the iced surface by the concept of the freezing fraction. Thus with the help of energy and mass conservation equations the problems of ice accretion and anti-ice design are presented and solved.

Chapter 5 classifies and describes the main ice prevention systems (IPSs). It starts by proposing a procedure of ice prevention system assessments. Then, IPS concepts are presented and systematically compared. Advantages and disadvantages of current wind turbine IPS are then discussed. Emerging technologies are reviewed, which include the pneumatic de-icing system (already in use in aerodynamic field), microwave, low adhesion coating materials, the intermittent (cyclic) hot gas heating, the regenerative ice prevention system and finally the film heating technology. Some simple calculations have been made to set up a comparison of the capabilities of such systems. From this discussion, a proposal of the energetic efficiency of an IPS is presented together with a synthetic model for estimating the anti-icing power and energy requirement. A worked example explains practically the theory.

The chapters include the detailed calculation of the design of a hot-air thermal anti-icing ice prevention, developed on the basis of the knowledge developed in the previous chapters. It describes how the blade can be geometrically discretised, the thermo-aerodynamic model, and the conjugate heat transfer model. Results are given and the simplification discussed.

Acknowledgments

I would like to thank Prof. Jens N. Sørensen and Prof. Martin O.L. Hansen of Risø DTU (DK).

Part of this work was developed and discussed during my visiting periods at DTU.

My sincere thanks also go to my former students and now appreciated collaborators, Dr. Alessandra Brighenti, Dr. Luca Zanne, Dr. Roberto Fedrizzi, Dr. Enrico Benini, the technicians of the Turbomachinery Lab of the Department of Civil, Environment and Mechanics, Eng. Sergio Dell'Anna and Filippo De Gasperi for their contributions to my researches.

Contents

Acronyms

AEP	Annual Energy Production
BoP	Balance of Plant
CC	Cold Climate
CL	Contamination Level
CW	Cold Weather
DLM	Damage Level Matrix
e.e.	Engineering Experience
EFL	Event Frequency Level
HAWT	Horizontal Axis Wind Turbine
IEC	International Electrotechnical Commission
IPS	Ice Prevention System
ISO	International Organization for Standardization
LB	Langmuir and Blodgett
LE	Leading Edge
LWC	Liquid Water Content in Air
MM	Meteorological Model
MVD	Median Volume Droplet Diameter
NWM	Numerical Weather Model
PETD	Pulsed ElectroThermal De-icing
RPIM	Reduced Parameter Icing Model
SLD	Super Large Droplets
TE	Trailing Edge
VAWT	Vertical Axis Wind Turbine
WECS	Wind Energy Converting System
WT	Wind Turbine

Chapter 1
Effects of Cold Climates on Wind Turbine Design and Operation

Abstract This chapter, after introducing the characteristics of cold climates, reviews the special equipment needed to safely exploit wind energy power systems in these locations and gives an updated picture of the current installations in the world. Then the analysis points at the effects of site elevation on inland plants. For these installations some aspects are often neglected or even ignored either in turbine design or in wind park development. Deviation from the standard density has a series of consequences both on power curve and loads, and if counteracting means are not adopted, the turbines suffer from poor performance and ruptures. Overview of offshore icing is also given. A brief introduction to issues concerning operations in icing conditions is given, in particular, the strategies and the special equipment are indicated, together with real indications of the penalties in annual energy harvest of existing wind parks.

1.1 Introduction

Wide cold regions are recently planned to become available for wind energy exploitation in the high latitudes of Eastern and North Europe, North America and Asia in general. In recent years, more and more turbines have been erected in mountainous sites. The increasing number of single wind turbines and wind parks in these locations, and the associated failure and maintenance reports have boosted recently the need for studying the effects of cold climates on plant operations. Cold climates are part of a wider compass of sites called *not conventional sites* opposed to the so-called *conventional sites*. A general classification of these sites is given in Fig. 1.1.

Conventional sites refer to sites located at open and windy areas, characterised by temperate climate, comprehensive knowledge of actual meteorological data and lack of obstacles in the proximity of turbines. Such sites have primarily been selected considering the compromise of level of energy production, closeness with the electrical grid and adequate distance to populated areas.

Nonconventional sites sites refer instead to hostile climate areas, leading the turbine to operate in extreme environmental conditions, and requiring special equipment for safe and continuous operations.

L. Battisti, *Wind Turbines in Cold Climates*, Green Energy and Technology,
DOI 10.1007/978-3-319-05191-8_1

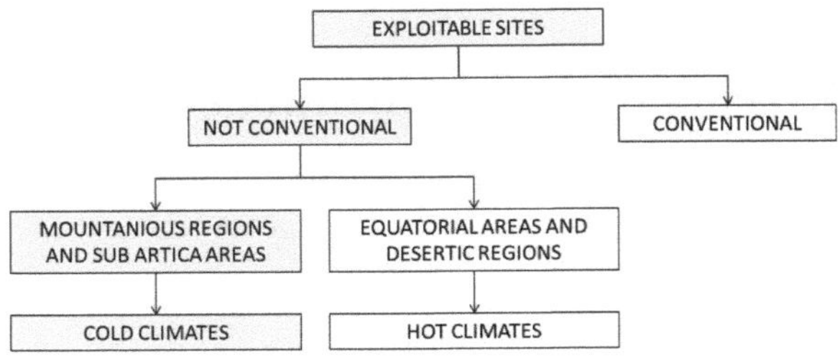

Fig. 1.1 Classification of sites

Among nonconventional sites, this book focuses on *cold climates sites*. Such sites exhibit the following characteristics:

- air temperature $T_a < 0\,^{\circ}\mathrm{C}$ for long periods during the year;
- complex terrains;
- site elevations above sea level (more than about 700–800 m a.s.l.);
- clouding in proximity of the ground surface;
- water content from atmosphere and/or sea water sprays;
- extreme conditions (high turbulence, extreme gusts, hail, lightning).

Besides average air temperatures below zero degree celsius for large parts of the year, the wet environment leads potentially to ice formation and icing persistence on structures exposed to wind and on the access paths to wind farms. Generally speaking, icing refers to both atmospheric icing, due to precipitation of water, and sea spraying and consists of accretion of ice shapes over stationary and moving parts, thus determining an alteration of the fluid-dynamic behaviour of aerodynamic profiles and an increase in the masses of the ice-contaminated components.

Certification Guidelines, i.e. GL Wind, arbitrarily define low temperature as an hourly averaged temperature of less than $-20\,^{\circ}\mathrm{C}$ which happens in an average year on the WECS site on more than nine days per year and/or the yearly average temperature is below $0\,^{\circ}\mathrm{C}$ [1]. About the characteristics of the wind turbine, the same certification guideline reports: *The nine-day criteria are fulfilled, if the temperature at the site remains below* $-20\,^{\circ}\mathrm{C}$*for one hour or more on the respective days. In this case, it has to be counted on special requirements for the wind turbine and the wind turbine shall be designed for cold climate conditions.*

The IEA XIX Annex, Wind Energy in Cold Climates, more consistently defines cold climates as: *sites that have either icing events or low temperatures outside the operational limit standard wind turbines* [2].

The retrieval of reliable data on characteristics of cold climate sites is still crucial. Although actually several maps of average site temperatures and frost are available (in Fig. 1.2 an example of typical maps is given), and a list of icing maps have been produced in the years on the basis of different criteria, there is a substantial lack of information that can be used to assess the severity of the phenomenon for preliminary design of wind parks in cold regions. It is not enough stressed the fact that the common icing evaluation methodologies for meteorological purpose are of

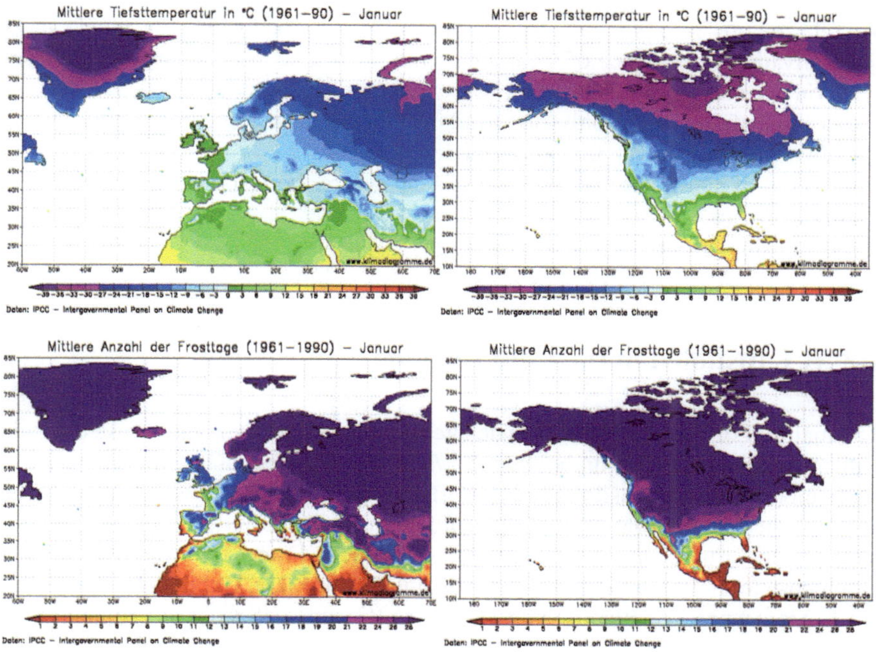

Fig. 1.2 Mean minimum temperature in °C in January (1961–1990), and average frost days for Europe and North America (*source* www.klimadiagramme.de)

limited help in forecasting the severity of the icing process on the wind turbine parts. The most common approach for icing forecast is carried out by coupling meteorological information with an ice accretion model. The former information derives from either numerical weather models or from analysis of measurement data from weather stations. The latter uses the ice accretion model of Makkonen [3] with a 50 cm long, freely rotating cylinder (diameter 3 cm) as reference body. This model was originally developed for icing on overhead power lines (here the conductors behave as rotating cylinders collecting ice, referenced as ISO 12494: Atmospheric Icing on Structures [4]. This model can hardly be stretched to simulate the behaviour of a rotating wind turbine blade. In fact, as it will be explained in detail in the following chapters, it is the dimensions of the body (i.e. the blade) and its relative velocity with regard to the water droplet dimension and speed that drives the ice formation on the surface. The conclusion is that there is currently no model able to convert the ice load modelled on a cylinder into an ice load on a wind turbine blade. Therefore, indications of icing maps can be regarded as indicative of the presence of conditions favourable for icing, but direct measurements of icing parameters to be used in models are necessary for the safe design of wind farms in cold climates.

Despite these shortcuts, cold climate sites show some potential for developers because of:

- a relevant presence of wind, also in macro areas classified with a low wind energy density, due to existence of local high wind spots (due to peculiar terrain shapes as, ridges, etc.);
- this kind of site is the only one available for the land (i.e. sub-Arctic regions, China, Russia, Finland, Canada, cold desert regions);
- this kind of site is the only portion of land where the resource is available (i.e. alpine regions);
- low housing density (relevant for safety).

1.2 WT Installed in the Cold Regions

The majority of potential cold climate sites are located in open and forested terrain with average wind speeds higher than 7 m/s. The total potential is estimated to be 10 times more than is for easily accessible offshore sites (personal evaluation based on not published market analysis). This potential is located in Sweden, Finland, Norway, Iceland, other European mountainous areas (Pyrenees, France, Austria, Switzerland, Liechtenstein, Italy, Germany. Slovenia, Romania, Slovakia, Ukraine, Hungary, Serbia & Montenegro, Scotland), North America (Canada, USA), Asia (Himalayas in China, India, Nepal, Bhutan), a part of South America and non-Himalayan parts of China.

According to the International Energy Agency (Task 19—Wind Energy in Cold Climates) the cold climates wind installations located in Northern Europe, Canada and Asia (North China and Russia) amounted to 3 GW at the end of 2008 and reached 10 GW at the end of 2011, when the total installed worldwide wind capacity had grown to 239 GW in 2011. In Table 1.1 the wind power capacity of some countries are given (primarily onshore plants).

Atmospheric icing in Northern Europe is very much a local phenomenon. Icing may occur at all existing wind farm sites in Finland, Sweden and Norway but the icing climate of different regions varies considerably. Despite this, wind power represents one of the fastest growing industries in Sweden and the installed wind power capacity reached 3,745 MW at the end of 2012. In Norway, wind power installations have grown up to 715 MW (2012) and the major part of these are built in areas where there is a significant risk of icing and actually several companies have their own programs for finding solutions to this problem. The installed wind power capacity in Finland was 197 MW at the end of 2011 and 288 MW at the end of 2012. The amount of installed wind power capacity in Finland is rather low compared to that of other European countries, but new projects are expected (some of them planned in the north of the region), thanks to the new long-term subsidy concerning the production of electricity from renewable sources of energy.

Cold weather occurs in Canada's vast areas and the best wind resources are often located in severe icing areas. Actually, in these regions there are many remote communities (i.e. not connected to the grid) entirely powered by diesel generators.

Table 1.1 Wind turbine installed capacity in some of the Nordic countries [5] and in cold climates [6]

Country	Production capacity installed (MW)		
	2010	2011	2012
Finland	194	197	288
(in cold climate areas)	(197)	n.a.	n.a.
Norway	436	520	715
(in cold climate areas)	(48)	n.a.	n.a.
Sweden	2,163	2,798	3,745
(in cold climate areas)	(124)	n.a.	n.a.
Germany	27,191	29,075	31,332
(in cold climate areas)	(1,000)	n.a.	n.a.
Canada	4,008	5,265	6,200
(in cold climate areas)	(1,823)	n.a.	n.a.

Authorities, governments and companies are currently working for the development of wind power integrated plants in such areas.

In Germany, atmospheric icing has been observed and reported from all site categories, i.e. from coastal sites, the plains of northern Germany and from low mountain regions.

1.2.1 WT Installed in the Alps

A map of the wind farms installed in the Alps, updated as at 2012, is shown in Fig. 1.3, and built up by a personnel site scouting. Among the European countries in the Alpine region, Austria and Switzerland were the first countries investing in wind energy exploitation in the mountains. By the end of 2002, the Swiss had an installed wind power amounting to 5 MW.

As part of an action by the Swiss Federal Office of Energy (SFOE) and the Association for promoting wind energy in Switzerland (Suisse Eole), several promising locations have been identified at moderately high elevations and the objective was to reach the target of 80 MW installed power by the year 2010, using modern, medium-to-large scale systems.

The highest wind park of Europe is sited in the Swiss Alps (Gtsch), near Andermatt, at 2,350 m a.s.l., with three Enercon Wind turbines, a 600 kW E40 (erected in 2004) and two 900 kW E44 erected in 2012 and 2012, capable of generating yearly over 4 GWh of electric energy. All turbines are equipped with a blade anti-icing system, the first erected being one of the first pilot plants in the world. After a temporary dismantling due to structural problems identified in the rotor of the first turbine, the wind plant has been powered with two new turbines. The largest Swiss wind farm

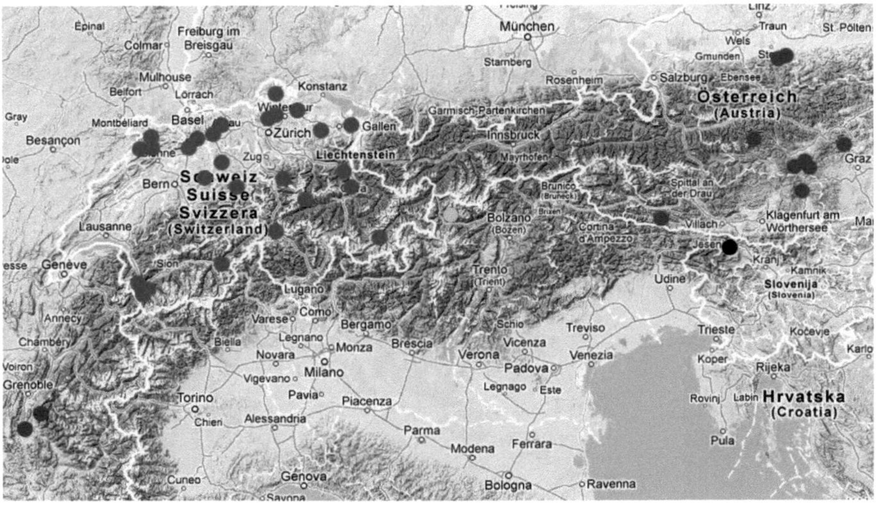

Fig. 1.3 Wind farms installed in the Alps updated as at 2012: Swiss (*red dots*), Austrian (*orange dots*), South Tyrol (*yellow dot*), France (*green dots*), and Slovenia (*blue dot*) sites (Color figure online)

is installed at 1,100 m a.s.l. on Mont Crosin, in the Bernese Jura alpine region. It comprises 16 Vestas turbines (from 600 kW to 2 MW), which generate annually 45 GWh. It is also worth adding that this installation has become an attraction, with thousands of visitors walking the nature trail that leads to the farm, which is equipped with stations providing visitors with information on energy-related topics, and on solar and wind energy in particular. It should be emphasised that studies have already been conducted in Switzerland to establish guidelines on the exploitation of wind energy in the mountains and an ice cover map has been drawn up to identify the icing risk.

Austria began exploiting wind energy in 1994 and, by the end of 2012, the situation is shown in Table 1.2 where it is compared to Italy and Switzerland. Although the majority of its windfarms are currently located in low-lying areas or on gentle hillsides in the eastern parts of the country, interest in installing wind turbine systems in alpine sites has grown considerably in recent years, numerous feasibility studies have been carried out, and several wind farms have already been installed. Austria hosts the “Tauernwindpark Oberzeiring” wind farm, which stands at the highest elevation for a wind farm in Europe (1,835 m a.s.l.) with an installed capacity of more than 20 MW.

Italy has a few plants in the Alpine regions, while many more wind turbines and wind parks are located in the Apennines. Such sites, although being at relatively low elevations and in the ridges of Central and Southern Apennines result to be particularly affected by ice. This is due to the condensing wet air rising from the Adriatic and Tyrrhenian seas.

Table 1.2 Wind turbine installed capacity in some of the Alps countries [5, 6]

Country	Production capacity installed (MW)		
	2010	2011	2012
Austria (in cold climates areas)	1,014 (200)	1,084 n.a.	1,378 n.a.
Italy (in cold climates areas)	5,797 (3)	6,737 (3)	8,124 (3)
Switzerland (in cold climates areas)	42 (35)	46 n.a.	52 n.a.

1.2.2 General Forecasts on Potential Development

According to BTM World Market Update 2012 [7], which made a forecast in the short term (2012–2017), between 45 and 50 GW of wind energy will be exploited in cold climates by 2017, which would mean an increase of as much as 72 % since the end of 2012 and investments amounting to approximately EUR 75 billion. This market has indeed the potential to compete with offshore wind power. The latest published prediction is given by BTM World Market Update 2012, and the complete picture of the actual situation and potential exploitation by interest area is given in Fig. 1.4.

The Chinese Wind Energy Association estimates that about 40 % of the 253 GW onshore exploitable power will be located in cold desert areas.

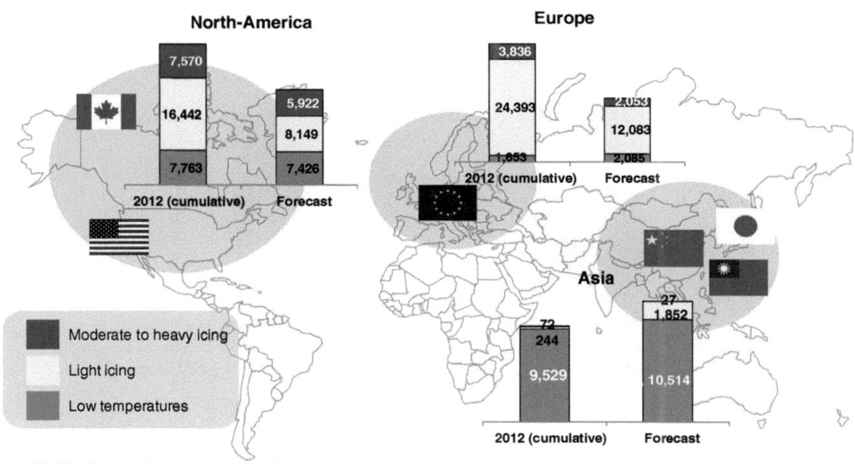

Source: BTM World Market Update 2012, Navigant Research, 2013.

Fig. 1.4 Capacity in cold climates (up to end 2012) and forecasted (2012–2017) in MW

1.3 Operation of Wind Turbines in Cold Climates

In general, if special precautions are not taken, operation of WTs in cold climates will lead to a reduction in energy harvest. Figure 1.5 shows a typical drop of energy yield compared to sea level operation for a WT installed in mild climate at 1,000 m a.s.l., and in a harsher climate at 2,000 m a.s.l. The combined effect of extreme events, air density reduction and ice can cause energy harvest reduction as large as 55 %.

Although a thorough definition will be given in Chap. 2, with the purpose to understand the graph in Fig. 1.5, the icing phenomena can be seen by the summation of *direct icing* and *indirect icing*. The former refers to ongoing icing on the body during bad weather conditions, while the latter consists of the persistence consequent to direct icing that hampers normal operation, causing additional mechanical or electrical sources of malfunctioning, and power adsorption by auxiliary equipments, as confirmed by infield observations.

The phenomena leading to potential reduction of the WECS technical availability are schematically shown in Fig. 1.6. Extreme events cause additional load and fatigue, damage and sudden failure, energy losses caused by precipitation events, or prolonged WT standstill and reduced WT availability. The air density alteration (low temperature, high elevation) changes the energy harvest and has a major impact on the control strategy. Low temperatures affect physical properties of materials and normal operation on electronic devices. Finally, icing determines additional load and fatigue, vibrations, reduced availability and energy losses. To mitigate these effects additional ancillary equipment (anti-icing and/or de-icing systems) are required.

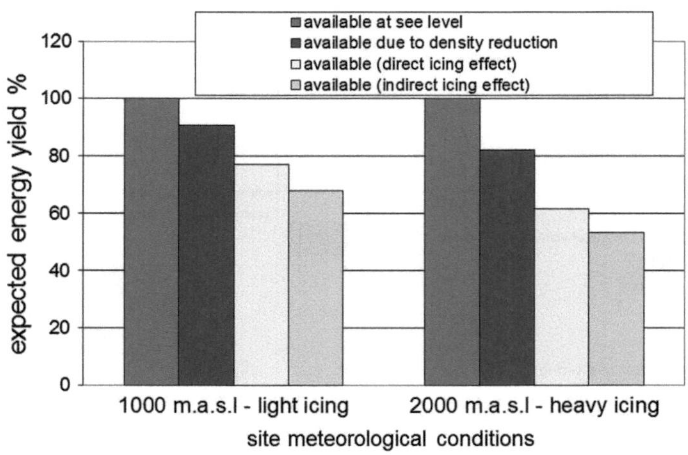

Fig. 1.5 Reduction in energy harvest in cold climates compared to conventional installation

Micro-climatic factors			
extreme events	**air density**	**low temperature**	**icing**
• additional loads and fatigue; • damages and sudden failures; • energy yields losses; • reduced availability.	• Alteration of power curve; • Inefficacy of on the control strategy.	• effect on physical properties of materials; • effect on electronic devices; • Need of CWP	• reduced energy harvest; • Inhibition of the control systems; • additional loads and fatigue; • damages and sudden failures; • vibrations; • Ancillary equipment (anti-icing de-icing systems); • reduced technical availability; • safety problems.

Fig. 1.6 Effect of cold microclimate on wind turbines

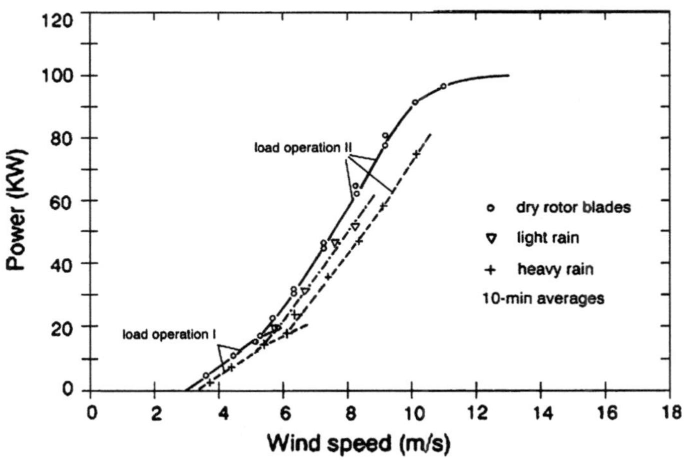

Fig. 1.7 Effect of heavy rain on WT power curve (Energiewerkstatt 1995) [8]

1.3.1 Heavy Rain

Slight or moderate rain does not influence the performance of a WT. However during heavy rain massive drop impacts occur at the rotor blades causing the flow around the profile to be considerably disturbed. Instantaneous measurements indicate power losses up to 30 % could be found based on the intensity of rain.

Figure 1.7 shows an alteration in the power curve due to rain precipitation. The occurrence of hail represents a rare event and therefore hardly measurable losses in the annual energy yields are observed. Much more important is the damage that can occur by the impact of hailstones on the leading edge at terminal speeds exceeding 100 m/s.

1.3.2 Lightning Strikes

Lightning strike is usually a fatal event for WTs. Mountainous regions are particularly affected by lightning. Density of yearly strikes per unit square kilometer as high as 5–10 can occur in certain areas. The strike can heavily damage or even destroy the rotor generator and the electrical parts. Two types of lightning impact can be distinguished:

- In direct strike event the WT is directly hit, usually on a rotor blade, and high electric currents are routed from the discharge point to the rotor hub, the bearing, the tower and foundations into the ground. Serious damage can occur to the rotor blades and the electric components even if lightning receptors are embedded into the blade tip.
- In indirect strike event the lightning falls in the vicinity of the WT. The effects are felt over the medium voltage grid, and excess voltage waves can propagate along the electrical distribution lines. Damage usually involves components which are insufficiently protected against overvoltage.

Table 1.3 shows the regional distribution of lightning damage in Germany, over the period 1992–1994. One can notice the high recurrence of the phenomena in the sub-Alpine regions.

1.3.3 Cold Weather Packages

Specific cold weather packages are used to allow WTs to survive in cold climates. Besides general certification assessing that the design and construction of a given turbine/tower assembly conforms to the accepted standards (in terms of design load

Table 1.3 Regional distribution of lightning damage in Germany, over the period 1992–1994 [9]

	Coast	North-German plains	Sub-Alpine mountains	All
WEC numbers	584	455	263	1,302
Years of operations	1,251	936	465	2,652
Observation (all)	85	64	109	258
Direct strike	25	16	22	63
Observation/WEC-year	7 %	7 %	23 %	10 %
Direct strike/WEC-year	2.0 %	1.7 %	4.7 %	2.4 %

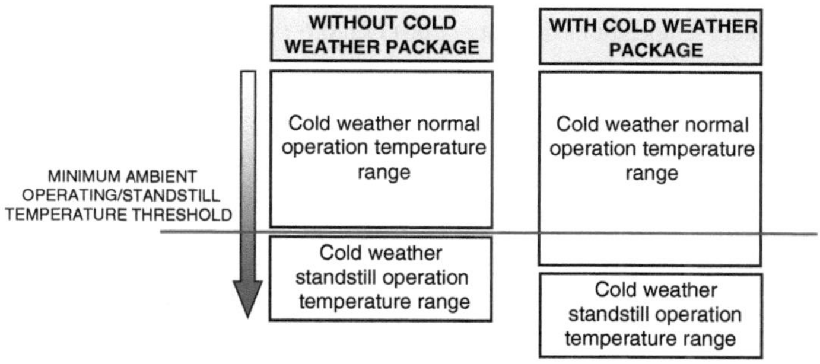

Fig. 1.8 Definition of minimum and standstill temperatures

assumptions, construction materials and methods, control systems and safety margins), operations in cold climates need further caution toward the applicability of the system design and construction to the site-specific conditions. All wind turbine manufacturers specify the temperature operating thresholds for their equipment. Materials and lubricants are designed to withstand temperatures within their specified ranges and when ambient temperatures exceed these ranges, increased maintenance shall be scheduled, as an accelerated decay of equipment life will occur. The most important of these operating limits is the *minimum ambient temperature* below which operation is stopped, according to the characteristics of the materials and lubricants used in the wind turbine. *Standstill temperature* refers to the temperature the turbine can withstand while not operating.

This temperature reflects the limit on the turbine materials' ability to withstand stress without exceeding normal or acceptable wear and tear. According to the specifications of many manufacturers, most turbine models are designed for operation in ambient temperatures down to −20 °C, although some companies indicate operational ranges up to −30 °C, and structure threshold limits up to −40 °C. Special versions exist for such applications (cold climate version, Arctic version, cold weather packages, etc.), allowing to widen the operating temperature range of the turbine (see Fig. 1.8).

Table 1.4 provides background information regarding cold weather operation of utility-scale (1–3 MW) wind turbines, gathered from manufacturers' brochures and data sheets (updated 2012).

Small and micro WT are rarely equipped with cold weather packages. The preferred operating philosophy is the stop of the turbine when extreme conditions occur or even during the whole hostile season period. Some examples of earlier cold weather packages equipping small wind turbines are listed in Table 1.5.

Cold weather packages are associated with a cold station service providing power for preventing damage during the turbine standstill or shutdown periods as schematically shown in Fig. 1.9.

Table 1.4 Minimum temperatures of utility-scale (1–3 MW) wind turbines, without and with cold weather packages (2012)

Manufacturer	Minimum operating/standstill temperature	
	Without CW packages	With CW packages
ENERCON	−10 °C/−20 °C	−30 °C/−40 °C
Gamesa	−10 °C/−20 °C	−30 °C/−40 °C
GE wind turbine	−15 °C/−20 °C	−30 °C/−40 °C
Nordex	−10 °C/−20 °C	−30 °C/−40 °C
REpower	−10 °C/−20 °C	−30 °C/−40 °C
Siemens	−10 °C/−20 °C	−30 °C/−40 °C
WinWinD	−10 °C/−20 °C	−30 °C/−40 °C
Vestas	−10 °C/−20 °C	−30 °C/−40 °C

Table 1.5 Cold weather (CW) packages for small wind turbines (update 2004)

Manufacturer	Size (kW)	Number	Date of installation	Location	Minimum operating temperature (°C)	In service
Bonus	150	1	1993	Yukon	−30	Yes
Vestas V47 LT II	660	1	1999	Yukon	−30	Yes
AOC	66	10	n/a	Alaska (Kotzebue)	−40	Yes
Northwind	100	1	n/a	Alaska (Kotzebue)	−46	Yes

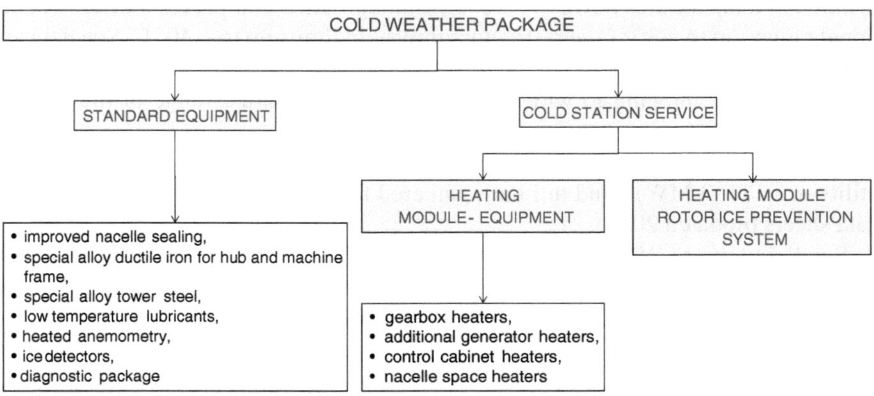

Fig. 1.9 Special requirements for cold climates

1.4 Operations on Mountainous Regions

1.4.1 General Effect of High Elevation

The mountainous site is part of the wider group of nonconventional sites and is characterised by areas of land with complex orographic features, irregular winds, low temperatures and extreme weather conditions. In addition, the existence of microclimatic phenomena such as thermal currents introduces further effects on the temporal and spatial distribution of the wind's velocity, in both its deterministic and stochastic aspects. With the term *mountainous regions* one more specifically refers to areas at elevations ranging between 800 and 2,500 m a.s.l. Authoritative sources, such as the European Wind Energy Association (EWEA), have estimated that 20–25 % of the approximately 60 GW expected to be installed in Europe between now and 2020 will be situated in cold climate areas, and a part of them will be on hills and mountains. Clearly, by comparison with conventional sites, such locations are bound to be more critical, not only in landscape and environmental terms and as regards their acceptance by the community, but also because the more severe microclimate conditions will have adverse effects on the functional capacity of the wind turbines and on the electricity generated. In fact, while numerous studies (see for all [10]) have demonstrated that the wind resource in the mountains is generally greater than on the valley or in the neighbouring lowlands, it was also shown that this resource is strongly influenced by the peculiar features of the land, being characterised by a lower air density, complex turbulence mechanisms and—in winter—very cold temperatures and conditions favouring ice accretion. In addition, the particular logistics of certain mountain sites demand a careful assessment of their real accessibility. The installation of wind farms in the mountains consequently demands an in-depth analysis, both in the stage of assessing the resource and in the design of such plants, as procedures and technologies developed for conventional sites cannot be directly used. For the time being, the IEC standards (originally developed to provide a reference picture relating to conventional sites) fail to provide recommendations on this type of site, where the structure of the flow field is substantially more complex in terms of its effect on the stresses involved. It is therefore worth to outline the main features of mountain wind farm sites and critically discuss the effects of some of the said features on the structural assessment of the turbines in light of the IEC (International Electrotechnical Commission) standard requirements. IEC 1400-1 and, subsequently, IEC 61400-1 [11, 12] deal mainly with providing guidance for classes I to IV, delegating the certification of special turbines operating at unconventional sites to class S. Although they recommend to ascertain the effects deriving from the complex texture of the land and the combination of extreme events into account (in the case of turbulence, for instance), they do not specify what kind of procedures to adopt for it. Some aspects, such as structural effects due to ice, are treated exclusively as part of the rules for certification contained in the Germanischer Lloyd Wind Energie Rules and Guidelines [13].

1.4.2 Feature of the Mountainous Environment

At elevations higher than sea level the air pressure is lower leading to a lower air density. The low air temperatures are a prerequisite for ice accretion on the structures exposed to winds.

In addition to the characteristics of the wind resource (which are quite distinctive and substantially different from those of conventional sites), there are also other site-related factors to consider, as distance from the electrical grid, site accessibility, road communications, and in general, the logistic for transport, erection and maintenance. Moreover, there are constraints related to the landscape and to land usage that may limit the size and the positioning of the turbines (passage of birds' migratory routes, and site specific plant species).

Mountain sites also suffer from a high yearly rate of lightning strikes per unit of land surface area.

These factors have a decisive effect on the definition of the wind resource and on the functional and structural features of the wind turbines to use in mountain installations.

1.4.3 Wind Resource

Mountain sites have Weibull shape parameter values that are unlikely to exceed 1.8. This means the presence of high energy levels in the high-intensity wind bins. The first consequence is to adopt the actual Weibull distribution instead of the Rayleigh distribution established by the IEC standards, specifying the actual value of the shape parameter k, which (as mentioned earlier) is well below two. This is fundamental to the calculation of the extreme loads and the assessment of turbine fatigue. The parameters used for this purpose are variable quantities of deterministic and stochastic type, such as the mean wind intensity V_{ave}, vertical wind profile, reference velocity U_{ref}, extreme velocities V_{e1} and V_{e50}, and parameters that describe the instantaneous changes in velocity and direction, like the intensity of point turbulence measured at the nacelle level.

The extreme events are used to determine the critical stress inducing structural damage or functional loss in the components. These are typically gusts of wind and rapid changes in wind velocity and direction. Hence the interest in predicting the number of times the said critical states occur, or their probability of occurrence, during the life of the turbine. The IEC standard 61400-1 [11] distinguishes between normal and extreme wind conditions during the life of the turbine. The latter (averaged over 3 s) are evaluated for a return time of 1 and 50 years (EWM—Extreme Wind Speed Model), for use in calculating the ultimate stress in installed turbine conditions. Both are based on the velocity U_{ref} and calculated as a function of the height z, respectively, using the following formulas:

$$V_{e1}(z) = 0.75\ V_{e50}(z) \tag{1.1}$$

$$V_{e50}(z) = 1.4\ U_{ref}\left(\frac{z}{z_h}\right)^{0.11} \tag{1.2}$$

Since the return time is specified, the probability of the event's occurrence is also known. The reference velocity U_{ref} is defined as the velocity averaged over 10 min at the height of the nacelle with a 2 % probability (i.e. once every 50 years) of this intensity being exceeded. For the turbine classes from I to IV, the IEC standards define a constant ratio of five between the reference velocity and the annual mean velocity on a level with the nacelle:

$$\frac{U_{ref}}{V_{ave}} = 5 \tag{1.3}$$

Assuming that the statistical models of the Weibull distribution and normal turbulence models are adoptable, the probability of an extreme event can be deduced. The probability of extreme annual mean velocities on the basis of 10 min average is given by:

$$F(V) = 1 - \exp\left[-\left(\frac{V}{C}\right)^k\right] = 1 - \exp\left[-\left(\frac{V}{V_{ave}}\frac{V_{ave}}{C}\right)^k\right] \tag{1.4}$$

Having defined U_{ref} as the extreme annual mean velocity (averaged over 10 min) with a return time T, and assuming [14] the following relationship between the return time and the probability of the event for a number N of annual independent observations (subset of the 52,596 annual 10-min observations):

$$F(V) = \frac{T}{1 - F(U)} \tag{1.5}$$

the extreme annual velocity probability results to be:

$$F\left(U_{ref}\right) = \exp\left[\frac{ln\left(1 - \frac{1}{T}\right)}{N}\right] \tag{1.6}$$

By writing Eq. (1.4) in terms of U_{ref} and equating the result to (1.6), the ratio U_{ref}/V_{ave} can be obtained:

$$\frac{U_{ref}}{V_{ave}} = \frac{1}{\frac{V_{ave}}{C}} \ln\left[1 - \exp\left(\frac{\ln\left(1 - \frac{1}{T}\right)}{N}\right)\right]^{\frac{1}{k}} \tag{1.7}$$

Recalling the Gamma function:

$$\frac{V_{ave}}{C} = \Gamma\left(1 + \frac{1}{k}\right) \tag{1.8}$$

Equation (1.7) can be expressed as a function of the return time and the Weibull shape factor only:

$$\frac{U_{ref}}{V_{ave}} = \frac{1}{\Gamma\left(1 + \frac{1}{k}\right)} \ln\left[1 - \exp\left(\frac{\ln\left(1 - \frac{1}{T}\right)}{N}\right)\right]^{\frac{1}{k}} \tag{1.9}$$

Similar expressions can be calculated using the Gumbel distribution instead of the Weibull one. Due to high figures assumed by N (number of wind 10 min averaged yearly records) the equation is actually independent of N.

While for flat ground at temperate latitudes, where the value of k (averaged over 10 min) can vary from 1.8 (inland areas) to 1.9 (coastal areas) and 2.1 (offshore), a value of five can be conservatively applied to the ratio of the reference and mean velocities. Mountain and complex sites (where $k < 1.8$) appear having ratios higher than five as established by the standards. Taking a site where $k = 1.5$ for example, the curve in Fig. 1.10 obtained by Eq. 1.9 indicates a value of 6.5 for this ratio, i.e. 25 % higher than the one specified by the standards. This is also tantamount to saying (see curve with dotted line) that, for the U_{ref} value given in the standards, a turbine certified as class I can be used only at mountain sites with an annual mean velocity below about 7.8 m/s (instead of 10 m/s at hub height), and below about 6.6 m/s for class II (instead of 8.5 m/s at hub height). This increase in the value of U_{ref} also affects the calculation of V_{e50} and V_{e1}, which are proportionally higher. The lower air density of mountain sites attenuates the impact of these considera-

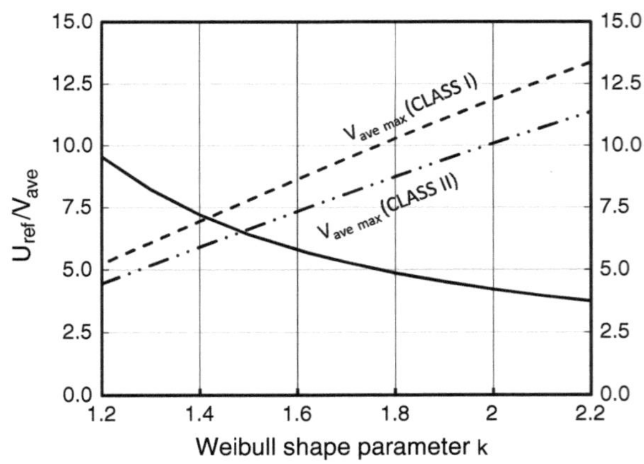

Fig. 1.10 U_{ref}/V_{ave} ratio and maximum V_{ave} for class I turbines as a function of the Weibull shape parameter

tions on the resulting mechanical stresses (albeit only marginally), given the linear vis-a-vis quadratic dependence of the velocity on the aerodynamic forces. Stochastic effects have a dominant effect on fatigue loads. Experimental and numerical studies [15, 16] indicate that the parameter with the greatest effect on fatigue is the standard deviation of the longitudinal velocity.

Typically, the turbulence intensity has values ranging from 0.08 (flat land in conditions of stable stratification of the air) and 0.5 (soil with complex terrain and unstable weather conditions). However, it can reach values close to 0.8 in complex textures (urban centers). The IEC 61400-1 indicates average reference values I_{15} of 0.18 or 0.16.

Under conditions of atmospheric stability and land with homogeneous values of z_0, one can use the following simplified expression:

$$I(z) \approx \frac{1}{ln\,(z/z_0)} \tag{1.10}$$

where z is the generic height from the ground. This expression indicates that the turbulence intensity decreases with increasing height from the ground level.

The standards assume that intensity of turbulence is used as a parameter to describe differences in the level of turbulence at different sites for the purpose of calculating the related stresses:

$$\sigma_u = I_{15}\frac{15 + aV_u}{a + I} \tag{1.11}$$

I_{15} (characteristic value of the turbulence at 15 m/s) and a are both coefficients charted for turbine classes I to IV.

Measures and theoretical analysis indicate that the levels of turbulent kinetic energy are greater in mountain sites and even more significant is the fact that there are complex mechanisms of distribution in the three components compared to conventional sites. Unlike conventional flat sites, the mountain sites shows typically the following ratios:

$$\sigma_u : \sigma_v : \sigma_w = 1 : 0.9 : 0.8$$

Alternately, the *kinetic energy of turbulence* K_I parameter is defined as:

$$K_I = \frac{\sigma_u^2}{2}\left[1 + \left(\frac{\sigma_v}{\sigma_u}\right)^2 + \left(\frac{\sigma_w}{\sigma_u}\right)^2\right] \tag{1.12}$$

The expression of K_I depending on σ_u requires the knowledge of the relations σ_v/σ_u and σ_w/σ_u.

Consequently, the kinetic energy of the turbulence results for flat sites:

$$\sigma_u = \frac{\sqrt{K_I}}{0.97} \tag{1.13}$$

and for complex sites:

$$\sigma_u = \frac{\sqrt{K_I}}{1.1} \tag{1.14}$$

which indicates how in complex terrain, the content of turbulent energy in the longitudinal direction is smaller than in flat ones.

At constant standard deviation, the intensity of turbulence in mountain sites is often lower, compared to conventional sites, for local effect of the accelerations of the average longitudinal velocity (*speed up effect*).

For this reason the standard deviation of velocity in the longitudinal direction or the kinetic energy of turbulence, rather than the intensity of turbulence, are the parameters to be referenced to explain the increase in the fatigue damage detectable in the complex sites.

The equivalent fatigue load R_{eq} [11] is defined through the common Palmgren-Miner damage rule according to specific Woehler curve with slope (1/m):

$$R_{eq} = \left(\frac{\sum_{i=1}^{N} R_i^m N_i}{N_{eq}} \right)^{\frac{1}{m}} \tag{1.15}$$

where N_i is the number of cycles in the i_{th} load range, R_i is the maximum value assumed by each load level bin, N_{th} is the equivalent number of constant amplitude cycles and m the slope of the material S–N curve, which can be modelled by the following:

$$N = kS^{-m} \tag{1.16}$$

As the Weibull coefficients C e k are known, the average velocity, and the variance, can be expressed as:

$$V_{ave} = C\Gamma\left(1 + \frac{1}{k}\right) \tag{1.17}$$

$$\sigma_u = C\left[\Gamma\left(1 + \frac{2}{k}\right) - \Gamma^2\left(1 + \frac{1}{k}\right)\right]^{1/2} \tag{1.18}$$

With the help of these relationships, the standard deviation results to increase as a constant site scale parameter C is considered, giving the plot of Fig. 1.11.

Therefore, by assuming a constant value of the intensity of turbulence with variations in the site average velocity, we can write the following functional relationship [17], which illustrates how the equivalent fatigue load R_{eq} can be represented as a function of the standard deviation:

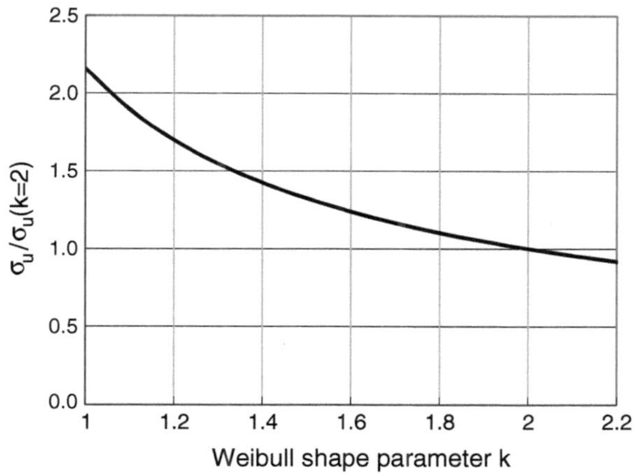

Fig. 1.11 Relationship between standard deviation and Weibull shape parameter *k*

$$R_{eq} = \alpha V_{ave}^2 + \beta V_{ave} = \beta(\varepsilon V_{ave}^2 + V_{ave}) \tag{1.19}$$

$$R_{eq} = \alpha \sigma_u^2 + \beta \sigma_u = \beta(\varepsilon \sigma_u^2 + \sigma_u) \tag{1.20}$$

where α and β are suitable quadratic and linear parameters modelling the turbine's response. The equivalent fatigue load R_{eq} is a quadratic function of the standard deviation σ_u. This equation can be converted by using the relationships between variance or standard deviation, Weibull shape parameters and gamma function to obtain a functional relationship of the following type:

$$R_{eq} = f(\varepsilon, \beta, k, m, \Gamma, C) \tag{1.21}$$

or alternately:

$$R_{eq} = f(\varepsilon, \beta, k, m, \Gamma, V_{ave}) \tag{1.22}$$

The equivalent load spectrum R_{eq} causes a damage D:

$$D = \frac{N_{eq}}{q(pR_{eq})^{-m}} \tag{1.23}$$

where q is a component constant and p takes into account a particular location and load. Equation (1.20) becomes:

$$R_{eq}^m = \beta^m (\varepsilon \sigma_u^2 + \sigma_u)^m \tag{1.24}$$

Equation (1.24) can be expanded as exponent series, and after discarding terms containing second order factors of ε (ε is $<<1$), it becomes:

$$R_{eq}^{m} = \beta^{m}(\sigma_{u}^{m} + m\varepsilon\sigma_{u}^{m+1}) \tag{1.25}$$

After introducing the analytic expressions for the standard deviation and average velocity as function of gamma function, one can define the RR_{eq} parameter (equivalent fatigue loads ratio), normalising equation (1.24) to the equivalent fatigue load calculated according to the IEC standards (k= 2). Then one obtains:

$$RR_{eq} = \frac{f(\varepsilon, k, m, \Gamma, \sigma_{u}, V_{ave})}{f(\varepsilon, 2, m, \Gamma, \sigma_{u}, V_{ave})} \tag{1.26}$$

This parameter indicates the increment in the equivalent fatigue load of a generic site with respect to a site with Rayleigh distribution (k= 2), in accordance with the recommendations of the IEC standards.

Figure 1.12 shows the trend of this relationship with respect to the shape parameter, referring to the assumptions given in Table 1.6. The results indicate that for sites characterised by about the same average velocity, higher equivalent fatigue loads arise with lower k values. This dependence is less evident if materials with a lower m value are used. It is also evident that the effect is more severe as more the component behaviour (i.e. the blade) departs from linearity.

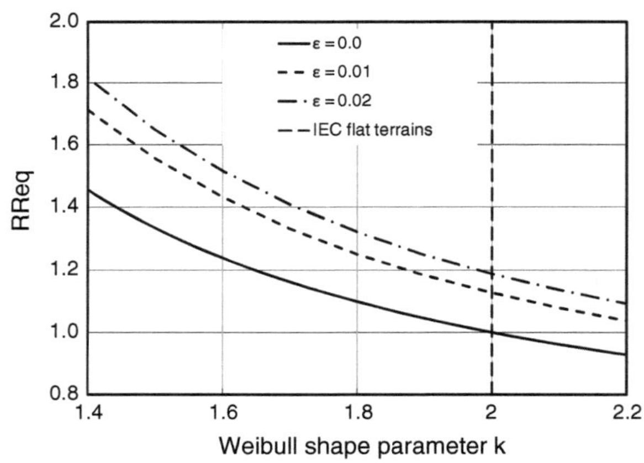

Fig. 1.12 Relationship between equivalent loads with variations of Weibull shape parameter *k*

Table 1.6 Assumptions for the estimation of the fatigue loads ratio [17]

Parameter	Value
k	1.4–2.2
ε	0–0.01–0.02
m	10
V_{ave}	8.5

For a k value of 1.4, within the range of values considered, RR_{eq} varies between 1.45 and 1.80. The installation of a turbine at sites with a k below 1.8 should consequently be conservatively classified in special class S.

The figure also shows (see curve with dotted line) the effect of the lower density due to the location of a site at approximately 1,000 m a.s.l. The mitigating effect of the lower air density is negligible, however, on equivalent fatigue loads.

1.4.4 Variation of Air Density with Site Elevation

Thermodynamically, the wind behaves like a technical gas and obeys the law of ideal gas. By considering the air as a mixture consisting mainly of dry air and vapour, the density can be expressed as:

$$\rho = \frac{p}{\bar{R}_{air} T} \left[1 - \phi \left(1 - \frac{\bar{R}_{air}}{\bar{R}_{vap}} \right) \right] \tag{1.27}$$

where ϕ indicates the specific humidity, with $\bar{R}_{air}$ and $\bar{R}_{vap}$ the gas constant for air and vapour respectively. Equation (1.27) shows that the density is a function of climatic variables pressure, temperature and relative humidity.

Standard conditions for air are defined at $T_{st} = 15\,°\mathrm{C}$, $p_{st} = 101{,}325\ Pa$, and the resulting standard density of dry air is $\rho_{st} = 1.225\ \mathrm{kg/m^3}$.

Table 1.7, shows the qualitative effects of pressure and temperature of typical world locations on air density, Reynolds number, power and aerodynamic forces. Generally speaking, we can define a *direct effect* and an *indirect effect* of the density on power and forces. Power (and energy) and forces are directly proportional to the air density (direct effect), while the Reynolds number (via the modified density and dynamic viscosity of the air) acts in a more subtle manner on the aerodynamic efficiency of the blade (indirect effect) as will be evidenced in the following. For wind installations in *not conventional* sites, such as, for example, the sites at high altitude (above about 750 m above sea level), the deviations in the air density from the standard value should be taken into account, as also the international standards of the *International Electrotechnical Commission* (IEC 61400) suggests.

Table 1.7 Effect of the location of the site on the density of the air, the Reynolds number, at work and on the aerodynamic forces

Site	T_{ave}	p_{ave}	ρ_{ave}	Re	Power	Aerodynamic forces
Deserts	+	st	−	−	−	−
Temperate plateau	+	−	≈	≈	≈	≈
Sub-artic	−	st	+	+	+	+
Mountaniuos	−	−	−	−	−	−

Note st: value of standard conditions, +: value larger than standard conditions, −: less than, ≈: value unchanged

By assuming that the air is dry and the temperature gradient varies linearly with altitude, one can adopt the following polytropic model to compute the air density at a given altitude z:

$$\rho(z) = \rho_{st}\left[1 - \frac{m-1}{m}\frac{z}{T_{st}}\frac{g}{\bar{R}_{air}}\right]^{\frac{1}{m-1}} \tag{1.28}$$

The polytropic exponent m depends on the local vertical temperature gradient (*Environmental Lapse Rate, ELR*) and, as such, the characteristics of the atmosphere which vary in a time period.

A volume of air moving in the atmosphere is subjected to the polytropic transformation:

$$pv^m = \texttt{const} \tag{1.29}$$

The polytropic coefficient m is defined as:

$$m = \frac{c_p - c}{c_v - c} \tag{1.30}$$

c is the specific heat of the generic transformation (while c_p and c_v are the specific heats at constant pressure and volume respectively). This relationship describes the process where the air volume considered exchanges heat with adjacent masses and generates heat internally due to dissipation caused by viscous stresses. The polytropic coefficient is thus related to the so-called *atmospheric stability*, parameter which, in coarse terms, indicates the attitude of the atmosphere to mix.

As provided by *U.S. Standard Atmosphere*, the vertical thermal gradient has an average value of −0.65 °C/100 m. Figure 1.13 shows the trend of the ratio $\rho(z)/\rho_{st}$ as a function of altitude and the degree of atmospheric stability.

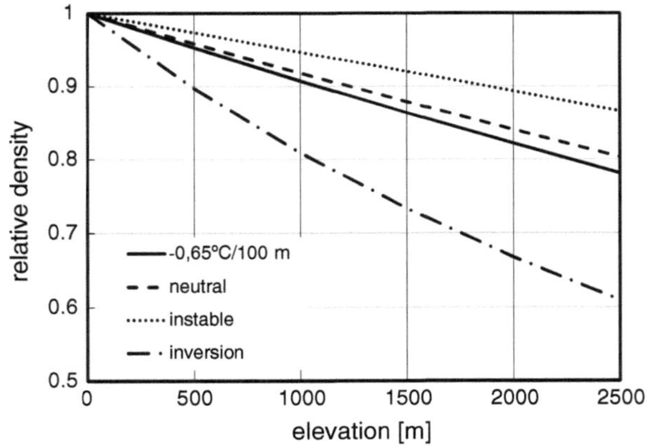

Fig. 1.13 Trend of the relative density as a function of elevation z for different conditions of atmospheric stability (standard density at sea level equal to 1.225 kg/m^3)

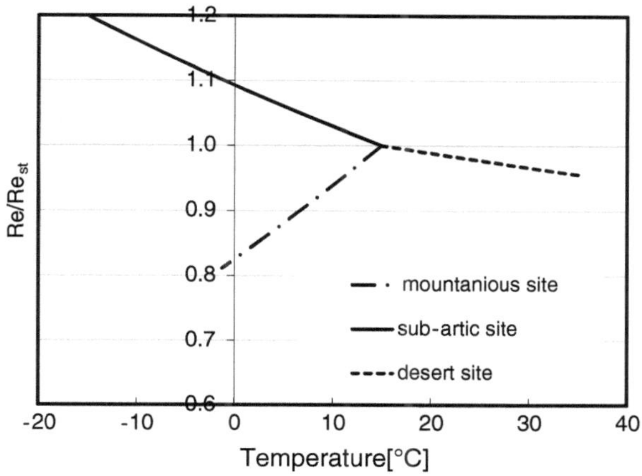

Fig. 1.14 Dependence of Reynolds number on site temperature

A dry atmosphere is in a neutral state when the temperature decreases with altitude at a rate of about 1 °C/100 m. If this gradient assumes higher values, the atmosphere becomes unstable, if it is lower or even becomes positive, the atmosphere becomes stable, up to generate thermal inversion. In the presence of neutral atmosphere, at 1,000 m above sea level, the air density is 1.125 kg/m^3, approximately 91 % of standard conditions and 83 % at 2,000 m above sea level. These values can change significantly in the case of inversion as shown in Fig. 1.13.

The dual dependence of the Reynolds number on the density and dynamic viscosity prompts the considerations of Fig. 1.14, where the dependence on the Reynolds number from site temperature at different locations is shown.

For turbines operating at sub-Arctic sites (considered at sea level) with average temperatures around 0 °C the Reynolds number increases by around 10 % leading to higher aerodynamic performances and stronger aerodynamic forces. The Reynolds number affects the drag and the lift coefficients of the profiles and has an effect on the expected power. This effect is particularly relevant for small blade chords, where changes in maximum lift and lift-AOA slope can be relevant as the Reynolds number decreases. In Fig. 1.15 the dependence of maximum lift on the Reynolds number for a common series of NACA profiles is given. Large blades operating at high Re ($>2 \times 10^6$) tend to be less sensitive to Re variation, while smaller blades suffer a more drastic reduction in aerodynamic performances.

Figure 1.16 shows the same results for the new DELFT profiles series DU. This family of profiles exhibits a much less sensibility of the lift coefficient with the Reynolds number.

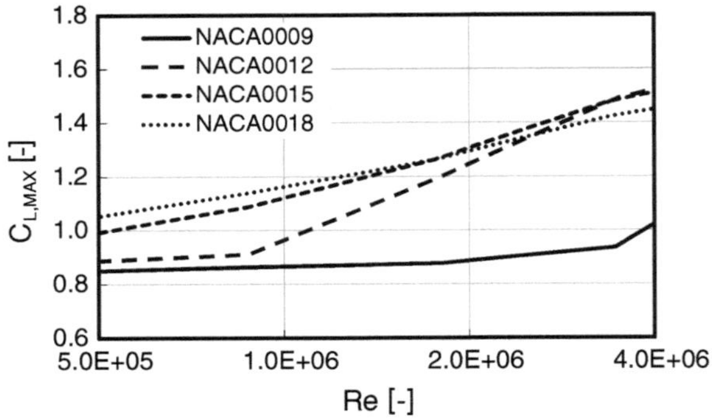

Fig. 1.15 Dependence of maximum lift coefficient of NACA profiles on Reynolds number [18]

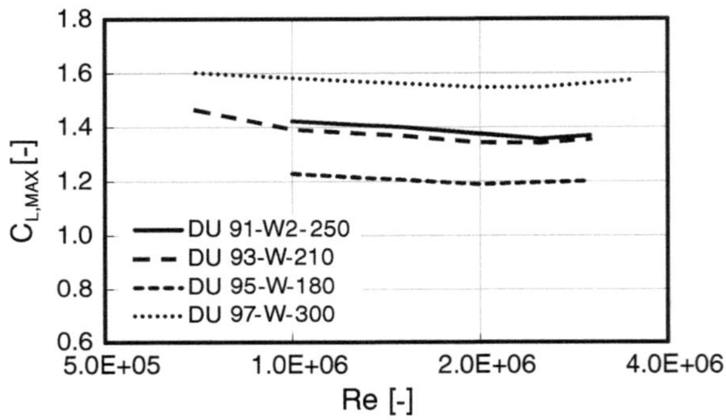

Fig. 1.16 Dependence of maximum lift coefficient of DU profiles on Reynolds number [18]

1.4.5 WT Power and Thrust in Different Air Density Environment

To analyse the effect of air density on power, power coefficient, thrust and thrust coefficient, three main typologies of wind turbines, namely stall, variable pitch and variable pitch and variable speed controlled wind turbines with the same rotor diameter have been designed. Their performance has been computed for standard and non-standard density conditions with the aid of WT_Perf BEM code [19], after a proper design of the rotor characteristics and operation data sets.

The wind turbines power curve and thrust have been computed for the cases that are described in Table 1.8. Case A stands for stall regulated, B for pitch regulated and C for pitch regulated full variable speed. Sub case 1 stands for design with standard density, 2 stands for design with reduced density, 3 for reduced density where the rotor diameter has been increased by means of hub extenders and keeping the same blade, and finally 4 stands for an increased density environment compared to the standard one. The magnification of the rotor size is a common procedure to recover part of the energy drop due to the reduced density. This operation is actually limited in efficacy for large rotors, since the maximum feasible root elongation for the blades

Table 1.8 Data and results of the operation of a 66 m diameter wind turbine in different environments

Case	Stall regulated				Pitch regulated				Full variable			
	A1	A2	A3	A4	B1	B2	B3	B4	C1	C2	C3	C4
ρ (kg/m^3)	1.225	1.00	1.00	1.35	1.225	1.00	1.00	1.35	1.225	1.00	1.00	1.35
D (m)	66	66	69	66	66	66	69	66	66	66	69	66
Ω (rpm)	20.0	20.0	19.27	20.0	20.0	20.0	19.27	20.0	–	–	–	–
Pitch (deg)	−4.6	−4.6	−5.1	−4.6	−4.5	−4.5	−5.25	−4.5	−4.6	−4.6	−5.1	−4.6
$V_{tip,max}$ (m/s)	69.1	69.1	69.6	69.1	69.1	69.1	69.6	69.1	106.6	114.1	112.4	103.2
$c_{p,max}$ (−)	0.53	0.53	0.526	0.53	0.53	0.53	0.52	0.53	0.53	0.53	0.52	0.53
V_r (m/s)	–	–	–	–	14.47	16.73	15.78	13.55	12.19	13.04	12.75	11.80
P/P_{st} (−)	1.00	0.82	0.86	1.102	1.00	1.00	1.00	1.00	1.00	1.00	1.00	1.00
λ/λ_{st} (−)	1.00	1.00	0.98	1.00	1.00	1.00	0.98	1.00	0.98	0.98	0.99	0.98

A—stall regulated, *B*—pitch regulated, *C*—full variable speed. Item *1*— standard density environment, *2*—reduced density environment, *3*—reduced density with enlarged rotor size, *4*—increased density environment compared to standard

is about 1.5 m. A diameter of 66 m has been considered for the computations with NACA 44XX profiles. The blade pitch and rotational speed have been optimised to deliver the maximum power coefficient for each operational case. The Reynolds number effect on blades lift and drag is also considered.

The summary of the main results of the simulations is shown in Figs. 1.17, 1.18, 1.19, and Table 1.8. For stall regulated wind turbines the reduction in density (case A2) brings to a generalised decrease in the power, while an increase (A4) results in extra power for each wind speed (many observations are reported in the past for several unexpected high power records for stall regulated wind turbines operating under very low temperatures). The same trend is confirmed also for the thrust. The wind speed at which the maximum power occurs remains almost unaffected. The extended diameter of the rotor (A3) allows recovering almost the whole power, but is has to mention that a structural verification on the tower for the bigger rotor size and thrust is necessary. In pitch regulated and full variable speed wind turbines (case B and C), the rotor will deliver the maximum power in all conditions, but the rated wind speed will shift to higher values for the reduced density sites conditions (B2, C2) and lower for the higher densities (B4, C4) ones. The peak thrust is also reduced as the density drops.

Note that for higher density cases the control lets the thrust to increase well above the maximum setting given for the standard conditions. This shift of the power curve also has an adverse effect on the optimum turbine-site (siting) matching, which can be partially mismatched if this effect is not considered. The general conclusion is that without any correction on the control settings, the power curve results generally altered with a suboptimal siting.

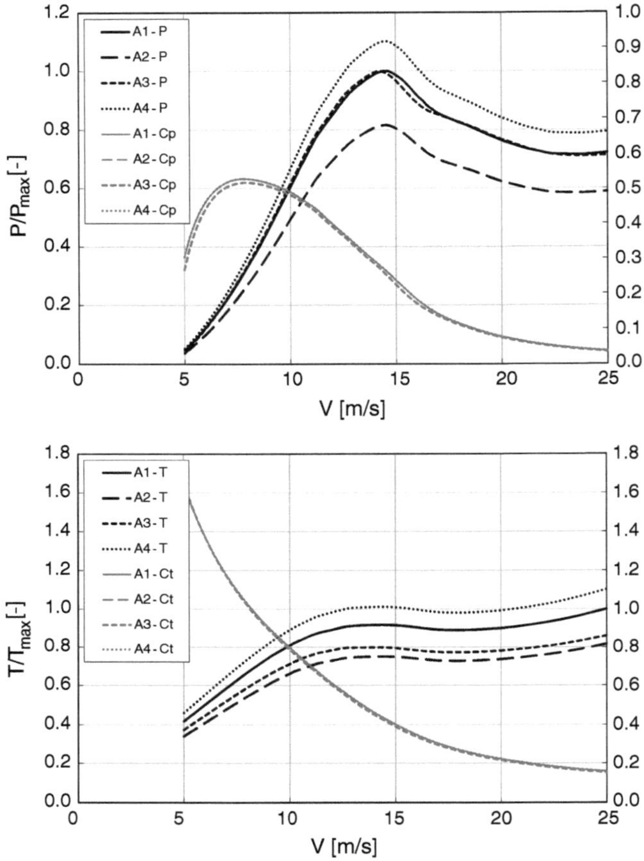

Fig. 1.17 Effect of density variation on power, power coefficient, thrust and thrust coefficient for the stall regulated (case A) wind turbine

1.4.6 Site-Power Curve Mismatch at Non-standard Air Density

In the above paragraphs the direct and indirect effects of density variations on the energy yield have been discussed. It has been proved that the density reduction causes a shift in the rated velocity toward higher values. This effect produces, if not corrected by the control setting, a reduction in the capacity factor of the plant. To analyse this situation, a design power curve is generated, called *optimum power curve*, (according to the procedure outlined in [18]) with the task to analyse the best match of the power curve with a given site.

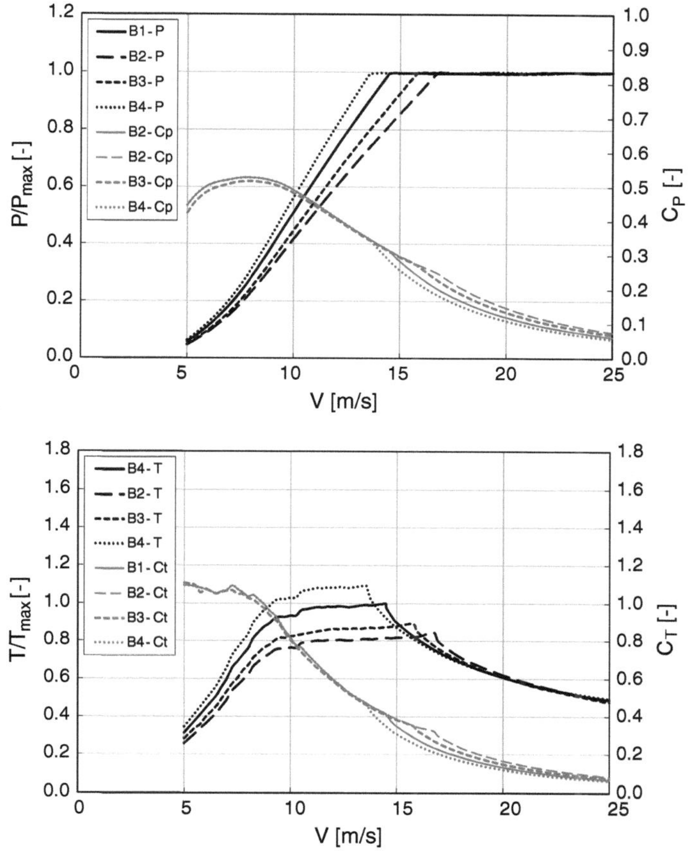

Fig. 1.18 Effect of density variation on power, power coefficient, thrust and thrust coefficient for the pitch regulated (case B) wind turbine

1.4.6.1 Optimum Power Curve

Let us consider the power curve of a wind turbine with variable speed and pitch control, such as the ideal one depicted in Fig. 1.20.

As below $V_{cut,in}$ and above $V_{cut,out}$ there is no power production, the integral of the average power can be expressed as:

$$\bar{P}(V) = \int_{V_{max}}^{V_{cut,out}} b(V)V^k f(V)dV + \int_{V_{max}}^{V_{cut,out}} P_{el,max} f(V)dV \qquad (1.31)$$

The maximum power $P_{el,max}$ corresponds to the maximum available electrical power equal to:

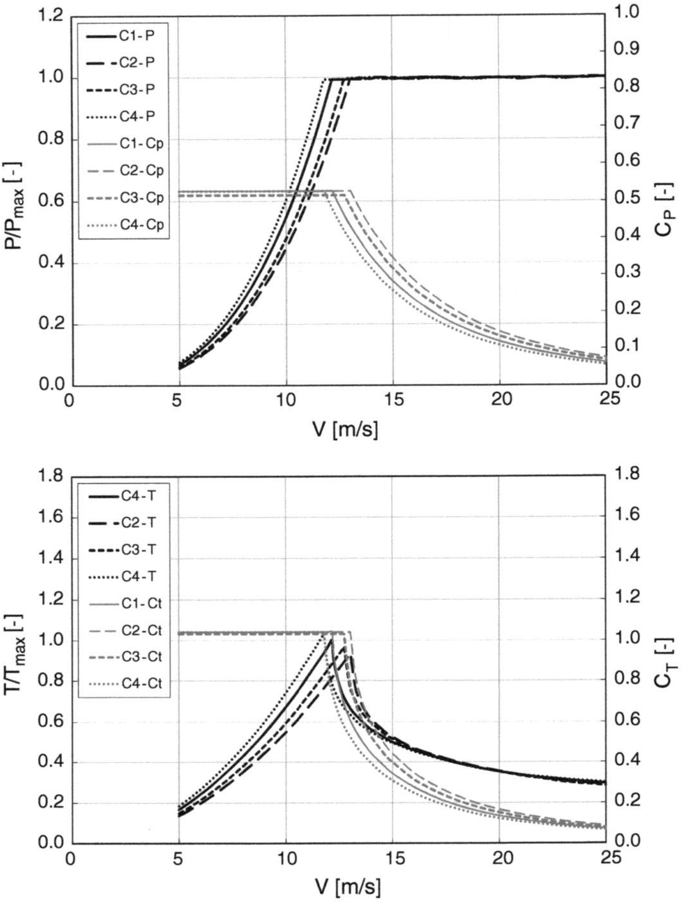

Fig. 1.19 Effect of density variation on power, power coefficient, thrust and thrust coefficient for the pitch regulated full variable speed (case C) wind turbine

$$P_{el,max} = \frac{1}{2}\rho V_R^3 A\ C_{P,R} \tag{1.32}$$

being V_R the wind speed where the maximum electrical power is achieved, while $C_{P,R}$ is the result of the various efficiencies at that wind speed. By multiplying and dividing Eq. (1.32) by the Weibull scale factor C, one obtains:

$$P_{el,max} = \frac{1}{2}\rho C^3 \left(\frac{V_R}{C}\right)^3 A\ C_{P,R} \tag{1.33}$$

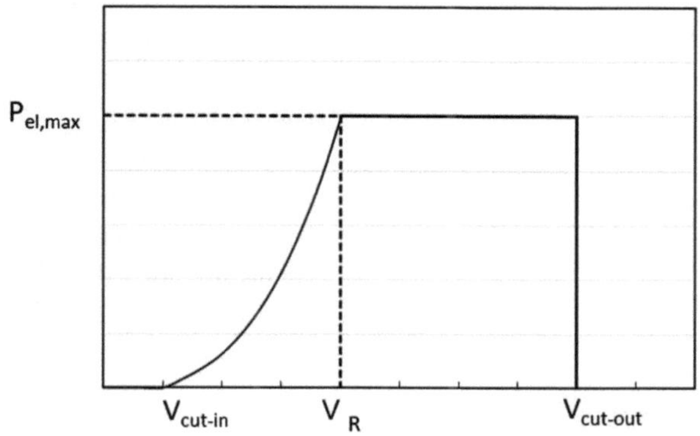

Fig. 1.20 Idealised power curve

since

$$V_{ave} = C\Gamma\left(1 + \frac{1}{k}\right) = C\Gamma(k) \tag{1.34}$$

and by setting that

$$\frac{V_R}{V_{ave}} = k_v \tag{1.35}$$

it follows:

$$P_{el,max} = \frac{1}{2}\rho C^3 (k_v C\Gamma(k))^3 A\ C_{P,R} \tag{1.36}$$

From the above equations, the distribution of $b(V)$ between $V_{cut,in}$ and V_R giving the maximum site energy yield is obtained as:

$$b(V_R) = \frac{1}{2}\rho C_{P,R} V_R^{3-k} \tag{1.37}$$

and

$$b(V) = \frac{1}{2}\rho C_{P,R} V^{3-k} \tag{1.38}$$

while from V_R to $V_{cut,out}$ it holds that

$$b(V) = \frac{1}{2}\rho A C_P V^{3-k} = \frac{1}{2}\rho A \frac{P_{el,max}}{\frac{1}{2}\rho a V^3} V^{3-k} = P_{el,max} V^{-k} \tag{1.39}$$

On the basis of this model, the power curves of three turbines with the same diameter, rated power and maximum power coefficient, but working at different air densities, have been calculated. The design variables are listed in Table 1.9.

Table 1.9 Input data for the calculations

Weibull scale factor (m/s)/average wind velocity (m/s)	6.1/5.51
Turbine SRO (W/m^2)	0.25
$C_{P,R}$ (−)	0.30
Cut in velocity (m/s)	4
Cut out velocity (m/s)	25
Air density case standard (kg/m^3)	1.225
Air density case 1 (kg/m^3)	1.100
Air density case 2 (kg/m^3)	1.000

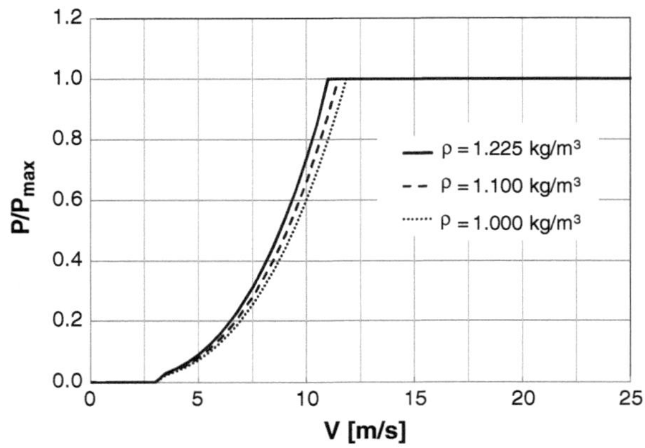

Fig. 1.21 Alteration of the optimum power curve due to air density variation

The resulting power curves are shown in Fig. 1.21 and the resulting capacity factors in Fig. 1.22.

The capacity factor is computed as:

$$F_u = \frac{\int_0^{8760} P(t)\,dt}{8760\, P_{el,max}}$$

Figure 1.21 shows that the rated velocity shifts from 11 to 11.5 m/s and to 11.86 m/s as the density drops from 1.225 to 1.1 and to 1.0 kg/m^3 respectively. This shift causes different reductions in the capacity factor depending on the annual average site velocity. Lower the velocity, lower the capacity factor. For 5.5 m/s average wind speed, the site-turbine mismatch at relative density of 0.9 (elevation of about 1,000 m a.s.l) causes a reduction as high as 7 %, while at a relative density of 0.84 (elevation of about 2,000 m a.s.l) the drops can achieve 13 %. Again, adjustment of the control curve setting is crucial for operation at high elevations.

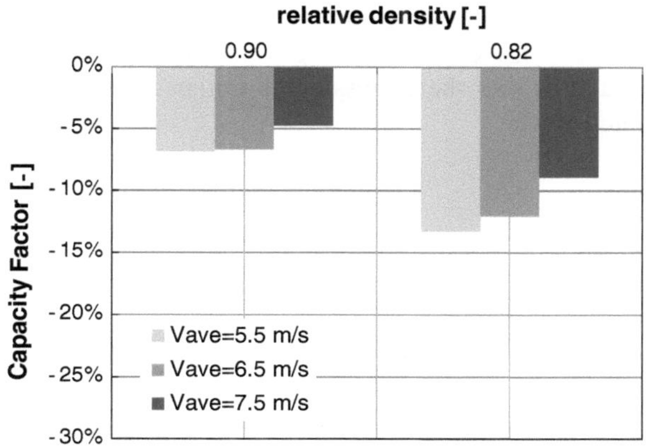

Fig. 1.22 Expected percent capacity factor drop compared to standard air density

1.4.7 Strategies to Mitigate the Density Reduction Effects

Different strategies can be adopted to adapt the turbine to non-standard site characteristics. The case is relevant in reduced density environment in order to partially recover the energy drop caused by the aforementioned direct and indirect effects of air density on the rotor performance. IEC standards [20] give recommendations about the corrections due to non-standard density on the power curve (but nothing is specified for the Reynolds corrections).

For stall regulated WT it holds that

$$P_n = P_T \frac{\rho_0}{\rho_T}$$

while for pitch controlled (above 70 % of the rated power) the correction is made on the wind velocity:

$$V_n = V_T \left(\frac{\rho_T}{\rho_0}\right)^{\frac{1}{3}}$$

The pedex T designates the actual conditions.

For stall regulated rotors the lower density environment badly affects the peak power control because the tip airfoil sections can no longer restrain the maximum lift coefficient designed for standard conditions. Thicker airfoils could help in restraining higher C_L, when density reduction is of concern.

For pitch controlled rotors with blades that pitch toward stall to control peak power (active stall), restrained maximum lift coefficient is desirable. This strategy is eligible for compensating the decrease in density and hence in lift performance.

It is worth to mention that the resulting gain in energy yield of such strategies is not sufficient to lead to a correspondent decrease in the cost of energy, because the latter has to be compared with the higher investment costs, as some design changes are necessary. Limited density variations can be nevertheless compensated by inexpensive control setting adjustments.

Let us start the discussion on mitigation strategies from the formulation of the rated power:

$$P_R = \frac{1}{2}\rho\, A\, V_R^3\, C_{P,R}$$

where

$$C_{P,R} = C_{P,aero}\, \eta_{m,R}\, \eta_{el,R}$$

$\eta_{m,R}$ is the mechanical efficiency at rated power and $\eta_{el,R}$ is the electrical generator efficiency at rated power.

If the same electric generator is maintained, the strategy leads to the following condition:

$$P_R = \frac{1}{2}\rho\, A\, V_R^3\, C_{P,R} = \frac{1}{2}\rho'\, A'\, V_R'^3\, C'_{P,R}$$

or

$$\rho\, D^2\, V_R^3\, C_{P,R} = \rho'\, D'^2\, V_R'^3\, C'_{P,R}$$

where the apex $'$ denotes the density conditions lower than standard.

Two scenarios are possible:

1. V_R is kept at the same value as in the case at standard density. In this scenario, the optimum, cost-effective design is maintained, but the size of the electric generator has to be changed. A site-specific design of the electric generator can reduce the economic benefit of a largely standardised production of the components, so a unique compromise design is advisable.
2. V_R is tuned in order to achieve the standard rated power anyhow. In this scenario, the cost-effective design is no longer maintained, but the penalty introduced could be overtaken by adopting a standard electric generator capacity.

The first scenario determines a linear downscale of the power curve of the WECS and this does not require a detailed analysis from the aerodynamic point of view.

The second scenario requires to compute a new value of V_R as well as a new pitch regulation set.

V_R constant

Here since the rated velocity is locked not to drift to higher values to compensate the density drop, an increase in the rotor diameter is necessary according to the following relationship:

$$\frac{\rho D^2}{\rho' D^2} = K = 1$$

$C_{P,R}$ is considered not to be altered or only moderately altered in the different density conditions. This hypothesis is valid if the blade is sufficiently large to keep the aerodynamic behaviour invariant to the Reynolds number changes.

Table 1.10 Diameter increase to accomplish for the density drop

z (m)	T (°C)	ρ (kg/m^3)	Density drop (%)	Diameter growth (%)
0	+15	1.225	0	0
250	+13.4	1.196	2	1
500	+11.8	1.167	5	2
750	+10.1	1.139	7	4
1,000	+8.5	1.112	9	5
1,250	+6.9	1.085	11	6
1,500	+5.3	1.058	14	8
1,750	+3.6	1.032	16	9
2,000	+2.0	1.006	18	10
2,250	+0.4	0.9815	20	12
2,500	−1.3	0.9572	22	13
2,750	−2.9	0.9331	24	15
3,000	−4.5	0.9093	26	16

Table 1.10 shows the diameter increase needed to compensate for the density drop, according to this strategy.

Such operation can be accomplished also by using root extenders to increase the blade length, without changing the blade set. This solution has some limits in the maximum length of the extender, which would not be longer than about 1.5 m for a 30 m blade (personnel communications from companies). A larger diameter will keep the thrust theoretically unchanged in reduced density environment. Actually since the tip parts result to be also extended, a moderate increase in the thrust has to be expected. If the rotational speed is not increased the tip speed ratio increases lightly and the peripheral speed accordingly.

V_R variable

This case admits a partial increase in the rated velocity by a change in blade pitch settings to compensate the density drop, and to limit the diameter increase. Therefore:

$$\frac{\rho\, D^2}{\rho'\, D'^2} = K'$$

with

$$\left(\frac{V_R'^3}{V_R^3}\right)^3 \frac{C'_{P,R}}{C_{P,R}} = K'$$

In this case, the increase in the rated velocity is accompanied by a reduction in the capacity factor and, thus, the potential recovery of energy yield compared to a corresponding sea level installation is no longer achieved.

It is worth to remark that each solution will bring about a more or less important increase in the investment costs, either throughout a separate increase in rated velocity, rotor diameter and hub height or throughout a combined increase of two or three of them. The diameter can be considered independent of the last two terms and can be discussed as before. Its increment can be more relevant if the rated power has to be increased markedly compared to the standard conditions, and results to be feasible only for small rotors when the length of the extenders can be a relatively high fraction of the diameter. The increase in the rated speed has a counteracting effect on the energy yield, because it increases the rated power while reducing the capacity factor.

The full examination of the data of Table 1.10 has allowed to obtain an optimum relationship between diameter and density. An objective function was set for the optimisation problem with constraints of not to violate the design power and the design power coefficient:

$$F_{ob} = \{min\mathrm{D}|\mathrm{P} = \mathrm{f}(\rho, \mathrm{V}, \theta, \mathrm{D}), (\mathrm{P} - \mathrm{P}_{\mathrm{el,max}} = 0, \mathrm{C_P} - \mathrm{C_{P,R}} = 0)\}$$

and the resulting relationship between diameter and density resulted as:

$$\frac{D}{D'} = \left(\frac{\rho}{\rho'}\right)^{-0.7}$$

This conclusion is directly applicable for full-variable and for pitch regulated WT if the equations provided that small variations around optimum tip speed ratio (TSR_{opt}), and optimum pitch setting (θ_{opt}) are involved. As stall regulated WT are of concern, the equation is valid near the optimum C_P only.

1.5 Operations During Icing

Where icing is expected during operations, anti-icing or de-icing systems, synthetically classified as IPS (Ice Prevention Systems), are necessary. These systems require usually a source of power and energy to be driven, which is taken from the online production of the turbine or from the electrical net. As a consequence, operation in cold climates needs a cold station service to provide energy for both the cold weather package and the ice prevention system and to prevent damage during the turbine standstill or shutdown periods.

All such issues need to be examined in the design phase preceding the installation of the turbines in their working environment. Neglecting these issues would mean a lower energy production to that expected, and prolonged periods of inactivity required for safety purpose or because of the turbine inability to perform satisfactorily. Equipment resulting from Fig. 1.6 are usual for large WT, while they are usually not provided to small ones, due to the relevant additional investment costs and en-

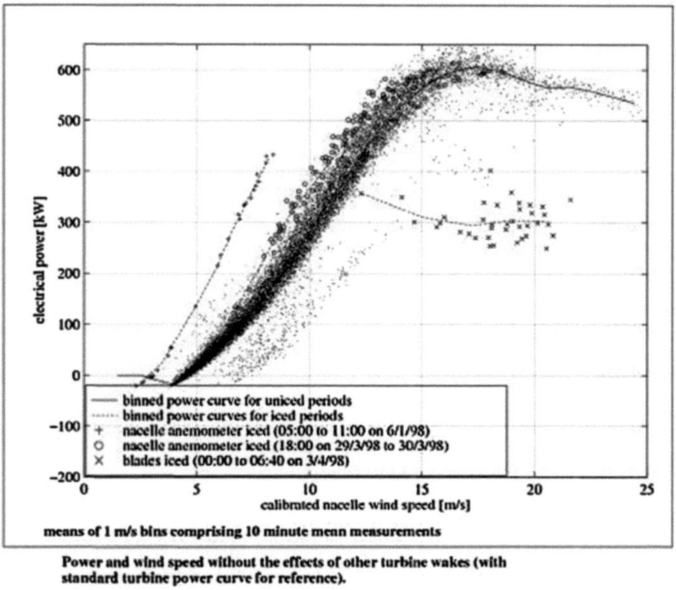

Fig. 1.23 Ice effects on Neg/Micon 637 wind turbine (original in poor quality) [21]

Table 1.11 Anti/de-icing systems on commercial wind turbines (2013)

Manufacturer	Anti/de-icing technology
ENERCON	Anti-icing hot air system, but a new hybrid hot air-electrical system is under development
Leitwind	De-icing heating with electrical system
Nordex	Anti-icing heating with electrical system on N100/2500 and N117/2400 models
REpower	Passive de-icing with special coating
Siemens	De-icing heating with electrical integrated system
WinWinD	De-icing heating with electrical integrated system
Vestas	De-icing heating with electrical system, but a new hot air system is under development

ergy consumption associated, which make them economically not viable for small WECSs.

Icing is a major source of additional downtime and loss of availability for WT operating in cold climates. Substantial decay of the power curve is usually reported in case of icing. Figure 1.23 shows the effect of icing on a Neg/Micon 637 [21].

Table 1.11 provides background information about the main WT manufacturers proposing commercial anti-icing systems.

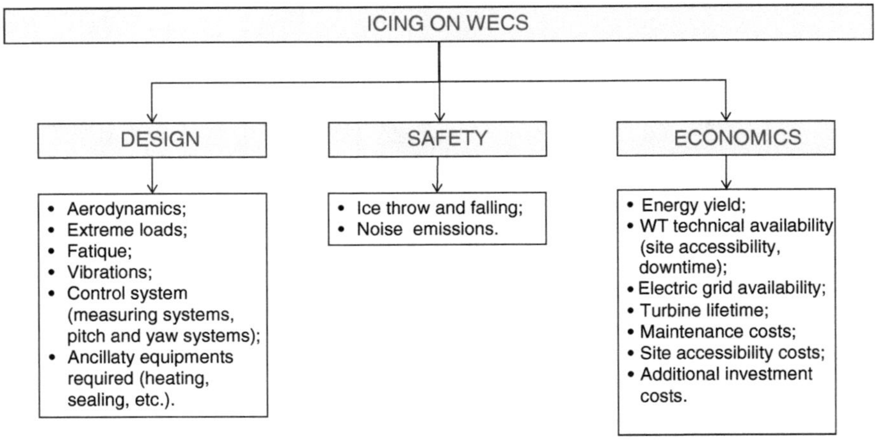

Fig. 1.24 Schematic of ice effects on wind turbines

Figure 1.24 synthesises the basic effects of ice on the turbine, grouped into three classes of design, safety and economics.

Data from ISET 250 MW Wind-Programme [9], and its accompanying Scientific Measurement and Evaluation Programme (WMEP) analysed approximately 55,350 reports with about 2,000,000 h of total turbine downtime. In more than 880 cases (1.6 %) the reports refer to icing events with a total downtime of approximately 64,200 h (3.1 %).

The resulting frequencies of events were:

- plant stoppage (89 %)
- reduced power output (13 %)
- noise (2 %)
- vibration (5 %)
- overspeed (1 %)
- overload (1 %)
- causing follow-up defects (1 %)
- other consequences (4 %).

In most cases (90 %) icing of turbines resulted in plant stoppage. In some cases the turbines remain in operation but effects like noise (2 %), reduced power output (13 %), and vibration (5 %) have been reported to ISET [9].

Figure 1.25 illustrates the results in terms of duration of operation disruption and distribution of icing events over Germany [9].

Turbines operating in cold climates have to be conceived for safe operation: dedicated strategies and special equipment could be considered after the scheme shown in Table 1.12 showing the climate characteristics and the suggested strategy, although the final decision needs to be accomplished by business plan including real costs of additional devices and their benefits.

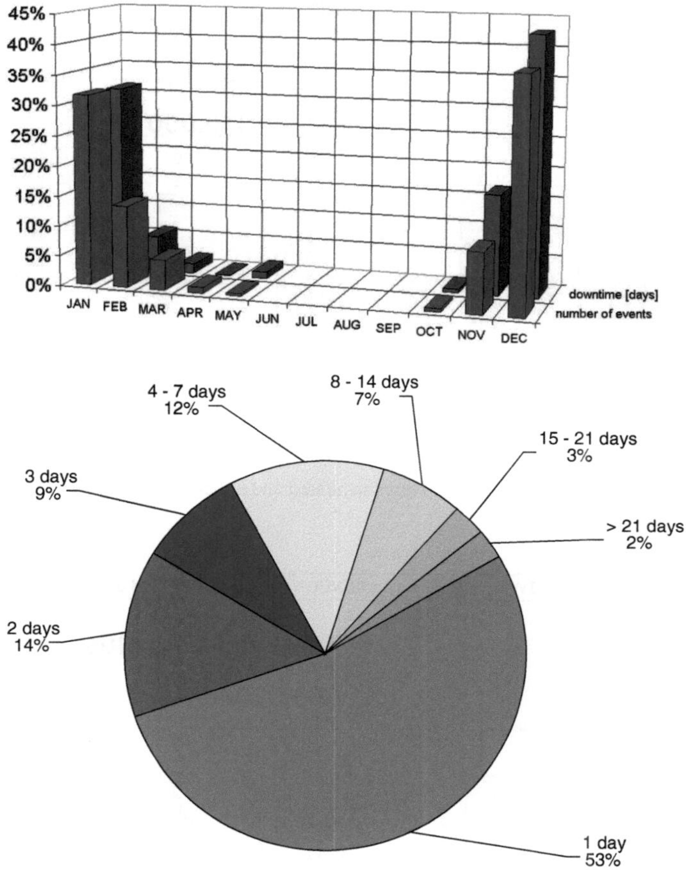

Fig. 1.25 Duration of operation disruption and distribution of icing events in Germany [9]

Table 1.12 Strategies suggested for icing operations in cold climates

Climate characteristics	Strategies
Low temperatures (0 to −3°C) and light icing	None and occasionally stopped of WT
Very low temperatures (less than −3 °C) and moderate icing	Cold weather packages
Real icing risks	Ice prevention systems

Generally, the adoption of icing mitigation strategies (as the IPS, increased preventive maintenance, pre-stocking replacement parts in the site or near each turbine) aims to increase the WT availability and performance. As a consequence in cold cli-

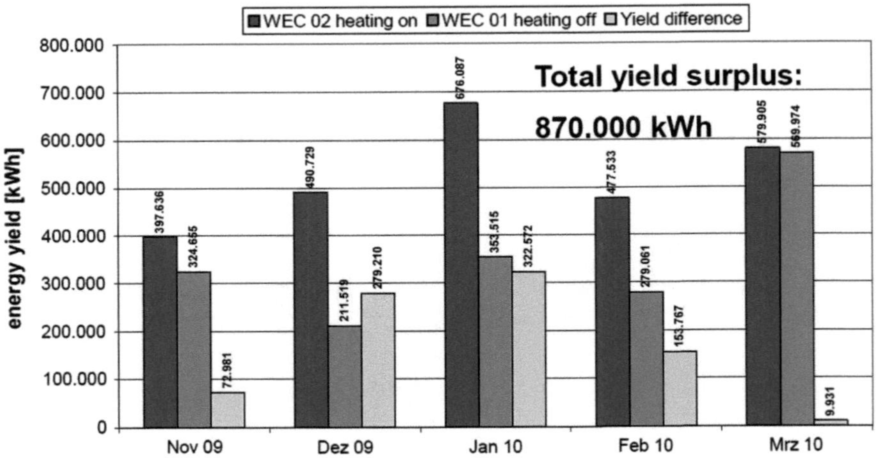

Fig. 1.26 Difference in yield per month between heated and unheated WEC E-82 2MW at location in Dragaliden (SE) [22]

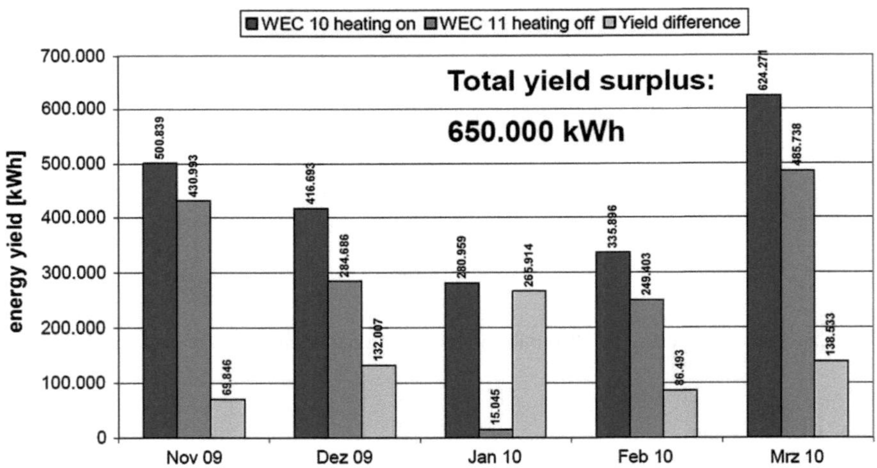

Fig. 1.27 Difference in yield per month between heated and unheated WEC E-82 2MW at location in Krystofovy-Hamry in Czech Republic [22]

mates, additional costs and the different performance scatter with respect to a wind project in conventional site have to be taken into account. The major economic risks arise from:

- increased initial costs (limited installation schedules, higher equipment costs, higher installation costs);

- additional costs for installation of dedicated equipment (as cold weather packages or anti-deicing system) and their operational costs;
- increased periodic (e.g. after snowfalls and icing events) and unscheduled (premature failures due to increased fatigue loading) maintenance costs;
- increased downtime or power penalty due to icing events;
- increased downtime due to extreme low temperatures (only in very cold weather);
- increased downtime between repairs due to turbine inaccessibility;
- increased turbine downtime for public and labour safety (blades and tower ice throw).

1.5.1 Energy Harvest During Icing

Open data on IPS equipped WT in icing conditions is quite rare. Efficiency of the rotor blade de-icing system has been tested for five months in winter 2009/2010 in locations at Dragaliden in Sweden and Krystofovy-Hamry in Czech Republic by Enercon technical service. On both locations two WECs E-82 2MW have been compared. Both WECs are located next to each other. On one WT the rotor blade heating was activated, while on the other it was deactivated.

The percentage of energy yield surplus in relation to WT without de-icing system in the five testing months was 54 and 48 % for the Czech Republic and Sweden sites, respectively, as declared from Enercon technical service reports [22]. In Fig. 1.26 the monthly detail on the difference in yield between heated and unheated WT E-82 2MW at location in Dragaliden (SE) is shown, while in Fig. 1.27 the analogous situation for the E-82 2MW at location in Czech Republic is given. Unfortunately, no indications were available on the site meteorological conditions and ice severity.

1.6 Offshore Icing

Icing has been recognised as a serious long-lasting problem for ships and sea structures.

Icing on the sea is caused by in order of importance from:

- sea sprays originating from sea water;
- atmospheric icing originating from atmospheric freshwater.

Both phenomena can occur simultaneously and the respective prevalence is essentially dependent on the height above the sea surface of the structures. Sea sprays do not usually reach higher than 15–30 m depending on the relative velocity of the wind.

The water content in air drops with height, and at 4 m of level above seal level and wind speeds less than 25 m/s [23] is one or two orders of magnitude smaller than the typical maximum values observed during atmospheric icing.

Fig. 1.28 Iced sea close to WT [25]

Sea ice (flows, driving ice, land-fast ice) is another important issue related to wind turbine performances in offshore sites located in cold climates.

In offshore conditions ice pack or floating blocks on the sea surface cause additional static and dynamic forces on the turbine structure. The effects of sea ice occur as mechanical shocks and increased vibrations that may result in additional operational loads.

It is well recognised from field experiences that offshore wind turbines suffer from sea ice actions much less at areas where sea ice is mostly land fast ice compared with that of drifting ice. For land fast ice the structure is typically surrounded by more or less uniform ice. The ice sheet interacting with the turbine structure produces a wide range of deformation states, each generating different reactions on the structure [24]. Static loads are induced by a stationary contact of the ice with the turbine tower, and the surface forces arise from loads applied by a combination of winds, currents drags and thermal expansion, which slowly push the ice cover against the structure. The tower behaves as a single isolated pinning point resisting the applied driving force, which can be more or less distributed over the tower surface. Weather conditions, applied force level and icing-deicing cycles of the interface determine the uniformity of the mutual ice-structure contact. Thick ice in cold sea waters may sometimes induce the pile-up phenomenon, as a result of irreversible damage of the offshore cantilever structure systems. Some amount of this "pack ice" occurs every winter, typically in spring time when sea ice starts to move, as shown in Fig. 1.28.

Dynamic loadings arise from pieces of floating ice or even ice fields which can cover several square kilometres (see Fig. 1.29), hitting against the structure with appreciable velocity (even higher than 1 m/s). The duration and the forces exchanged with the tower depend on the kinetic energy of the ice and on its features. Floating and pack ice on the water surface and atmospheric icing induce the wind turbine

Fig. 1.29 Floating sea ice pushed from wind to shore at Bay of Bothia [26]

to excessive vibrations. Ice drift and hitting against the foundation might trigger structural vibrations or even damage it by exciting the tower, while structures icing will excite flapwise the blades but the main effect is felt on the tower [27].

Sea ice accumulation on the tower could possibly modify the tower weight and aerodynamic, thus modifying loads on foundations. Moreover, as different researchers showed in [28], ice accumulation could accelerate the corrosion speed process of the tower and support structure, if current offshore corrosion protection systems are not adopted.

References

1. Germanischer Lloyd Industrial Services GmbH (2005) Business segment wind energy, guideline for the certification of offshore wind turbines
2. Laakso T, Holttinen H, Ronsten G, Horbaty R, Lacroix A, Peltola E, Tammelin B (2010) State-of-the-art of wind energy in cold climates. http://www.vtt.fi/publications/index.jsp
3. Makkonen L (2000) Models for the growth of rime, glaze, icicles and wet snow on structures. Philos Trans R Soc Lond 358(1776):2913–2939
4. International Standard ISO 12494:2001. Atmospheric icing on structures. ISO/TC 98/SC3
5. The wind power: wind turbines and wind farms database (2013) http://www.thewindpower.net, last upload: May 2013
6. Peltola E et al (2012) State-of-the-art of wind energy in cold climates. http://www.vtt.fi/publications/index.jsp
7. BTM wind report (2012) World market update 2012, Navigant Research. http://www.navigantresearch.com/research/world-market-update-2012

8. Dobesch H, Kury G (2006) Basic meteorological concepts and recommendations for the exploitation of wind energy in the atmospheric boundary layer. Zentralanstalt fur Meteorologie und Geodynamik. Vienna
9. Durstewitz M (2005) A statistical evaluation of icing failures in Germany - 2050 MW wind-programme. Institut fr Solare Energieversorgungstechnik e.V. (ISET). http://renknownet2.iwes.fraunhofer.de
10. Botta G, Cavaliere M, Casale C (2006) Exploitation of wind energy: ENEL's first experience at a mountain test site. In: Proceedings of the EUROSUN conference. Glasgow
11. International Electrotechnical Commission (2005) International standard IEC 61400-1. Wind turbine generator systems - part 1: safety requirements, 3rd edn
12. International Electrotechnical Commission (2001) International standard IEC 61400-13. Wind turbine generator systems - part 13: measurement of mechanical loads, 1st edn
13. Germanischer Lloyd Industrial Services GmbH. Business Segment Wind Energy (2010) Guideline for the certification of wind turbines
14. Spiegel MR (1975) Probability and statistics. Schaum's outline series in mathematics. McGraw-Hill
15. Mounturb (1996) Load and power measurement program on wind turbines operating in complex mountainous regions, vol I–III. CRES, Pikermi
16. Winterstein SR, Kashef T (1999) Moment based load and response model with wind engineering applications. Wind energy symposium AIAA/ASME, p 346
17. European commission non nuclear energy Joule-III RD (1998) European Wind Turbine Standard - II, ECN Solar & Wind Energy Publishing
18. Battisti L (2012) Gli impianti motori eolici. Lorenzo Battisti (ed) ISBN: 978-88-907585-0-8
19. Buhl M (2012) NWTC design codes WT_Perf a wind-turbine performance predictor. National renewable energy laboratory, official web site: http://wind.nrel.gov/designcodes/simulators/wtperf/. Accessed 6 Nov 2012
20. International electrotechnical commission (2005) International standard IEC 61400-12-1. Wind turbines - Part 12–1: power performance measurements of electricity producing wind turbines, 1st edn
21. Tammelin B, Seifert H (2000) The EU WECO-project wind energy production in cold climate. In: Proceedings of an international conference BOREAS V. Finnish Meteorological Institute, Levi
22. Jonsson C (2012) Further development of ENERCONs de-icing system. Winter wind. Skelleftea
23. Makkonen L (1984) Atmospheric icing on sea structures. Army Cold Regions Research & Engineering Laboratory, CRREL Monograph, 84–2. US
24. Mróz A, Holnicki-Szulc J, Karna T (2005) Mitigation of ice loading on off-shore wind turbines, feasibility study of a semi-active solution. II ECCOMAS thematic conference on smart structures and materials. Lisbon, 18–21 July 2005
25. Eranti E, Lehtonen E, Pukkila H, Rantala L (2011) A novel offshore windmill foundation for heavy ice conditions. In: Proceedings of the 30th international conference on ocean, offshore and arctic engineering OMAE 2011 Rotterdam, The Netherlands 19–24 June 2011
26. Battisti L, Fedrizzi R, Brighenti A, Laakso T (2006) Sea ice and icing risk for offshore wind turbines. In: Proceedings of the OWEMES 2006. Civitavecchia, Italy 20–22 April 2006
27. Battisti L, Hansen MOL, Soraperra G (2005) Aeroelastic simulations of an iced MW-class wind turbine rotor. In: Proceedings of the VII BOREAS conference. Saarisalkä, Finland 7–8 March 2005
28. Morcillo M (2004) Atmospheric corrosion of reference metals in Antarctic sites. Cold Reg Sci Technol 40:165–178
29. Tammelin B, Cavaliere M, Holtinnen H, Morgan C, Seifert H (2000) Wind energy in cold climate - final report WECO (JOR3-CT95-0014), Finnish Meteorological Institute, Helsinki. ISBN: 951-679-518-6
30. EWEA (2004) Wind force 12. http://www.ewea.org

Chapter 2
Relevance of Icing for Wind Turbines

Abstract The chapter addresses specifically the topic of the effect of icing on wind turbines. General icing characteristics are discussed, the prerequisite for icing occurrence and the ice growth on wind turbine. The problem of ice detection and the main ice detection system, the behaviour of iced sensors is then discussed. An experimental campaign have been undertaken with a dedicated wind measurement station at 2,000 m a.s.l. with heated and not heated anemometers to enlighten the phenomenon of direct icing and icing persistence, and to present a procedure to deduce the number of icing days in the year. The problem of meteorological icing data forecasting is twofold. Not only are historical data and spatial extrapolation tools needed for wind farm design, but continuously up-to-date forecasts of icing events, their intensity and duration are also essential for wind turbine operation programmes. In fact, the latter enable a prediction of the energy available for dispatching to the grid in a given period and a rational use of the IPS. Short-term forecast approaches are therefore reviewed and discussed. A probabilistic-based methodology has been further proposed to evaluate the icing period on a site with few information. The problem of ice throw is analysed as it involves also safety of people and goods. The deriving icing risk has been approached with a dedicated, fully explained model based on a Monte Carlo method. Finally, the economic risks of adopting or not adopting ice prevention system are discussed by the help of a dedicated break-even model. The analysis aims to asses the minimum number of icing days that makes the IPS investment viable.

2.1 Effect of Ice on Wind Turbines

As shown in the scheme of Fig. 1.24, icing of rotor blades or other wind turbine components can impact on the design (aerodynamics, loads, control system, material response), the safety (ice throw, noise emission) and the economics (energy yield, design life duration, wind turbine ancillary equipment). Sites characterised by heavy icing can result in a total stop of the turbine and the duration of ice on the blades can be considerably longer than the meteorological icing period. Without any ice prevention systems at these types of sites, the turbine would be full of ice over long periods. Since in some sites the icing season can be relatively high, long downtimes

L. Battisti, *Wind Turbines in Cold Climates*, Green Energy and Technology,
DOI 10.1007/978-3-319-05191-8_2

due to iced rotor blades will cause severe losses of production, as indicated in the Figs. 1.26 and 1.27.

BOREAS IV Conference in 1998 ended with an estimation that wind turbine operations in severe icing would yield to an energy loss of 20 % raising to 50 % on harsh sites on a yearly basis [1–4].

It is not clear at the moment how severe the problem of icing will be for large offshore plants, since only occasional observations are reported in open literature, not sufficient to be statistically usable.

Icing conditions can therefore occur at great variety of sites, from mountainous to inland and sea sites, as the pictures collection of Fig. 2.1 shows.

In these locations the ice structures can be so important to block completely the nacelle operation, or alter the aerodynamic surfaces depleting their efficiency. How relevantly the ice growth can affect turbine structures is shown in Fig. 2.2. Here a compilation of iced blades and nacelles in various location, taken from the web, are shown.

The principles of operation of wind turbines under icing conditions have been listed in several projects starting from the pioneer and fundamental NEMO (1988–1992), NEMO 2 (1993–1998) and EU-funded WECO project *Wind Energy Production in Cold Climates* [3]. However, since finishing the research work, many wind turbines of larger size have been installed and more experience has been gained. Almost yearly updated reports on the topic are appearing on the open literature to confirm the raise of the interest in the matter. Among them, the Elforsk Reports, the publication of the VTT Technical Research Centre of Finland and two dedicated international conferences, The Boreas conference and the Winterwind conference can be cited. A quite exhaustive compilation of the Research and Development projects on various subjects can be found in the appendix of the report on wind power in cold climate edited by WSP Environmental, and commissioned from the Nordic Energy Research in fall 2011 [5].

In detail, all cited sources agree on identifying the following consequences of icing on wind turbines:

- inefficient or inoperative wind measuring equipments (both during wind assessment and turbine operating phase);
- rapid WT performance degradation;

Fig. 2.1 Picture collection of typical iced wind turbine sites (*Source* IEA)

Fig. 2.2 Picture collection of iced wind turbines (Maissan, J.F. "Wind Power Development in Sub-Arctic Conditions with Severe Rime Icing", Technical Services Yukon Energy Corporation—Circumpolar Climate Change Summit and Exposition 2001, web photo: Kent Larsson, ABVee)

- increased noise level;
- increased fatigue on wind turbine and foundations;
- WT downtime due to excessive vibrations;
- disconnection from the grid because of rotor deceleration below critical threshold;
- ice throw and falling (critic for public, maintenance personnel and near properties safety);
- damages of power lines;
- additional power supply to heat up the measuring instrumentation, cold weather devices and ice prevention system on board;
- additional troubles (site accessibility, site data communication).

In particular, hazard of ice shedding and snowfall from wind turbines is relevant towards the people, objects and operator personnel safety. The area on the ground covered by ice fall around the wind turbine depends on the mass and size of the ice fragments, as well as on wind turbine size and prevailing wind directions [6].

From the maintenance point of view, as the turbines are usually located at remote sites, the access may be difficult or even impossible during long periods of the harsher seasons. Sometime the access may be limited to motor sledge carrying light repair instruments. The resulting reduced site accessibility impacts on wind turbine technical availability.

For these reasons de-icing and anti-icing systems became subject of a fast development. From a single commercial anti-icing system available in 2004, today more companies are proposing systems based on different operation principles, even if some of them are still in the prototype phase or in small serial production lines. Blades thermal heating systems are currently used at new wind power plants in Finland, Sweden, Switzerland and other countries outside Europe. Only little data of operations with anti-icing and de-icing system are openly available despite of the large number of turbines being erected worldwide. At sites with a high probability of icing—e.g. several weeks per year—an active or passive de-icing or anti-icing system for the rotor blades is suggested [2]. However, it is worth to remark that it is not always recommended, from the economical point of view, to blindly install ice prevention system in ice risk areas. In fact, if icing occurs only a few days to a week per year at the site, the turbine should be more efficiently switched off at icing onset or has to be designed to withstand the additional loads [6]. The anti-ice thermal power $\dot{Q}_{IPS}$ to be supplied to the blades (Fig. 2.3) increases with the turbine size as a consequence of the increase of the area to be heated. This figure anticipates the results of the model developed in the next chapters to estimate the thermal power required to keep the blades free from ice as function of turbine size and blade setting (twist and chord taper). However, the ratio of anti-icing power to WT rated power output (Fig. 2.4) is unfavourable for the smaller turbines, growing from 4–10 % for the largest (three bladed) to 11–16 % for the smallest (three bladed) ones [7]. This conclusion is relevant since the thermal energy supplied by ice prevention systems has to be subtracted from the annual energy yield of the turbine.

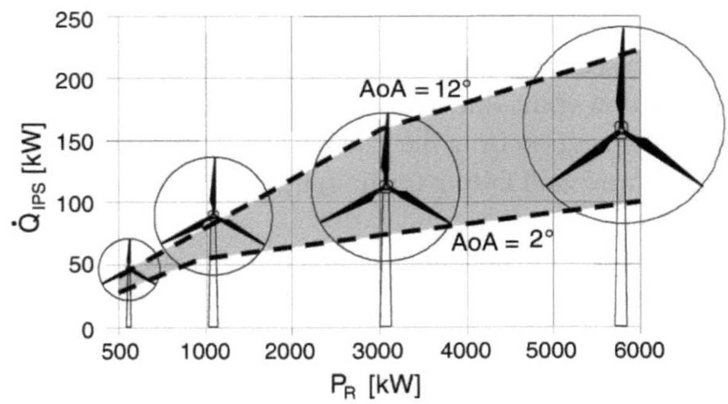

Fig. 2.3 Comparison of the total expected electric power $\dot{Q}_{IPS}$ [kW] demand for anti-icing purpose of the turbine as the size varies [7]

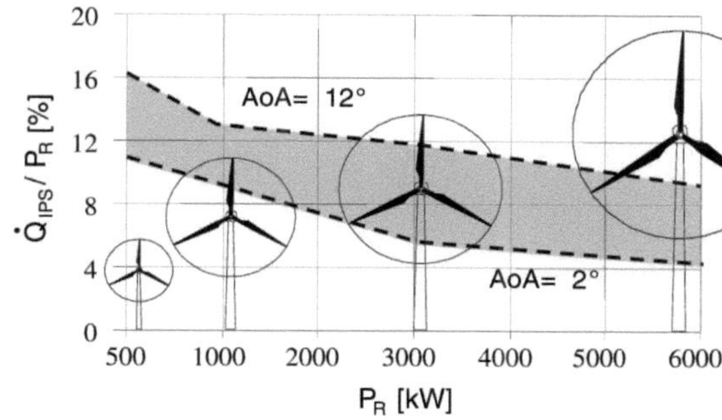

Fig. 2.4 Comparison of the total expected electric power demand to the rated power of the turbine $\dot{Q}_{IPS}/P_R$ as the size varies [7]

Therefore, it is crucial for site developers to have reliable tools to ascertain the benefit of installing ice mitigation systems. One of the most difficult problems is to assess the icing risk of a given site. Meteorological and climatic available data are unfortunately not sufficient to determine how the wind turbine will be affected by ice, and can serve only preliminarily as a warning for potential risks.

A review of ice detection status of the art indicates that several systems to detect and measure icing on wind turbines are potentially available on the market, but there are currently no sufficiently reliable ice detection systems suitable for wind turbines. As will be discussed later, several reports assess that none of the available sensors works yet satisfactory for this specific field of application.

2.2 Ice Growth on Wind Turbine

Generally, there are three situations in which icing on WT can occur, power production, idling (the blades are rotating but the wind turbine is not connected to the grid) and standstill (the rotor is at rest).

Figure 2.5 shows the turbine in different working conditions. Qualitative ice accretions are depicted on the blades.

During power production the rotational speed causes the onset of centrifugal forces on the ice at the blade leading edge. These forces combine with the aerodynamic forces causing shear forces and bending moments between the ice structure and the blade body, resulting in the early break off of the cups of ice. Ice growth is generally dependent on the adopted wind turbine power control strategy: pitch controlled wind turbines or stall controlled. On a pitch-controlled WT, also leading edge (LE) ice accretion of up to 100 % (of the chord length) could be observed [8, 9] during idling in icing conditions, since the pitch angle is reduced and the rotor speed is low. As it will be evident, in horizontal axis wind turbines, due to the conjugate effect of increasing spanwise relative air velocity along the radius and chord reduction, the ice accretion builds up more at the outer part of the blade with an approximately

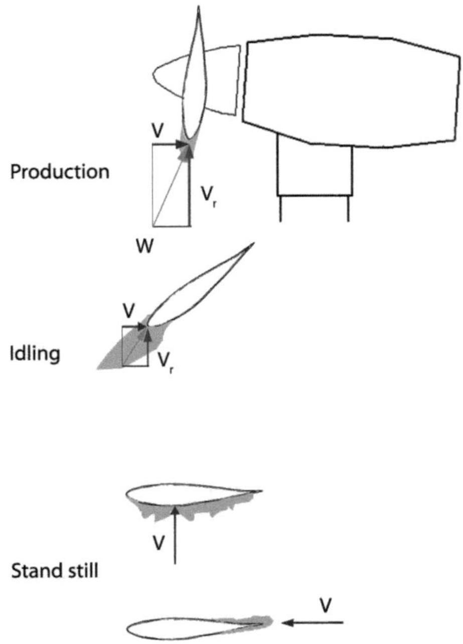

Fig. 2.5 Example of ice accretion on a pitch controlled blade for different working conditions

linear increase. The ice at the blade outer part breaks off and grows again during storm operations forming a typical saw-tooth distribution (see Fig. 2.6).

Idling and standstill cause a more relevant problem, since, depending on wind direction, large areas of the blade can be exposed to icing, also in moderate winds. As a consequence, de-icing applied on small blade leading edge region can be totally ineffective. Also trailing edge ice accretion is possible.

Based on field observations Germanisher Lloyd GL Wind 2010 (Chaps. 4—4.2.4.2.2 Ice) [10], prescribes that "the mass distribution (mass/unit length) shall be assumed at the leading edge. It increases linearly from zero in the rotor axis to the value μ_E at half of the radius, and thus remains constant up to the outermost radius". The following empirical rule is given:

$$\begin{aligned} \mu_{ICE} &= \rho_{ICE}\, k\, c_{min}\, (c_{min} + c_{max}) \\ \rho_{ICE} &= 700\,\text{kg/m}^3 \\ k &= 0.00675 + 0.3 \exp(-0.32\, r/r_1) \\ r_1 &= 1\,\text{m} \end{aligned}$$

where μ_{ICE} is the ice mass distribution per metre of blade at the half of the blade radius, c_{min} is the chord length at the blade tip linearly extrapolated from the blade contour and c_{max} is the maximum chord length.

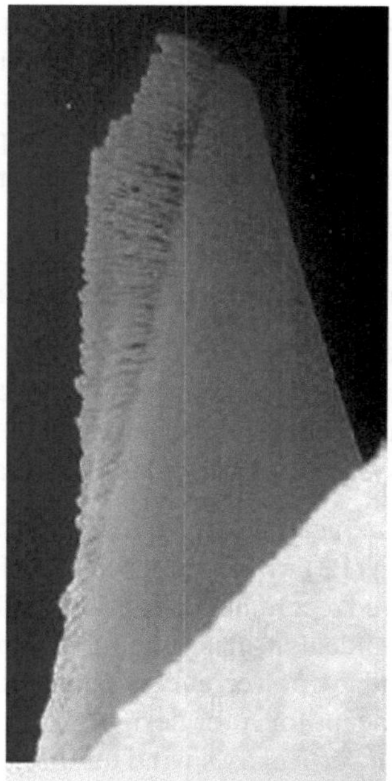

Fig. 2.6 Picture of the distribution of a typical ice accretion [10]

This relationship is quite simplified and do not consider neither the ambient nor the wind turbine operating conditions. This approach produces the mass distribution shown in Fig. 2.7 where the ice distribution for three rotors of different sizes is simulated. The mass inventory gives about 1,066 kg of additional mass for each blade

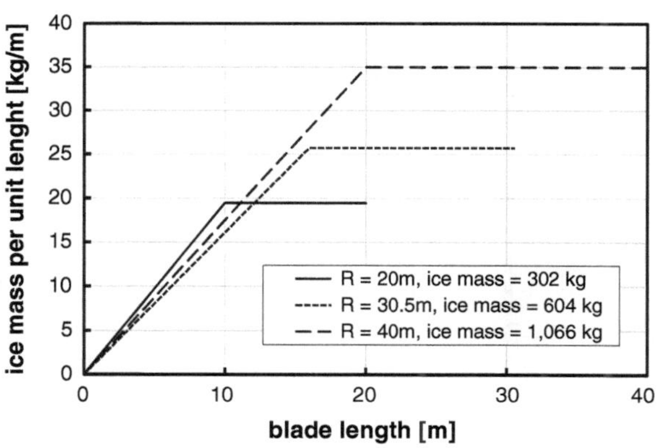

Fig. 2.7 Ice mass distribution for three rotors of different sizes computed to [10]

of the 80 m diameter rotor, and about 605 kg for the 61 m diameter, such additional weights being relevant for static and dynamic load analyses.

This model is open to some criticism, in fact, the mass inventory is unrealistic for large turbines due to the large dimensions of the blades preventing water catch and ice growth, as will be evident in Chaps. 4 and 5. Such quantities could be reasonable, although for standstill conditions.

In general referring to the rotor aerodynamic and aeroelastic behavior one could expect the following relevant modifications in the boundary conditions:

- the change of the mass per unit metre of the blades;
- the change of the aerodynamic performances (and forces) of the airfoils;
- the change of the material mechanical properties with temperature;
- the effect on the control system (inhibition of the actuators and or sensors, etc.).

Further information on the aeroleastic behavior of the iced rotor can be found in part of work package 3 (WP3) of the EC project NNE5 2001-00259 (New Icetools [11]), aiming at improved recommendations for load and design related to wind turbine in ice condition. WP3 deals with structural dynamics and turbine safety due to ice load conditions (standard power curves, standard frequency spectra of the vibrations) to set up new concept for rotor design.

Studies of ice accretion on vertical axis wind turbines are not reported yet in open literature due to scarce diffusion of this kind of turbine.

The joints between blades and arms can become a candidate area of accumulation of water, snow and ice, because the latter are centrifuged by the rotation as shown in the pictures of Fig. 2.8, reproduced by the kindly permission on Tozzi Nord Wind Turbines Company. At this location the snow or ice can compact and create high unbalance levels for the rotor causing turbine shut down. From infield observations, the relatively high centrifugal forces and the associated deformations of the blade seem to prevent risky ice accretion on this component during normal operations.

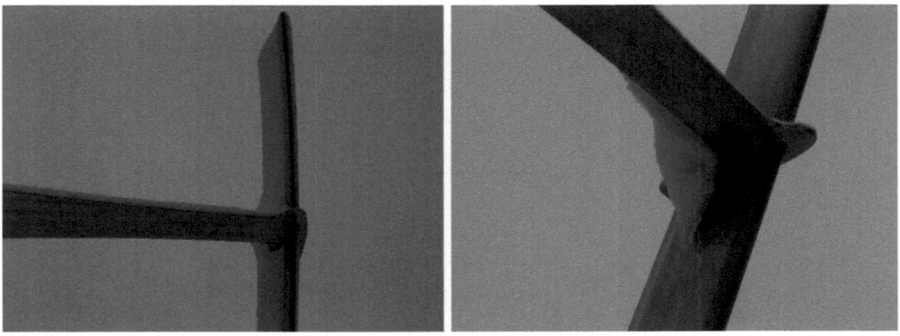

Fig. 2.8 Example of ice accumulation on the area between blades and arms for VAWT (kindly permission on Tozzi Nord Wind Turbines Company)

2.3 Prerequisites for Icing Occurrence

The term *atmospheric icing* collects all events due to rain, or snow precipitation, while by the term *sea spray icing* indicates the ice growth caused by sea water sprays splashing on the structure or sea water droplets carried by the wind and hitting the structure. The International standard ISO 12494-2001 *Atmospheric icing of structures* [12], describes the weather conditions leading to ice and different types of icing.

Atmospheric icing is generated by three different formation processes; precipitation icing, in-cloud icing and hoar frost (caused by sublimation when ice crystals precipitate from water vapour). Precipitation icing and in-cloud icing (see Fig. 2.9) are the most common situations for wind turbine operations.

Precipitation icing is formed as ice by cold (freezing) rain or wet snow. The accumulation rate of precipitation icing is usually higher than in-cloud icing. The resulting ice density and adhesion of freezing rain are high. Wet snow occurs for moderately high temperature, i.e. between 0 and −3 °C. Wet snow is usually dangerous for additional weight on standing structures as roofs, towers and power lines, and for wind turbines in standstill conditions. The density range is 200–990 kg/m^3.

In-cloud icing will occur when supercooled liquid droplets, typically cloud droplets, collide with moving structures and freeze on them. This leads usually to thick layers of ice. Two types of ice are observed during in-cloud conditions, rime and glaze ice. Since weather conditions can change during the precipitation event, also mixed (alternate glaze and rime) accretion can form. Predicting the type and shape of the ice accretion for a specified set of icing conditions is difficult because of the changes of atmospheric conditions occurring usually during a storm.

Depending on the range of ambient temperatures different types of ice are expected. Figure 2.10, taken from ISO 12494-2001 [12] indicates the empirical relationship between ice type and wind speed and temperature, although a more

Fig. 2.9 In-cloud operation of wind turbine (reproduced by kind permission of MX-Matrix s.r.l.)

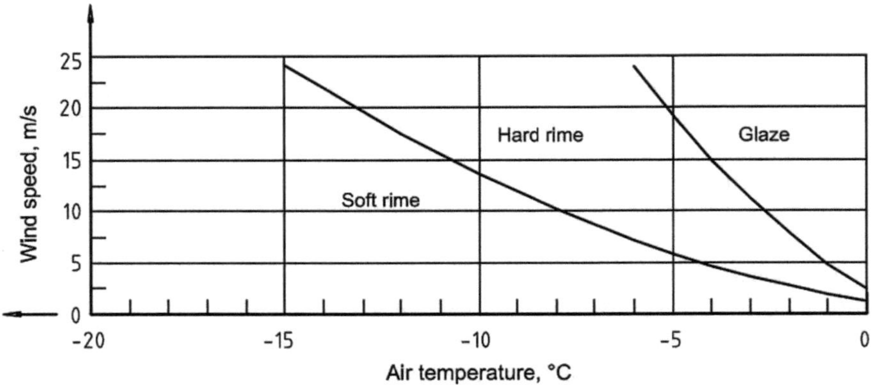

Fig. 2.10 Empirical relationship between ice type and wind speed and temperature taken form ISO 12494-2001 [12]

appropriate analysis should include also the heat balance occurring at the surface. Glaze or rime (either soft or hard) can result. Generally, icing is usually expected between 0 and $-15\,^{\circ}$C. Below this temperature, the liquid water droplets tend to freeze and form snow in the cloud, therefore the occurrence of freezing rain icing is less probable.

More precisely glaze and rime depend on the thermal conditions prevailing during ice formation, which lead to two different thermomechanical processes at the surface, as is illustrated in Fig. 2.11.

Glaze is usually associated with large values of liquid water content (LWC) of the air, droplet size 0–500 μm and temperatures between 0 and $-5\,^{\circ}$C.

It forms a transparent ice cap with a glassy surface. The ice cap adapts itself to the form of the enclosed object and can hardly be removed from it. Visually, this accretion appears hard, compact, almost transparent and bubble free. It exhibits strong adhesion, with density close to 900 kg/m^3.

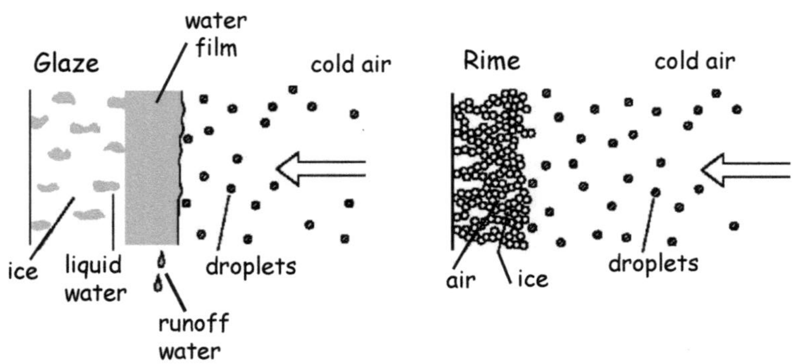

Fig. 2.11 Glaze and rime ice—mechanisms of formation [13]

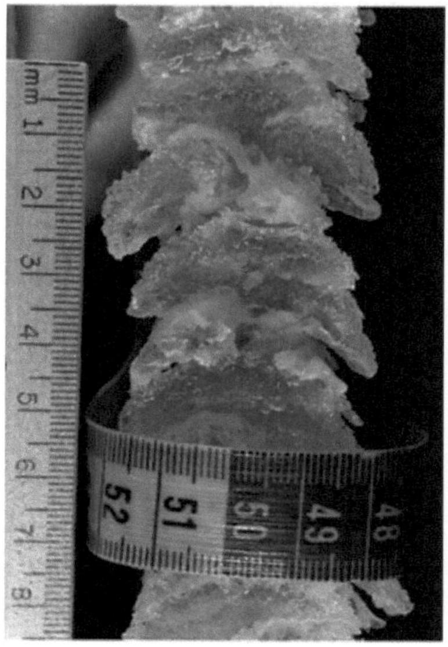

Fig. 2.12 Glaze ice [14]

Glaze ice is generally clear and is characterised by the presence of larger protuberances, commonly known as glaze horns, as shown in Fig. 2.12.

Rime is usually associated with freezing fog with droplet size 0–10 μm. When the air temperature is well below 0 °C (less than −5 °C), supercooled droplets freeze quasi-instantly on impaction.

As anticipated, there are two types of rime ice that can be visually identified:

- Hard rime: granular, white or translucent, density 600–900 kg/m^3 (Fig. 2.13);
- Soft rime: white or opaque, density: 100–600 kg/m^3.

The characteristics depend on the thermodynamic process at the surface. Faster the icing process, more air results trapped inside the ice structure (see Fig. 2.14) lightening the ice mass and making it more brittle.

A mixed ice accretion will have some of the characteristics of both glaze and rime ice. As shown in Fig. 2.14, the centre portion of a mixed ice accretion will have the characteristics of glaze ice accretion. This glaze centre will be surrounded by rime ice accretions, commonly called rime feathers because of their thin, feather-like shape.

Hoar frost is formed by a water vapour sublimation process. The density and persistence of hoar frost are very low, and the effects on wind turbines are negligible.

To conclude, whenever atmospheric or sea spray, two requisites are necessary for the appearance of icing on a body (i.e. the rotor blade):

1. the surface are exposed to impacts of supercooled water droplets (i.e. liquid water at temperature below 0 °C) and is therefore wet.
2. the surface temperature must lie beneath 0 °C;

Fig. 2.13 Rime ice [3]

Fig. 2.14 Example of a glaze ice shape and of mixed ice accretion

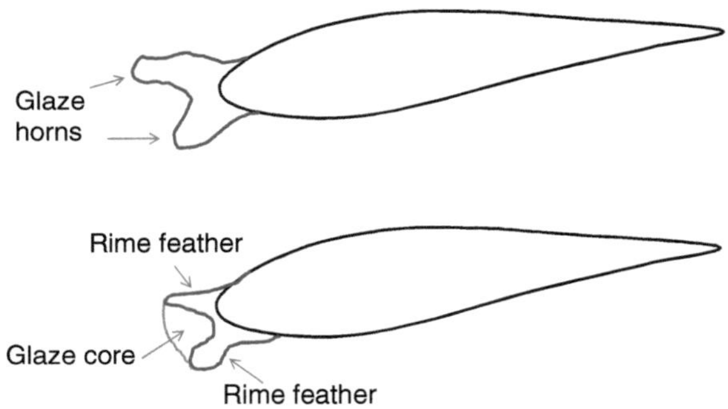

Note that due to the high relative velocity of the blade tip, condition 1 is satisfied also for very low wind speeds (in-cloud icing).

2.3.1 Physical and Mechanical Characters of Ice

The density of ice is one of the most important parameters to estimate the ice loads on structure, because:

- most of the observations of ice on structures report the ice thickness, so that the density is required to determine the accreted mass;
- the adhesive strength of ice and its mechanical properties are dependent upon the density;

The following empirical equation, due to Makkonen can be used [15]:

$$\rho_{ICE} = 0.11 \left(-\frac{d W_0}{2\, T_s} \right)^{0.76} \qquad \left[\frac{\text{g}}{\text{cm}^3} \right]$$

W_0 is the terminal velocity and it depends also on the hit object, with the consequence that for the same atmospheric conditions, the ice density will be smaller for larger structures than on smaller ones. As a consequence, the density is continuously decreasing during ice accretion as the deposit size increases.

In his observations, Makkonen [15] concluded that detachment of glaze and hard rime is usually due to a pure adhesive failure, while soft rime undergoes to a more cohesive strength. Since failure of ice is mostly adhesive, only a very thin layer near the substrate is relevant to ice rebounding, and this initial ice of layer is often formed closer to the wet growth limit than is the major part of the deposit. Therefore, it will have a well-defined density.

The adhesive bonding depends on the type of substrate on the temperature conditions during the icing process, and generally increases as the air temperature decreases and the wind speed increases [15].

2.4 Icing Variables

In the preceding paragraphs, it has been introduced the concept that the icing process, type and severity depends on many variables. One can therefore state the following icing paradigm: *Different turbines process identical weather conditions in different ways: the same conditions may produce different icing effects on a small wind turbine than on a MW-size one and on a stall–vs a pitch-controlled turbine. Identical weather conditions may even lead to different icing effects on the same turbine in different operating regimes, depending on the rotor, angle of attack, turning speed and other factors.*

The variables involved in icing can be conventionally classified in three different classes [16], and are listed in Fig. 2.15:

1. site-dependent variables;
2. turbine design-dependent variables;
3. shared variables.

The *site-dependent variables* can in turn be designated as: climatic variables (which include air pressure, air temperature and wind speed) and meteorological variables (such as air humidity, water content per cubic metre of air conveyed and water particle dimensions).

The *turbine design-dependent variables* group either turbine constructive (the number and length of blade, chord, twist and angle of attach radial distributions) and functional features (type of power curve, tip speed ratio and type of control system).

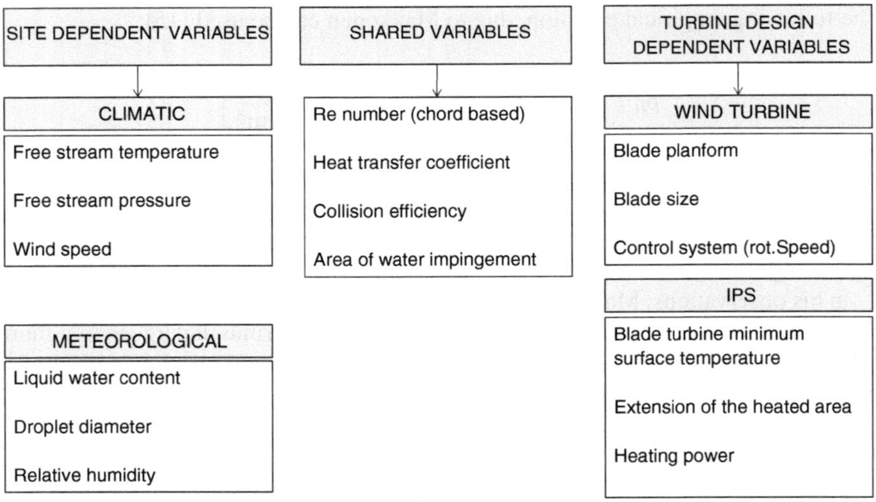

Fig. 2.15 Classification of basic variables involved in the icing process

Other turbine variables regarding more specifically the ice prevention system are the blade surface prescribed temperature distribution, the extension of the ice protected area and the anti-icing heat flux.

The *shared variables* depend at the same time on both climatic and turbine data and are synthesised for instance by the blade Reynolds number, the convective heat transfer coefficient, the local water collision efficiency and the area of water impingement over the blade.

The site-dependent variables govern the physics of ice accretion and are thus needed for the design of ice prevention systems. They have been very precisely recognised from over 50 years of experimental investigations in the aeronautic field. In-cloud measurements indicate that the liquid water content (LWC) ranges from 0 to 5 g/m^3. The LWC in stratiform cloud seldom exceeds 1 g/m^3, whereas much higher concentrations can occur in convective (cumuliform) cloud. Typical information about the relationship between droplet size, LWC and temperature is reported in the experimental founding of Cober [17], shown in Fig. 2.16. It emerges the tendency of LWC to decrease as droplet size increases. Note that the solid curves represent the potential accumulations of ice on a cylinder (diameter 3 cm) in g/cm^{2}h.

In icing-related applications, the actual droplet size distribution in-cloud is represented by a single variable called the droplet's median volume diameter (MVD). The MVD is the mid-point in the LWC distribution over the range of cloud droplet sizes that happen to be present at the time. The MVD therefore varies with the number of droplets in each size category. The MVD has proved useful as a simple substitute for the full droplet size distributions in ice accretion and anti-icing analysis. It ranges from 10 μm (in freezing fog) to 5,000 μm (in freezing rain). Over 1,000 μm we speak on super large droplets (SLD). Data reviewed by the FAA [18] indicate that

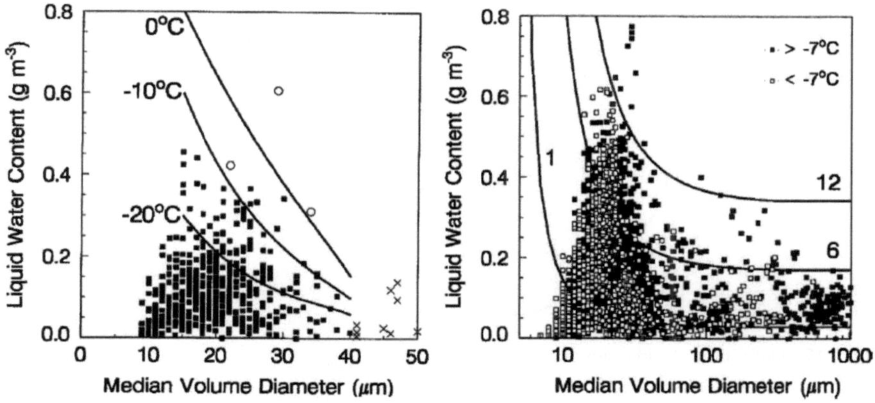

Fig. 2.16 Relation between droplet size, liquid water content and air temperature. Measured data from [17]

Table 2.1 Availability of information of the icing variables for aeronautic fields and wind energy field [19]

Parameter	LWC (g/m^3)	T (°C)	MVD (µm)	W (m/s)	p (Pa)
Aeronautic fields	Available	Available	Available	Available	Available
Wind energy	Not available	Available	Not available	Available	Available

the overall average is about 15 µm for stratiform clouds and 19 µm for convective clouds.

The availability of information on icing variables both in aeronautic fields and wind energy field is illustrated in Table 2.1 [19]. Although some parameters (e.g. LWC and MVD) are routinely measured and collected in the aeronautical field by specific instruments at airports (ground detection) or in-flight (aircraft-borne detection), their use is not common for wind turbines. Their extension in space (cloud extension) and time are also unavailable. Among them only temperature and pressure are routinely measured, while relative speed *W* can be easily computed.

2.5 Defining the Icing Event

The icing event is a critical variable for the assessment of the economic impact of icing on WECSs operating in sites with a given ice risk. The icing event is a probabilistic concept [19] depending on a subset of conditions defined by the shared area shown in Fig. 2.17.

The COST 727 project [20] stated, with particular concerns with ice detection instruments, the following definitions:

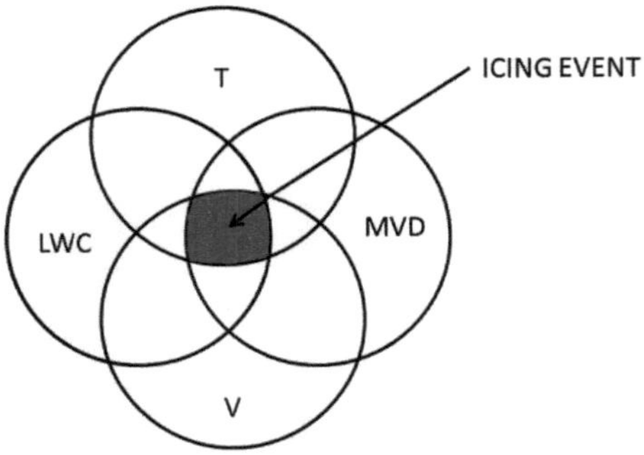

Fig. 2.17 Definition of direct icing event

1. *Meteorological icing*: the duration of a meteorological event or perturbation which causes icing (with time as unit).
2. *Instrumental icing*: the duration of the technical perturbation of the instrument due to icing (with time as unit).
3. *Incubation time*: time delay between the beginning of the meteorological icing and of the instrumental icing.
4. *Recovery time*: time delay between the end of the meteorological icing and the full recovery of the performance of the instrument.

These definitions are illustrated in Fig. 2.18 where the difference between the two different types of icing event, i.e. meteorological and instrumental icing is shown [21, 22]. The previous definitions allow to define without ambiguity the performance index *PI* of a meteorological instrument.

$$PI = \frac{I_{icing}}{M_{icing}}$$

being I_{icing} is the instrumental icing and M_{icing} is the meteorological icing. As the performance index approaches zero the meteorological instrument reflects increasing

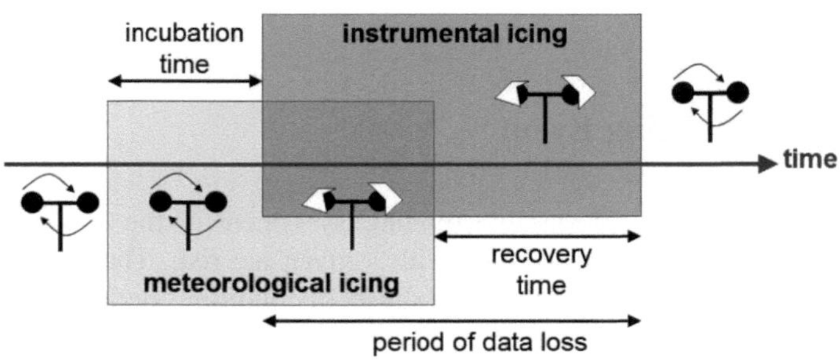

Fig. 2.18 Meteorological and instrumental icing [21]

insensitivity to icing, while figures larger than one indicate increasing sensitivity to icing. This parameter allows classification of ice detection systems.

This approach is quite relevant also towards icing on wind turbines, because, in a more scientific way it resembles the similar and intuitive concepts (see for instance [23]) of *direct icing* and *indirect icing*, having similar meaning. Direct icing is the stage where ice built up, whereas indirect icing indicates the stage of ice persistence on structures after direct icing causing a detrimental effect on their operations. Time durations are defined accordingly.

The author advises a future standard use of aircraft-based meteorological data. So, although such an approach suffers today of major shortage of data and a lack of cultural aptitude in the wind energy field, it will be adopted for the definition of the direct icing duration time.

A range of simultaneous combinations of meteorological variables, each of them having the same probability of being exceeded, will define the probability of icing.

Aeronautical models consider as icing events the atmospheric states defined by the variables LWC, MVD, V, T characterised in that their range are LWC > 0, MVD $> MVD_{min}$, T $< 0\,^{\circ}$C and V > 0. When such limits are simultaneously exceeded, ice will form. MVD_{min} is the minimum dimension of the droplets above which their trajectory will impact the structures, without being deflected from the aerodynamic field created by the body (the item will be treated in more detail in Chap. 4).

The direct icing duration (icing time) is defined as the minimum duration of the single event with a contemporary occurrence of LWC > 0, MVD $> MVD_{min}$, T $< 0\,^{\circ}$C and V > 0, and is expressed as:

$$t_i = t_{LWC>0} \cap \cdot t_{T<0} \cap t_{V>0} \cap t_{MVD>MVD_{min}} \tag{2.1}$$

Provided continuous observations and recordings of site meteorological data are available, the icing Table 2.2 can be constructed and the icing event duration can be calculated by Eq. (2.1).

The total icing duration depends not only on the direct ice accretion duration time, defined as icing event, but also on the time duration during which the turbine cannot produce because of persistence of ice on structures (indirect icing). Usually, indirect ice is not recorded by ice detection devices or can hardly deduced from

Table 2.2 Data for estimation of the number of ice events—icing table

Event number	V (m/s)	T (°C)	LWC (g/m^3)	MVD (m)	t_i (h)
1	...	...	...	...	...
...	...	...	...	...	...
i	...	...	...	...	...
...	...	...	...	...	...
I	...	...	...	...	...

it, but observations [24] indicate that it can be as high as 100 % of the direct icing time and cause till 50 % of increased unavailability. Disagreed information would be advisable.

Most of the data and provided by ice sensors are not sufficient to get a clear conclusion about the total icing days period (direct + indirect). There are not at the moment methods (either experimental or numerical) to help in the forecast of total icing duration, although after the author mind, this evaluation is of utmost importance for a reliable economic assessment and can at the moment be handled only by guessing a proper safety margin.

In Fig. 2.19 an example of direct icing and indirect icing is presented [23]. This level of information would be required for a correct economic analysis of the real relevance of the phenomenon.

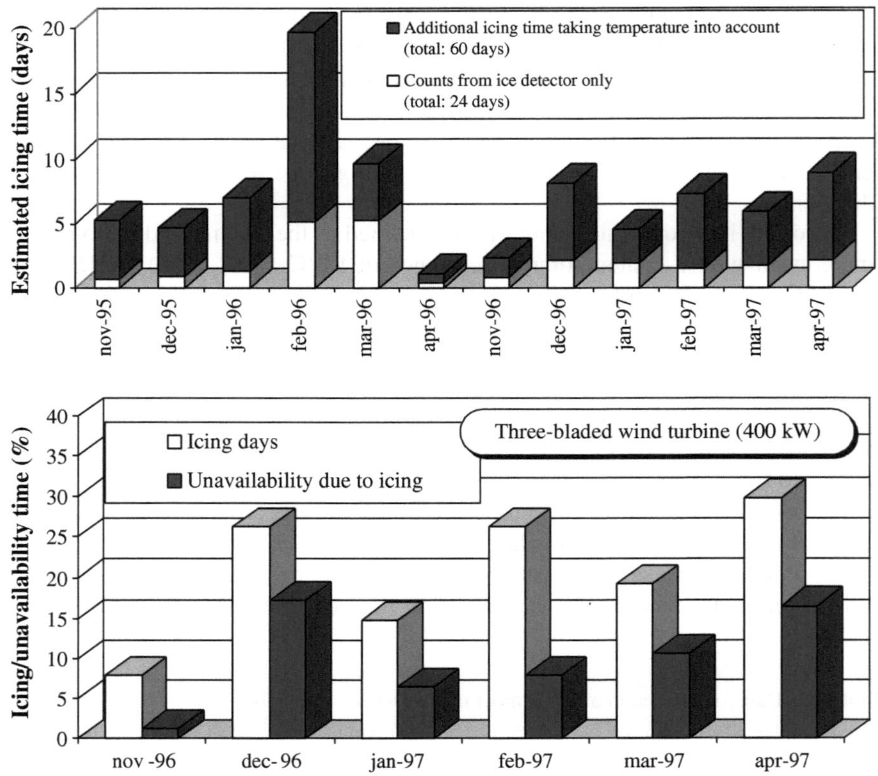

Fig. 2.19 Measurements of direct and indirect icing [23]

2.6 Ice Detection

As stated, the detection of ice formation on structures results complex because the accretion depend not only on the meteorological conditions but also on the body subject to icing. Compared to pure meteorological icing forecast, additional information are required to asses the icing risk for wind turbines as listed in Fig. 2.15. A reliable and really informative ice detection is a challenging task for wind energy purpose. Various sensor systems have been tested over the years but none of them has resulted yet sufficiently reliable or efficient.

A survey of scientific papers, patents and market technology suggests that the ice detectors can be classified as *direct icing sensing techniques* and *indirect icing sensing techniques*. A very comprehensive review can be found in the recent work of Homolosa et al. [25].

The *direct icing sensing techniques* detect changes of physical quantities caused by icing, and can be classified as *mechatronic systems* (integration of micromechanical and electronic devices) *electric systems*, and *optic systems*. The following compilation describes the main measurement principles. Most of them are subject of patents and proprietary knowledge.

2.6.1 Mechatronic Systems

2.6.1.1 Measurement of the Attenuation of a Signal

The method is based on the measure of the attenuation of a signal (usually ultrasonic or microwave) through a low acoustic attenuation mean. Typical waveguide can be made by steel or nickel tape. The signal generated at one end of the waveguide is collected at the other end, and its damping rate measured. Piezoelectric elements can be used for this purpose. The presence of ice causes an attenuation of ultrasonic signals, while the viscosity and density of liquid water causes much lower attenuation and can be so discriminated from ice presence on the surface.

2.6.1.2 Measurement of Shift of Resonant Frequency

The concept is based on the variation of the resonant frequency of a vibrating mass, based on the concept that the vibrating mass changes due to ice accretion. The natural frequency of the mass (typically a cylinder) is expected to drop as the mass increases, according to the laws of classical mechanics, and can be detected by piezoelectric and magnetostriction devices. The system allows melting the ice by joule effect heating in order to repeat the measuring procedure and to provide also information on the icing rate.

2.6.2 *Electric Systems*

2.6.2.1 Measurement of Change of Electric Properties

This methods use reading of the impedance/inductance/capacitance variation due the change of the electric properties of the water-ice layer (as for instance the dielectric constant). The capacitance and impedance technology is already successfully implemented in airplanes. The sensors can detect ice formation on a relatively wide area. The sensors are thin, letting them easily to be retrofitted on blades. The power consumption of the electronic is low.

2.6.3 *Optical Systems*

2.6.3.1 Direct Measurement of Reflected Light

According to this method a source of light either reflected or emitted by the body is measured. If ice accretes on the body the emittance/reflectance of the surface changes. The signal can be digitally sampled to associate the output with the absence/presence of ice. Also this technique allows performing icing rate measurements by the cyclic heating of the surface.

2.6.3.2 Infrared Spectroscopy

This method is based on the measurements of infrared light absorption and reflection by the ice. This technique cannot measure the thickness of ice since the type of ice affects the reflection rate. This method of ice detection on the surface of the rotor blade has the advantage that it needs no electrical wiring on the rotor blade surface being mounted on the hub. Possible drawback is that the sensor becomes inefficient if the reflection area is dirty, and that it is hardly retrofittable since the fibre optics need to be embedded into the blade.

2.6.3.3 Reflection from Inside

The technique is based on the Snell's law. As a light beam (e.g. a light emitting diode, LED) is directed through a medium with high refractive index towards a medium with a lower refractive index, the beam can be totally reflected back. As ice and water have an index of refraction of about 1.3 (at 0 °C, the refractive index of ice is 1.33049 and the one of water is 1.3354) compared to air which is very close to one, the light will be totally reflected as long as the surface is uncontaminated by ice, while the presence of water or ice will not totally reflect it. A drawback of the method is that it cannot discriminate ice from water since their indexes of refraction are about the same, and an additional measure of the ambient temperature is needed.

2.6.3.4 Web Camera Recordings

Web cameras pointing the blade coupled with image analysis techniques have the drawback to need a light source to record the images, therefore during scarce illumination conditions it cannot be efficient unless artificial lighting, (eventually outside of the visible range to avoid environment disturbances) could be employed.

The methods based on *indirect icing sensing techniques* use the contemporary measurements of parameters that can be combined as input for a diagnostic analysis. The contemporary occurrence of some signals can lead to a probable scenario of icing. These methods can be classified as following.

2.6.4 Wind Turbine-Based Parameters

An expected consequence of blade icing is the drop in the aerodynamic performance and hence in the power delivered. This measure, couple with the ambient temperature reading and blade accelerations can in principle discard the possibility of false signal due to mechanical failure. Also the measurement of the increased weight of the blade due to ice through the blade root bending moments have been proposed. A variant is the measure of the gyroscopic moment of the hub. The methods are based on the fact that ice accretion is naturally uneven on the blades, so a mass unbalance always occurs. Although the imbalance measurement can be integrated into a condition-monitoring system in the drivetrain, in order to provide any indication of imbalance, a substantial deviation in blade momentum is necessary. Typical rotor blade unbalances are computed in Chap. 3. A substantial mass and/or momentum deviation is necessary in order to assess that the imbalance is caused by ice. This seems to be achievable only statically at standstill or parked conditions where the additional ice weigh can be comparatively high (see Fig. 2.7). In fact any aerodynamic fluctuations can cause a misleading or false indication of ice build-up. The methods must determine whether the monitored meteorological conditions and operational conditions are consistent with blade icing. Thus, in configurations where blade mass imbalance is checked, these methods need also the measure of the ambient temperature, the humidity and the yaw, because blade unbalance can also result from yaw angle and a near zero yaw condition is necessary to avoid incorrect detection.

2.6.5 Noise Measurements

Seifert [2] proved that the noise level increases when the blade is iced, and in particular, the blade noise emission frequency is shifted to higher frequencies. The system needs customisation for any single site and turbine (or wind park) to clean out the response from background noises.

2.6.6 Thermodynamic Status of the Surface

The measurement of surface temperature and the contemporary detection of the presence of water allow to monitor the risk of icing. The presence of drops of water detected on a surface if the surface temperature is below a preset critical point, (i.e. $T_s \leq 0\,°\text{C}$), gives a precise indication of icing onset risk. The combination of the other conditions are not critical, as shown in Fig. 2.20 when the logic of operation of this method is given. Sensor built to accomplish these readings are simple, light and flash mountable and retrofittable on the blade. The signal (risk/no risk) emitted by the sensor on the blade surface can be received by a receiver installed on the tower as the blade passes in front of the tower.

2.6.7 Differential Reading of Heated and Not Heated Anemometers

The concept is based on the contemporary measurement of the reading of an heated and an unheated anemometer or wind vane. The unheated anemometer may experience ice accretion, and will be either blocked or retarded, while the heated one would continue measuring correctly during the ice event. This can be used as an indicator of potential ice hazard on the site during the wind measuring campaign, and indicates definitively if an heated anemometer (more costly compared to the standard one) is necessary for the turbine control on the nacelle. An example of the procedure is given in the next paragraph.

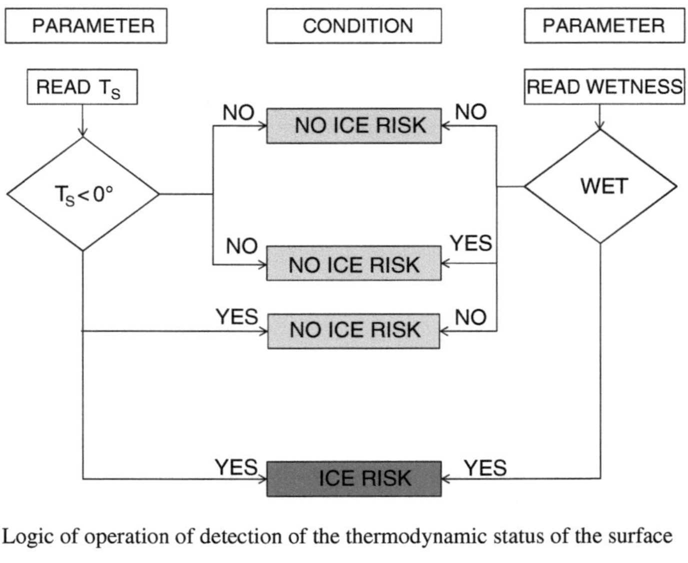

Fig. 2.20 Logic of operation of detection of the thermodynamic status of the surface

2.6.8 General Comments on Ice Detection Systems for Wind Turbines

The peculiarity of application of ice detection on wind turbines determines some not negotiable requirements:

1. *Efficacy on detection*. This is an essential property, because a few minutes of delay in the detection will create enough ice roughness to compromise the aerodynamics, as will be extensively discussed in the next chapter. The idling time in activation of anti-icing or de-icing devices if of the order of some minutes, and the ice mitigation strategy could result partially or completely ineffective.
2. *Reliability*. The system should prevent from false alarms. To achieve this goal, two prerequisites are necessary:

   - the area of detection should be sufficiently wide to avoid false detection.
   - the sensor should be rather insensitive to dirty.

   These conditions are satisfied if redundant sensors can be applied on the rotor, either on all blades or in multiple, jeopardised position on the blade tip area. Sensor redundancy overcomes the limits of some technologies having a small sensor area. Additionally, also runback ice in the chordwise direction can be detected. The contemporary measurement of other parameters, as the power curve, will offer the best strategy to make any technique completely reliable.
3. *Positioning*. Ice detectors used in stationary conditions (meteorology applications) or flying application (aeronautical) are heavy and installable only on wind turbine fixed parts as the nacelle. Unfortunately, since the icing process depends on the local heat transfer coefficients, the relative velocities at the blade tip are much higher than those occurring at the nacelle. For instance an absolute wind velocity of 12 m/s at the nacelle corresponds at least to a relative velocity of about 70 m/s at the blade tip. Most frequently, when the icing process is recorded to be incipient at the nacelle, the blade is experiencing full icing conditions and the maximum vibration sensors of the wind turbine are possibly already intervened. Therefore, the recommended position of ice detection sensors is in the outer third of the blade, possibly close to the tip.
4. *Easy of replacement*. Easy access to and substitution of the sensor in the event of failure should be guaranteed;
5. *Sensitivity to attract lightnings*. Although modern blades have lightning protection cables and collectors integrated, still in case of lightning the thin wires of the sensors will be the first to be burned, therefore a weakness of wire connected systems still remains. Additionally, there is the need to transfer the signals of the sensors through the rotating frame of the hub. If a wireless unit is used to overcome the problem, still exists the problem of duration of the battery for the emitting source.
6. *Lightness and intrusiveness*. As the optimal position is on the blades, a requisite of such devices is that the weights are not excessive and the aerodynamic is not disturbed by the presence of the sensor.

2.6.9 Measuring the Ice Occurrence on a Site

A typical use of the differential reading of heated and not heated anemometers to assess the relevance of direct icing consists on comparing the output of wind speed and direction measurements of heated and unheated sensors, together with the ambient temperature record. When temperature falls below 0 °C, and the heated and unheated wind traces diverge, and an icing event is considered in progress. The summation of the duration of these icing events in the year gives the total direct icing period hazard. This procedure was experimentally tested in a dedicated measurement station with a 20 m equipped mast situated in Castello Tesino (province of Trento) in the North of the italian Alpine Region (left Fig. 2.21). The meteorological station was installed close to the Monte Agaro (2,062 m a.s.l.) at the end of 2006 (see a picture of Monte Agaro in Fig. 2.22 with the detail of the met mast). The met mast of Agaro was located on an elevation of 2,018 m a.s.l. and at its top (20 m) were mounted the heated and the non-heated anemometers. Data have been delivered over time period from 20.01.2007 to 06.05.2010.

Table 2.3 summarises the main data of the installation. Both the heated sensors have a built-in electric element to fully heating the device and prevent icing up (class 0).

Figure 2.23 shows an image of the iced met mast during the winter season. The iced wiring of the tower is well visible.

In Figs. 2.24 and 2.25 some data recorded during a month characterised by a large snowfall events are shown. The recorded traces of the heated and unheated anemometers diverge during icing time after a short incubation time, as very clearly indicated in the sub-period of the bottom plot of Fig. 2.25. The snow up over the unheated anemometer and wind vane caused a rest period of 10 days (about 30 % of the whole month duration). These results were confirmed by visual observation on the Agaro site.

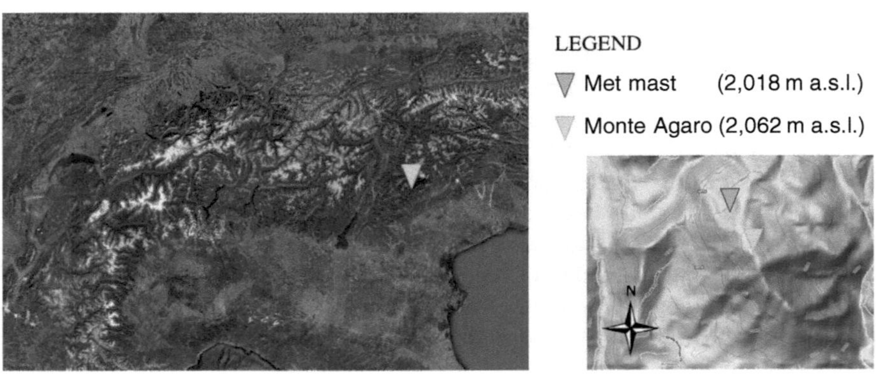

Fig. 2.21 Monte Agaro (*left*) and met mast (*right*) location

Fig. 2.22 Monte Agaro site

Table 2.3 Overview of the installed met mast of Agaro (Italy)

Met mast coordinates (UTM)	Easting: 705 221–Northing: 5108 587
Elevation	2,018 m a.s.l
Height heated anemometer	20 m a.g.l
Height heated vane	20 m a.g.l
Height non-heated anemometers	20 and 10 m a.g.l
Height non-heated vanes	20 and 10 m a.g.l
Height temperature sensor	4 m a.g.l
Measurement period	20.01.2007–06.05.2010
Averaging period	10 min

The method is sufficiently reliable to predict direct icing on a site. Since the very different size of anemometer cups and real dimension of a wind turbine blade, this prediction is not giving in absolute the number of days of potential icing of the turbine rotors. Anyway, since the small dimension of the cups makes the anemometer to be more prone to get wetted and iced (as will be evident from the theory outlined in Chap. 4) compared to the blades, the method can be considered conservative.

Fig. 2.23 Image of the iced met mast

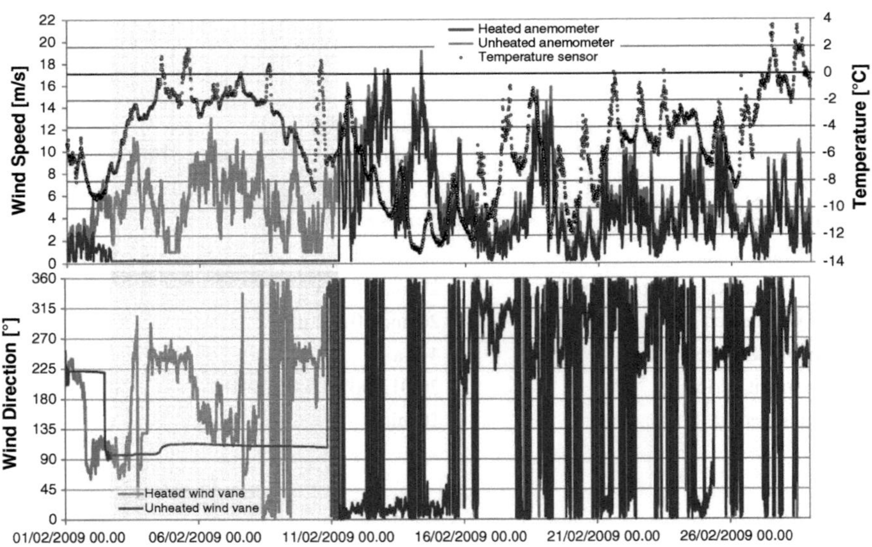

Fig. 2.24 Comparison between heated (*red line*) and unheated (*blue line*) anemometer and wind vane during February 2009 (Monte Agaro—Italy) (Color figure online)

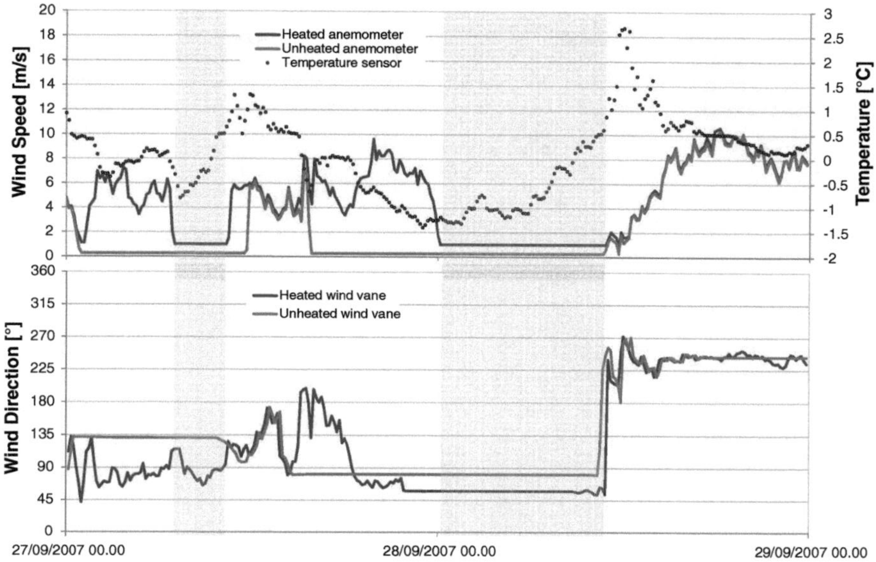

Fig. 2.25 Comparison between heated (*red line*) and unheated (*blue line*) anemometer and wind vane during 27–29 October 2007 (Monte Agaro—Italy) (Color figure online)

2.7 Wind Sensor Behaviour in Icing Climates

Cup anemometers and vanes are very sensitive to icing. Small amounts of ice significantly reduce the measured wind speed, and large ice accretion may even block the anemometer: small quantities of rime ice on the cups and shaft of an anemometer at 10 m/s lead to a wind speed underestimation of about 30 %. Figure 2.26 shows the occurrence of severe icing of an anemometer leading to complete stop. This can lead to turbine malfunctioning because of controller inadequacy to track the wind intensity and direction changes. If the wind turbine rotor during operations is not able to follow the wind direction changes, energy yield losses, vibrations, and finally also shutdown can occur. If the turbine is in standstill, as $V > V_{cut,in}$ it may start in the opposite rotational direction, or it does not even start due to wrong (underestimated) wind speed measurement. If finally the turbine is operating with a partially iced anemometer, large power fluctuations can occur.

Wind measurement during cold periods should be carefully checked because small amounts of ice or snow over the unheated cups could compromise the anemometer response. The Fig. 2.27 taken from the Eumetnet Sws II Project 2004 Final Report—Improvements Of Severe Weather Measurements And Sensors [26, 27], shows the effect of progressive icing of the cups. The ice causes a substantial variation of the aerodynamic behaviour altering the calibration curve and the anemometer response.

Fig. 2.26 Severe icing of an anemometer and wind vane

The reasons are the increase of the cup's mass, the increase of the cup's drag ratio (the high drag coefficient of concave side over the low drag coefficient of the convex side), and the different behaviour of lubricants into bearings.

Figure 2.28 shows the results of a model developed to analyse the effect of the latter parameters on the dynamic and static behaviour of a common wind vane anemometer. To obtain the result shown in Fig. 2.28, a simple model has been implemented based on the work of Pedersen et al. [28]. Ice thickness and ice mass are related via simple geometric relationships.

With reference to Fig. 2.29, the instantaneous aerodynamic torque produced from each cup is given by:

$$M(\vartheta) = \begin{cases} \frac{1}{2}\rho A_C C_{D,1}(V\cos\vartheta - \Omega R)^2 R & \Omega R < V\cos\vartheta \\ -\frac{1}{2}\rho A_C C_{D,2}(V\cos\vartheta + \Omega R)^2 R & \Omega R > V\cos\vartheta \end{cases} \quad (2.2)$$

R is the length of the arms, A_C is the frontal area, $C_{D,1}$ and $C_{D,2}$ are, respectively, the coefficients of resistance of the concave and convex side of the cups. The aerodynamic torque is counterbalanced from the resistant torque M_r due to mechanical friction and aerodynamic drag of the supporting arms. Assuming a uniform field of wind speed impacting the vane, one can determine the dynamic behaviour of the anemometer as plotted in Fig. 2.28 for different simulated icing conditions. Additional ice mass

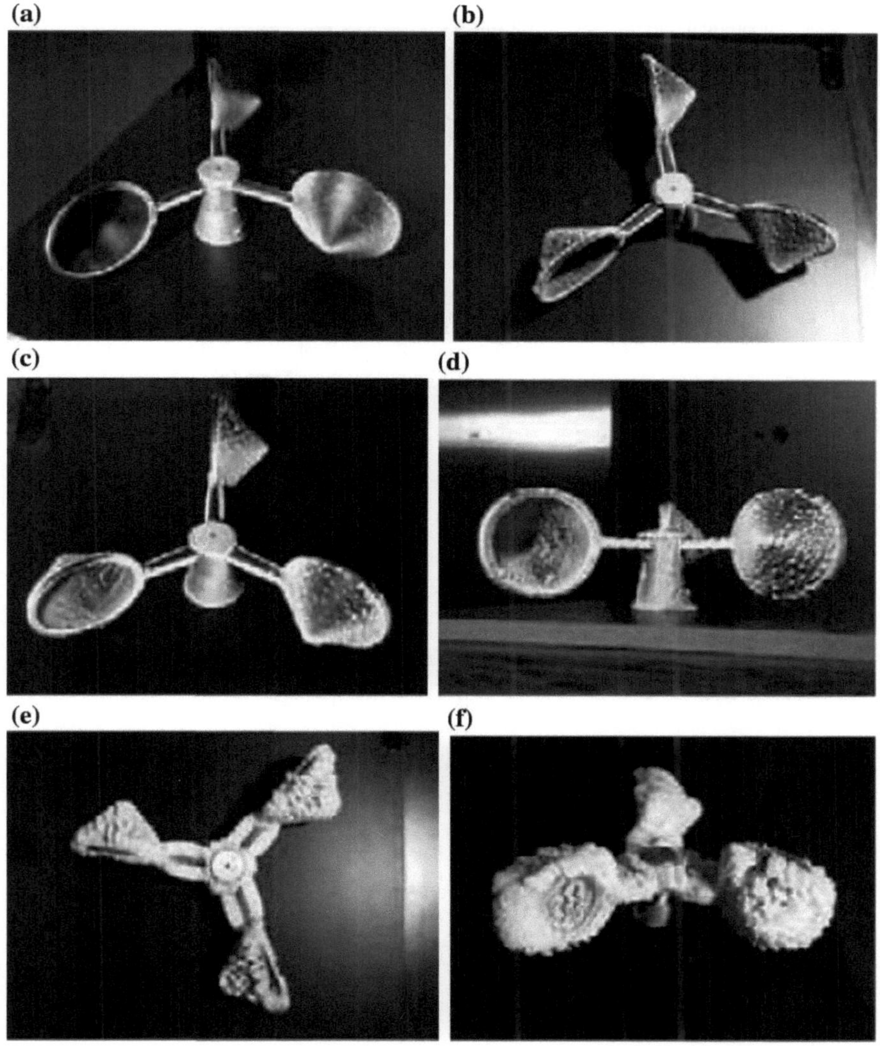

Fig. 2.27 Example of ice accretion on cup anemometer in dry icing conditions in cold climate wind tunnel simulations [27]. **a** 10 min. **b** 20 min. **c** 30 min. **d** 60 min. **e** 120 min. **f** 180 min

alters the acceleration and deceleration of the anemometer and the steady rotational velocity (at equilibrium) is lowered, giving a lower wind speed reading than the real one. A regularly dispersed ice roughness of 2 mm on the cups, generating a drag ratio of 1.3 on the cups causes a systematic drop of rotational speed of about 20 % compared to the clean one, with a corresponding drop of the some amount of the wind speed output (due to the linear calibration relationship between wind speed and rotational speed of this kind of sensor). This behaviour is confirmed by the

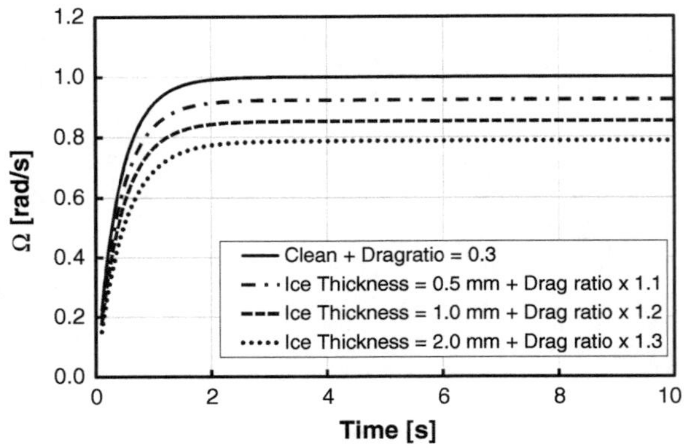

Fig. 2.28 Theoretical cup anemometer responses at 10 m/s: ice free (*continuous line*) versus iced cup anemometer for different ice thicknesses and masses. The drag ratio is defined as $C_{D,2}/C_{D,1}$

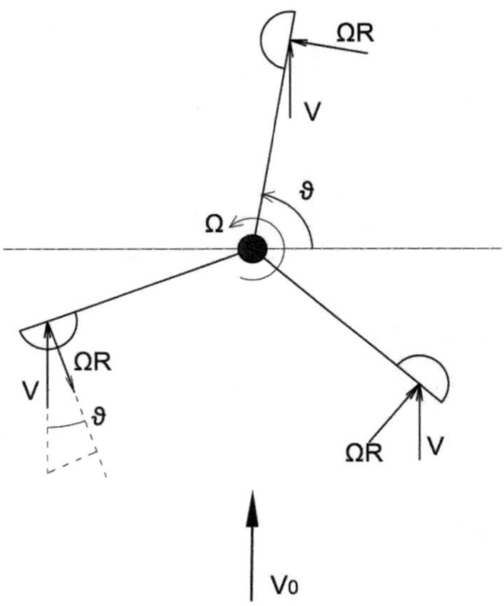

Fig. 2.29 Scheme of the kinematic model of the anemometer

recent experimental investigations of Fortin et al. [29], who propose a semi-empirical equation, taking into account air speed and temperature, LWC and MVD to predict a cup anemometer performance loss due to exposure time to freezing fog.

Wind sensors can be protected against icing by heating part of the components. Depending on the degree of protection, four classes are commercially available, from 0 to 4:

- class 0: totally free of ice (total heating);
- class 1: ice allowance on body and pole but not on the measuring elements;
- class 2: ice allowance also on measurement elements;
- class 3: sensor totally unheated.

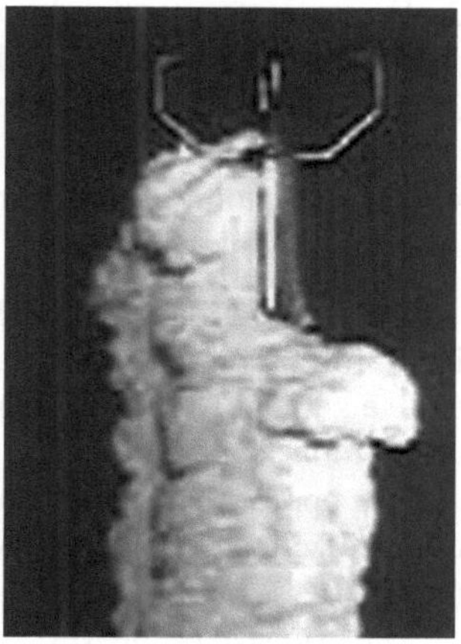

Fig. 2.30 Picture of large ice accretion on the wind sensor pole that can alter significantly the measurements [27]

Lower the class, higher the thermal (electrical) power required, and values as high as 200 to 300 W for a class 0 is common. This quantity makes wind measurements (especially in the wind scouting phase) technically difficult and very costly, because a relatively high power source has to be supplied to heat the sensors. This quantity largely exceeds a few watts required for conventional (not heated) measurement campaigns. Owing to this technical complexity, heated anemometers are usually not selected in routinely site assessments, even if risk of icing is under suspicion.

For this reason, as a general rule, heated wind sensors are recommended in icing environments or in sites with frequent snowfall. Not only the sensing part (i.e. cups or sonic arms) is important to be heated but also the supporting pole. In fact large ice accretion on the pole can alter significantly the measurements because of the local flow filed modifications induced by the ice structure (see for instance Fig. 2.30).

It is worth to remark that also fully heated sensors cannot guarantee to be free of ice or snow under particularly harsh conditions: for example the power supplied could suffer of blackouts (as shown in Fig. 2.25) or it could be less than the required one (in very cold events). In this case filtering techniques have to be used to remove samples that are affected by ice/snow.

2.8 Icing Forecast Models

2.8.1 Short-Term Icing Forecast

Short-term icing forecast for wind turbines is necessary to provide the following levels of information:

1. 2 to 3 days of forecast are important with concerns to energy production schedule for the spot market.

2. Forecast of potential drops in WT power curves and energy yield are essential also for BoP within WT clusters and keep the grid stable on larger areas.
3. 6 h to 1 day forecast is important for activation of anti-icing systems and to avoid waste of energy due to false alarm.
4. Determination of direct and indirect icing duration time to estimate the off-grid period of the wind park.

The most rational approach to icing forecast simulations is to couple meteorological information with an ice accretion model, according to the scheme of Fig. 2.31, that sets the need of a conjugate approach to the problem through the combination of the tools available:

The approach consists on using the output of a meteorological model (MM), coupled with an advanced physical icing scheme giving a Numerical Weather Model (NWM).

The grid size of the MM will mainly depend on the complexity of the topography surrounding the sites investigated. In fact higher resolutions are generally needed for more complex topography (high-resolution terrain grids down to 25 × 25 m) to adequately reproduce the physical processes affecting weather variables. However, the final resolution of the simulations is compromised between the need to reach an adequate fine resolution and the computational costs of the runs, which must produce real-time input variables for the icing model. In particular, the MM will supply the relevant meteorological and climatic variables (LWC, MVD, T, V, P). These variables will input in an icing growth model (IGM) that can have different levels of sophistication, as shown in Fig. 2.31.

Most icing data available derive from a meteorological network primarily designed for measuring purposes other than for assessing icing conditions [3, 30, 31]. The resulting maps are often inadequate to assess icing frequency and severity for wind turbine operations. A method suggested by Harstveit [32] enables LWC and droplet diameter above the cloud base to be obtained from routine meteorological data and cloud heights. The correlations are valid for very specific site conditions.

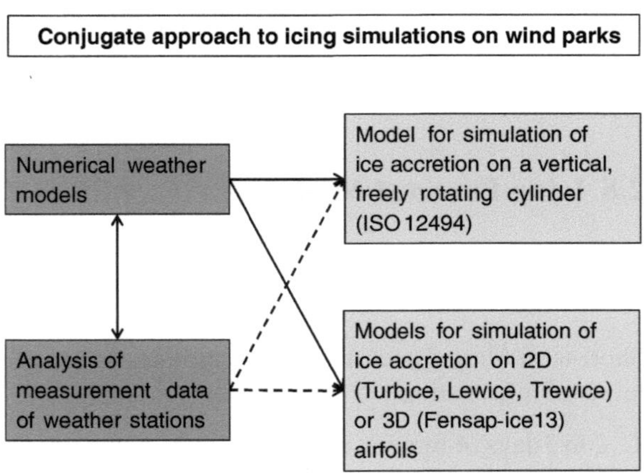

Fig. 2.31 Current approach for icing forecast

Today, Norway, Finland and Switzerland have produced their own national icing atlas based on coupling of a numerical weather model with the ISO 12494 [12] (read Makkonen on freely rotating cylinder) ice accretion formula.

Shortcomings of such approach are several. As first, model simulating icing for a vertical, freely rotating cylinder, although valid for power line icing, is proved to be not consistent for wind turbine rotors. By comparing stock numbers and blades relative velocity and blade planforms, it is apparent that the collision and collection efficiencies are completely different from that of resulting from a 3 cm thick rotating cylinder. Therefore, the ice accretion and ice risk forecast are far from the one gathered by observations. Additionally, in modern wind parks, a wind turbine blade tip moves at 100–200 m above ground level, where the climatic conditions are different from that recorded at ground level. In fact a higher probability that the structure will work inside clouds (icing frequency) and with higher water content levels (ice severity) is encountered. No simple algorithm leading to a straightforward conversion of the ice load modelled on a cylinder into an ice load on a wind turbine blade are available up now.

With concerns to dedicated numerical models for simulation of ice accretion to 2D and 3D airfoils, discussed in Chap. 4 (Fig. 2.31), the results are proved to be consistent as external conditions remain constant. This approach is rarely used for wind energy applications, because as LWC and MVD are not routinely measured and not immediately put into relation to usual data (i.e. relative humidity). Additionally, the runs are quite long and the rotating frame of the blade is not always considered, with the consequence that the amount of water collected is widely underestimated. Furthermore, only ice accretion is simulated, melting and sublimation effects are not simulated so far. Therefore, none of these ice accretion models has yet been coupled with the results of numerical weather models or with other time series of meteorological data during an icing event. An other deficiency arise from the fact that only short periods of ice accretion can be simulated with the 2D and 3D models (a few minutes). Longer model runs (2–6 h), typical of storm events, would lead to unrealistic results if variable operating conditions of the rotor are not considered during the ice event (rotational speed, vibration, etc.) leading to a break off of ice.

Combination of WNM with IGM alone are still not sufficient for a reliable forecast of ice risk and to select the proper ice mitigation strategy. A further module assessing the effect of ice accretion on the blades and sensors and therefore on power curve (potential energy yield losses), and onto structures (additional fatigue, unbalance, etc.) is needed (see Fig. 2.32).

In order to quickly run the conjugate model, a reduced parameter icing model (RPIM) would be more efficient. An example is given in Chap. 5. The output of this model are the synthetic variables as the ice Contamination Levels (CL) (mass,

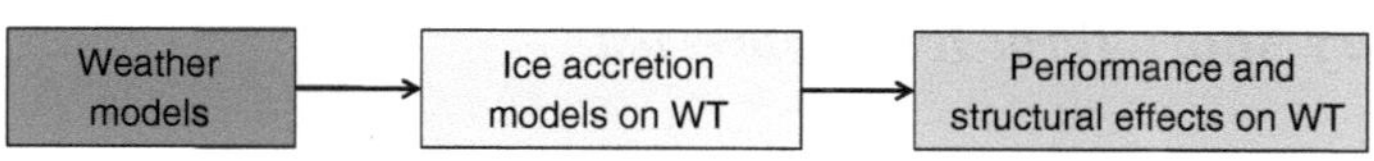

Fig. 2.32 Full ice forecast model on WT

EVENT FREQUENCY LEVEL - EFL	CONTAMINATION LEVEL - CL: NEAR ZERO (CL0)	MODERATE (CL1)	MEDIUM (CL2)	SEVERE (CL3)
N. ZERO (EFL0)	$D < D_{critical}$ $E_{IPS} = 0$ $\Delta E = 0$	$D < D_{critical}$ E_{IPS} = scenario ΔE = scenario	$D < D_{critical}$ E_{IPS} = scenario ΔE = scenario	$D < D_{critical}$ E_{IPS} = scenario ΔE = scenario
MODERATE (EFL1)	$D < D_{critical}$ E_{IPS} = scenario ΔE = scenario	$D < D_{critical}$ E_{IPS} = scenario ΔE = scenario	$D > D_{critical}$ E_{IPS} = scenario ΔE = scenario	$D > D_{critical}$ E_{IPS} = scenario ΔE = scenario
MEDIUM (EFL2)	$D < D_{critical}$ E_{IPS} = scenario ΔE = scenario	$D > D_{critical}$ E_{IPS} = scenario ΔE = scenario	$D > D_{critical}$ E_{IPS} = scenario ΔE = scenario	$D > D_{critical}$ $E_{IPS} = 0$ WT stop ΔE = scenario
SEVERE (EFL3)	$D < D_{critical}$ E_{IPS} = scenario ΔE = scenario	$D > D_{critical}$ E_{IPS} = scenario ΔE = scenario	$D > D_{critical}$ E_{IPS} = scenario ΔE = scenario	$D > D_{critical}$ $E_{IPS} = 0$ WT stop ΔE = scenario

Fig. 2.33 Damage level matrix is prebuilt (DLM), based on pre-evaluated rotor Contamination Levels (CL) and Event Frequency Level (EFL)

thickness and distribution of ice on the blades) and the Event Frequency Level (EFL). The knowledge of CL and EFL allow building up a Damage Level Matrix (DLM) as shown in Fig. 2.33.

The iced rotor power curve is computed by using profile aerodynamic database generated for the iced surface at different contamination levels and the resulting damage levels. Icing persistence is also considered. At this stage, a decision can be made on the basis of economic consideration (assessment of the structural integrity and life reduction of wind turbine components, IPS energy requirement, power curve drop). A possible logic procedure is shown in Fig. 2.34. It will deliver:

- probability of risk of occurrence of icing,
- ice loads (and structural damages) and anti-icing heat fluxes,
- penalties on power curve and energy yield drop,
- operation strategy (immediate stop if iced turbines, ice tolerating operation, etc.).

2.8.2 Very Short-Term Icing Forecast

Very short-term icing forecast can be obtained by on-time measurements of the meteorological conditions for a less than one hour temporal horizon. The aim is typically

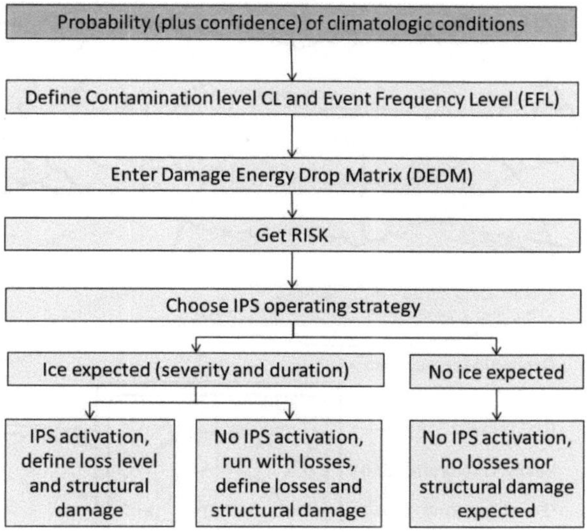

Fig. 2.34 Logic procedure for the ice prevention strategy assessment

to gather information for the actuation of the ice prevention systems. The method can be based on icing remote-sensing systems used in the aeronautical field. An icing remote-sensing system for wind turbines should be designed to monitor the environment, process the information collected and deliver it in useful form for the decision process, either at the farm or at a central control centre. In particular, it should detect conditions conducive to icing, including cloud and precipitation liquid water content, the droplet size spectrum and temperature. This can be done by scanning the airspace ahead of the turbine either from the ground stations or from the turbine. Radar and microwave radiometers are currently the most viable technologies. Ranging capability makes radar an attractive technology for detecting LWC, droplet size and possibly also temperature. Although radar can be used in a variety of directions, scanning vertically from the ground or horizontally from the air, its size, weight and power demands make ground-based radar a more viable near-term technology. Microwave radiometry is less well developed than radar technology, but the recent introduction of a radiometer that scans and profiles temperature, water vapour and cloud liquid water, and tests with methods using radiometers in a horizontal sensing mode (as well as the more traditional vertical or near-vertical modes) are promising [33]. Existing ground-based systems (airports) could provide an important basis for retrieving the land meteorological data needed, and together with the characteristics of the wind turbine, needed to build a DLM as shown in Fig. 2.33 of a given site. A possible near- to mid-term scenario is depicted in Fig. 2.35. Nacelle-based remote-sensing systems could provide basic information for the very short-term optimal operation of the farm, thanks to constant monitoring of the environment conditions [34].

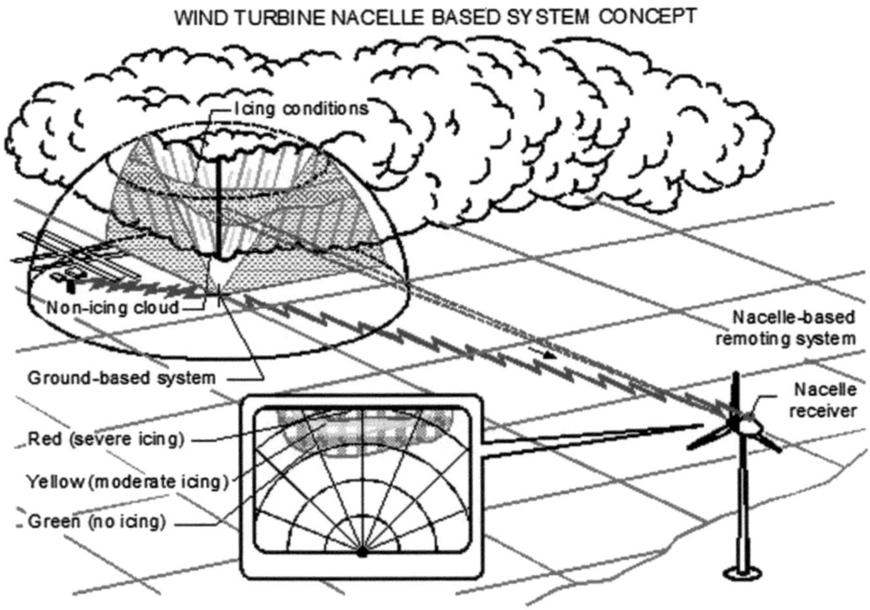

Fig. 2.35 Wind turbine nacelle remote-sensing systems

2.8.3 Evaluation of the Icing Risk on a Site with a Few Information

Table 2.4 by expanding the content of Table 2.1, resumes the list of meteorologic and climatic parameters needed to evaluate the icing risk together with their availability in aeronautical and wind energy fields. The relative humidity is here included as it is a parameter usually available in wind turbine site measuring equipment, although the author considers this parameter of less significance to assess icing severity. Field experiences with very harsh icing conditions were recorded with humidity levels below 70 % (large droplets freezing rain conditions).

Table 2.4 List of main meteorological parameters for input in ice prevention design (a = available, na = not available)

Parameter	LWC (g/m^3)	MVD (m)	V (m/s)	p (Pa)	T (°C)	% Rh (−)	Clouds hor. extent (m)
Aeronautic field	a	a	a	a	a	a	a
Wind energy	na	na	a	a	a	a	na

On the other hand the range of altitudes in which icing can occur is very condition-specific, so extrapolating from aeronautical data produces uncertain results. For mountain sites, interaction with the orographical conditions further complicates the issue.

One reason for the scarce use of these parameters is that characterising ice accretion conditions is expensive, typically requiring aircraft research quality. As a consequence, no quantitative data are available for assessing either icing frequency or icing severity at a given site.

When site developers attempt to assess icing severity, they are faced with the following problems:

1. establishing consistent simultaneous combinations of LWC, MVD, V and T, which can occur at a given site;
2. determining how these combinations will affect the rotor, leading to ice accretion (icing severity);
3. assessing the penalties (in terms of both energy yield losses and components reduced life) associated with icing severity;
4. assessing the energy and power requirement depending on the ice mitigation system adopted (thermal anti/de-icing, mechanical de-icing, etc.).

The following method, developed at the Fluid Machinery Laboratory of the University of Trento [19], gives a method for a rapid, but reliable estimation of the anti-icing heat requirement and on-board installed power needed to operate ice prevention systems on wind turbines installed in hostile climates. It attempts to overcome the difficulty in assessing the meteorological conditions at a given site, when scarce and incomplete data are available. Using typical probability distribution, the method enables a prediction of the icing frequency and severity at a site in relation to a specific type of wind turbine.

When available observations and recordings of the icing parameters are discontinuous, but sufficient to generate a distribution, the probability of occurrence of an icing event can be evaluated by the aid of the theorem of the compound probability. The probability density that a given potential icing event will take place when $V = V_1$, $T = T_1$, $LWC = LWC_1$, $MVD = MVD_1$ is given by the product of the conditioned density probability p(V) that V will exceed V_1, times the probability density f(T) that T will exceeds T_1 providing that $V > V_1$, times the conditioned probability density f(LWC) that LWC will exceed LWC_1 providing that $V > V_1$ and $T > T_1$, times by the conditioned probability density f(MVD) that MVD will exceed MVD_1 providing that $V > V_1$ and $T > T_1$ and $LWC > LWC_1$. This is expressed as follows:

$$f_1 = f_{V_1}\, f_{T_1}(V)\, f_{LWC_1}(V, T)\, f_{MVD_1}(V, T, LWC) \tag{2.3}$$

It is recognised that some relationships exist between these variables, such as between LWC and droplet diameter, or velocity and temperature distribution. Their practical evaluation is still uncertain for the wind turbine field, however. In the aeronautical field, though it emerged that droplet diameter was found only weakly correlated with either water content or temperature, the FAR regulation [18] provides charts relating LWC and MVD for intermittent and continuous cloud envelopes. But their extension to wind turbine sites is difficult to implement. Observations show that velocity and temperature distributions have a weak dependence as a function of site typology, but general trends are unlikely to emerge. Although the model shown in Eq. (2.3) can handle interdependencies among variables, the assumption is made here that the meteorological variables can be considered as independent. The inaccuracies introduced by this simplification are estimated to be of the same order of magnitude as those arising from the natural measuring uncertainty of meteorological variables. In this light, the composite probability of the icing event becomes the product of the single probabilities. Thus, Eq. (2.3) is written as:

$$f_I = f_{V_I} f_{T_I} f_{LWC_I} f_{MVD_I} \tag{2.4}$$

On the basis of Eq. (2.4), the anti-icing heat flux is given by:

$$\dot{Q} = \dot{Q}\left(\ldots,\ f(V)^{-1},\ f(T)^{-1},\ f(LWC)^{-1},\ f(MWD)^{-1}\right) \tag{2.5}$$

and the yearly anti-icing energy is written as:

$$E_I = 8760 \int_0^{\infty} \int_{-273.15}^{\infty} \int_0^{\infty} \int_0^{\infty} \dot{Q}\, f_V\ f_T\ f_{LWC>0}\ f_{MVD}\, dV\, dT\, dLWC\, dMVD \tag{2.6}$$

where $f_{LWC} > 0$ expresses the probability that LWC acquires values larger than zero. The duration curve of the elementary anti-icing power associated with each icing event is written as:

$$D(\dot{Q}) = 1 - \int_0^{\dot{Q}} f\left(\dot{Q}\right) d\dot{Q} \tag{2.7}$$

and depicted in Fig. 2.36, in non-dimensionalized form. The area under the curve represents the overall anti-icing energy when the ice-free condition is required for any environmental condition, (also the most extreme being considered).

The maximum installed thermal anti-icing power, for instance, having a less than 1 % probability of being exceeded, is given by:

$$\dot{Q}_{max} = \left\{\dot{Q} : P\left(\dot{Q}_{max}\right) < 0.01\right\} \tag{2.8}$$

It is worth bearing in mind that, if complete data from the site are available (see Table 2.4), Eq. (2.8) becomes:

$$\dot{Q}_{max} = (\ldots,\ V_{max},\ T_{min},\ LWC_{max},\ MVD_{max}) \tag{2.9}$$

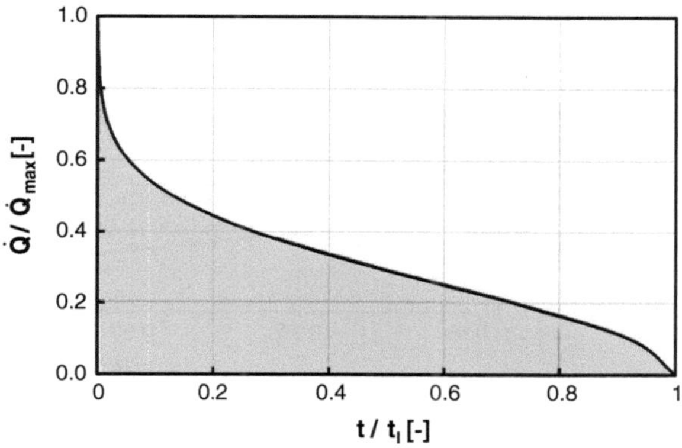

Fig. 2.36 Example of anti-icing thermal power duration curve. *Filled area* represents the anti-icing energy

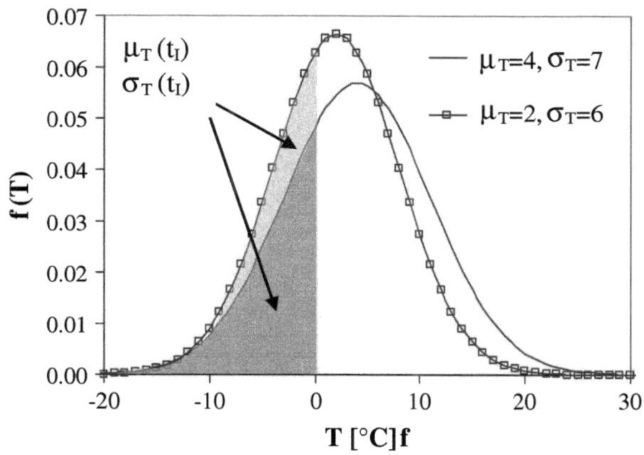

Fig. 2.37 Example of temperature probability density distribution for a trial site

These considerations can be generalised by working with a parameter distribution during icing events instead of discrete data. This approach enables considerations on generic sites.

In Fig. 2.37 a normal distribution is assumed for the temperature by assigning the mean value μ and standard deviation σ, while Weibull distributions are used for V, LWC and MVD, respectively, characterised by the parameters (k_V, c_V), (k_{LWC}, c_{LWC}), (k_{MVD}, c_{MVD}). The distribution of LWC refers to the events characterised by LWC>0. The probability of $f_{LWC} = 0.2$ has been assumed and is assigned on the strength of reasonable assumptions for humid sites and depends on general considerations on site severity. It might be evaluated, for instance, on the basis of visibility according to the method outlined in [31].

A family of site conditions was randomly generated and Eqs. (2.6) and (2.8) were solved using a Monte Carlo approach [35].

An exercise was conducted using the above procedure for a test case of a three-blade, full-span, pitch-controlled and full-variable speed MW-class wind turbine. A running wet ice prevention strategy is adopted and the characteristics of the main

Table 2.5 Main characteristics of wind turbine and anti-ice system

P_R (kW)	D (m)	Z (−)	Ω (rpm)	V_R (m/s)	Profiles (−)	$T_{s,min}$ (°C)	ε_{IPS} (−)	A (m^2)
1,200	62	3	22	13	NACA632-4xx	0.5	1	7.5

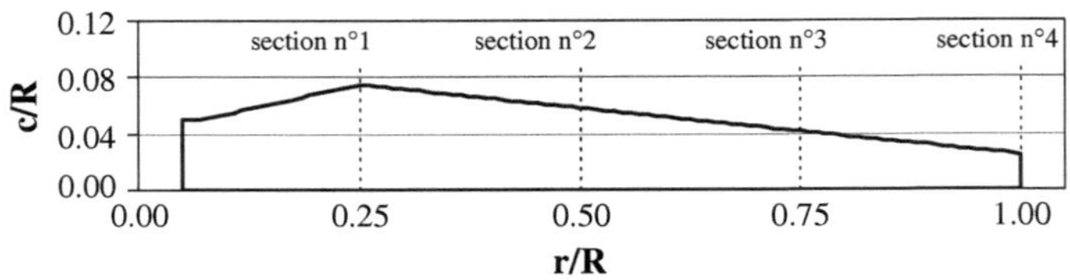

Fig. 2.38 Blade planform

Table 2.6 Samples of input (distribution parameters) and output (energy E_I and maximum thermal power $\dot{Q}_{max}$) for the blade described in Table 2.5 and Fig. 2.38

Case	μ_T (°C)	σ_T (°C)	c_V (m/s)	k_V (−)	c_{LWC} (g/m^3)	k_{LWC} (−)	c_{MVD} (μm)	k_{MVD} (−)	t_I (h)	E_I (kWh)	$\dot{Q}_{max}$ 99 % t_{ice} (kW)
1	6	6	7.5	2.5	0.6	1.5	30	2	288	2,273	28.0
2	0	7	6	2	0.9	2	25	2	876	11,582	40.8
3	2	7	9	2	0.9	2	22.2	1.5	679	10,318	42.3
4	4.3	7.9	5.7	1.8	0.8	1.5	20	1.8	518	6,101	39.2
5	8	4	7.9	1.5	0.7	2	30	2	40	180	16.5
6	…	…	…	…	…	…	…	…	…	…	…
…	…	…	…	…	…	…	…	…	…	…	…
N	…	…	…	…	…	…	…	…	…	…	…

system are listed in Table 2.5, according to the blade planform in Fig. 2.38. Table 2.6 lists typical data samples used as input for the simulation and the results obtained using Eqs. (2.6) and (2.8) in terms of required anti-icing power, icing time and power.

The data in simulation number 3 of Table 2.6 were used to obtain the ice frequency intensity charts of Fig. 2.39. Number of hours of direct icing and anti-icing energy E_I are given as a function of air temperature and wind speed. The graphs stress that the same number of icing hours can occur for the same temperature at different wind velocities. The maximum anti-icing energy requirements does not necessarily happen for the longest icing time due to the complex effect of icing parameters on both heat flux and icing duration.

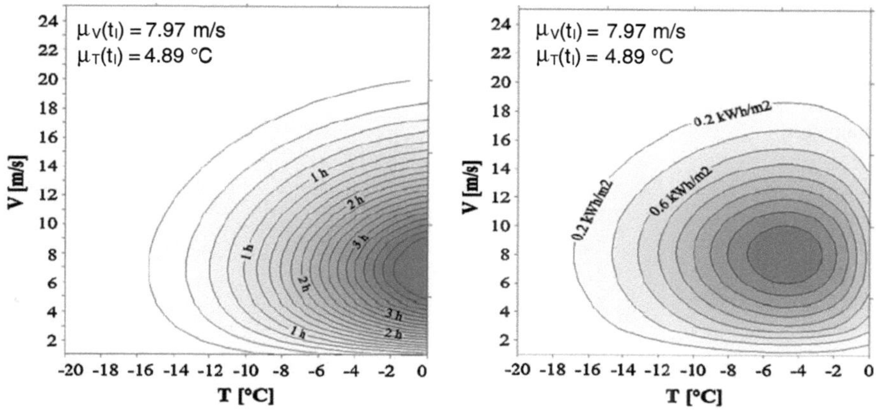

Fig. 2.39 Icing frequency (hours) and ice intensity (kW/m²) charts

2.8.3.1 Anti-icing Energy Requirements

The anti-icing energy can be determined using Eq. (2.6) only if the distribution of the icing parameters is known throughout the icing period. Since this is not the case for most sites, a simplified approach was investigated using discrete representative values of the above parameters to calculate both the anti-icing power and the energy. By adopting the mean values $\overline{V}$, $\overline{T}$, $\overline{LWC}$, $\overline{MVD}$ of the icing period, Eq. (2.6) is transformed into Eq. (2.10):

$$E_I = 8760\,\dot{Q}\left(\ldots, \overline{V}, \overline{T}, \overline{LWC}, \overline{MVD}\right) \int_{\infty}^{0} p\,(T)\; p_{LWC>0}\, dT \tag{2.10}$$

The solution provided by Eq. (2.10) is compared with that of Eq. (2.6) to evaluate the error introduced. The results, expressed as a percentage, are given in Fig. 2.40. An average overestimation of 5 % arises in the velocity range 5–10 m/s and in the temperature range −1 to −7 °C. This result is of practical importance for the preliminary evaluation of site ice severity. Site temperature appeared to be the most important variable. This parameter is often collected also by non-institutional observatories, and the average value approaches in most cases statistical significance.

2.8.3.2 Choice of Optimal Anti-icing Installed Power

As it will be evident in the discussion of Chaps. 4 and 5, designing IPSs based on the maximum thermal power demand would lead to high investment costs and a very low IPS load factor. The system would operate at maximum anti-ice power only for a few icing events a year and for some anti-icing technologies, such as the one based on hot air circulation inside the blade, this power level would also not be affordable.

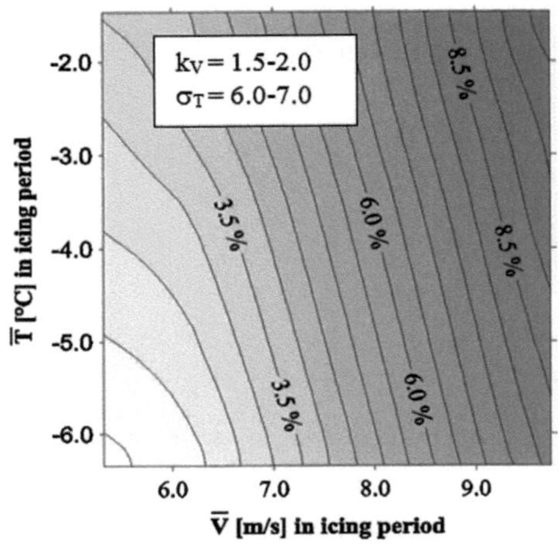

Fig. 2.40 Percentage error in anti-icing energy estimation when mean values are adopted for icing parameters

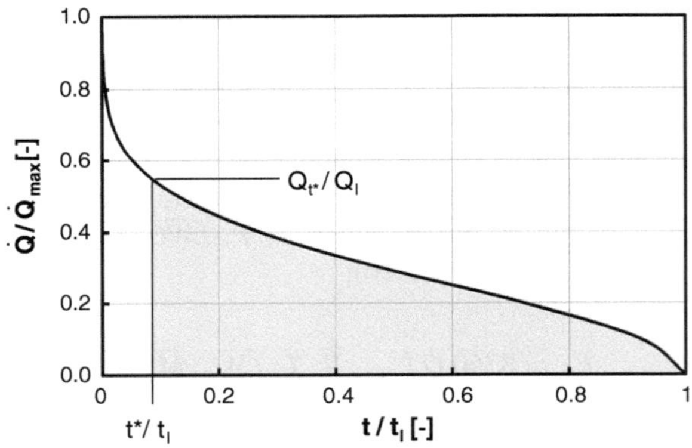

Fig. 2.41 Optimal anti-icing power level and energy

Moreover, this technology could prove ineffectual against very severe ice events. If a lower power capacity is selected, less than maximum power required, the duration curve of Fig. 2.41 shows that the power reduction $\Delta\dot{Q} = \dot{Q}_{max} - \dot{Q}_{t*}$ leads to a corresponding energy saving of $\Delta E^* = E_I - E_{t*}$. This implies that there is a time period t^*, during which the rotor is not protected against ice accretion. The turbine would have to be shut down, and no energy would be delivered to the grid during that period. Let us design $\Delta\dot{Q}$, ΔE^* e t^* as economic quantities used for the IPS break-even analysis.

Now, the standard deviation (accounting for the data dispersion around the mean) is relevant information, linking the energy and power dispersion to the mean value. The standard deviation is always associated with data series and can also be estimated when a historical data series is incomplete.

With reference to an arbitrary icing risk period t^*, shorter than the actual t_I, the reduced power $\dot{Q}_{t^*}$ is evaluated by the following expression, which replaces the previously shown Eq. (2.9):

$$Q_{max} = \left[\ldots,\ \overline{V} + \sigma(V),\ \overline{T} + \sigma(T),\ \overline{LWC} + \sigma(LWC),\ \overline{MVD} + \sigma(MVD)\right] \tag{2.11}$$

Icing parameters are expressed as a function of the means $\overline{V}$, $\overline{T}$, $\overline{LWC}$, $\overline{MVD}$, the respective standard deviations $\sigma(V)$, $\sigma(T)$, $\sigma(LWC)$, $\sigma(MVD)$. The novel *icing strength parameter* K is introduced here.

The following procedure allows to obtain a relationship between $\Delta\dot{Q}^*$, ΔE^* and t^* with the aid of the data in Tables 2.5 and 2.6. Trial case 1 of Table 2.6 is considered first. After guessing a value for K (=0 ± 1.4), Q_{t^*} is calculated using Eq. (2.11). Entering the Fig. 2.41 relating to case 1 with said value of $\dot{Q}_{t^*}$, gives us the corresponding t^* and E_t. If this procedure is repeated for K values from 0 to 1.4, and for a multitude of cases treated using a Monte Carlo procedure, the graphs of Fig. 2.42 are obtained. The quantity $(1 - t^*/t_I)$ represents the fraction of the icing period during which the IPS will protect the rotor, while $(1 - E_{t*}/E_I)$ represents the ratio between the energy saved by operating the IPS at a power $\dot{Q}_{t^*}$ less than $\dot{Q}_{max}$, for the reduced time period $(t_I - t^*)$. As the parameter K increases, the duration of ice protection and the energy required asymptotically approach t_I and E_I. Using values for the icing strength parameter K greater than about 1.4 does not add further information.

A possible example of the use of these graphs is given below. Let us suppose we have to install a wind turbine with the characteristics indicated in Table 2.5 and Fig. 2.38, using the sample case of Table 2.6.

The anti-icing energy is calculated using Eq. (2.10). If the maximum period when the IPS is unavailable is specified, i.e. $t^*/t_I = 0.2$, graph a of Fig. 2.42 indicates that K ranges between 0.4 and 0.7. Assuming that K = 0.55, the corresponding power $\dot{Q}_{t^*}$ is deduced from Eq. (2.11). Again, by entering K = 0.55 in graph b of Fig. 2.42, the ratio E_{t^*}/E_I furnishes a value ranging between 58 and 61 %. If we assume that $E_{t^*}/E_I = 0.59$, the actual predicted energy consumption for 80 % of the icing time is easily calculated. These figures can be associated with the investment costs of the installed power, the running costs for supplying the required anti-icing energy and the cost of the turbine's unavailability during the shutdown period t^*.

The method outlined allows linking the choice of a power level lower than the maximum (less than 1 % probability of being exceeded) with the duration of operation without anti-ice protection and the corresponding energy requirement. In view of the number and complexity of the possible sources of error, it is not possible to make a reliable estimate of the overall accuracy of the outlined procedure. The reliability of the probability analysis reported herein is affected by three factors, namely: (a) errors in icing parameters measurements; (b) amount and representativeness of available data; (c) simplification of the thermal model. In addition, if sites meteorological are not routinely measured, (d) limitations in extrapolation with respect to climate and

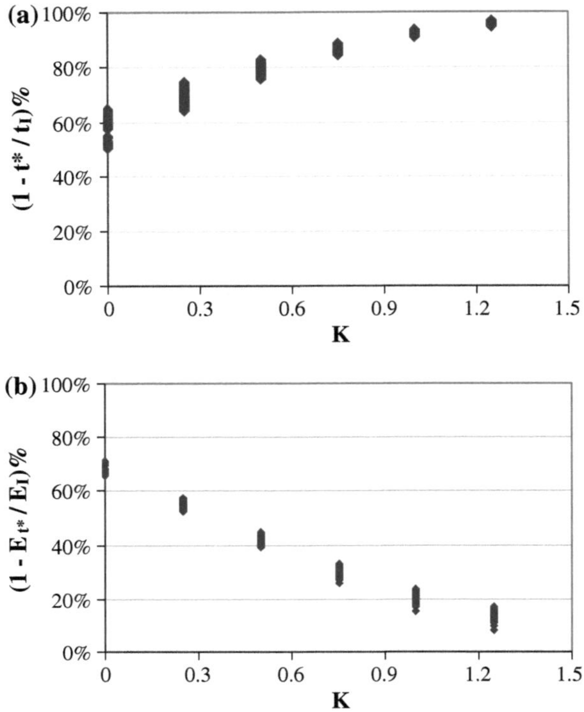

Fig. 2.42 Percentage duration of IPS protection $(1 - t^*/t_I)$ and % savings of the required energy $(1 - E_{t^*}/E_I)$ as a function of the parameter K

altitude. Actual field observations could be of great help in refining the method. The overall accuracy of the presented energy assessment method itself is estimated to be within 10–15 %.

2.9 Ice Throw and Icing Risk

The icing of wind turbine components, mainly the blades, affects not only the system performance because of ice accretion on blades, but also the safety due to ice shedding. Sudden detachment of ice pieces having different weights and shapes can occur from the tower, nacelle and rotor blades surfaces during standstill, idling and operation. Regarding the ice accreted onto the rotor blades, the detachment of ice pieces can be caused by natural conditions or de-icing operations. When protection from icing is accomplished by means of cyclical de-icing, some ice is permitted to form on the blades surface; during the turbine operation, the accreted ice is then periodically removed through short intensive warming or mechanical actions [36]. Gravitational, aerodynamic and centrifugal forces project the detached ice fragments away (see Fig. 2.43).

Fig. 2.43 Picture of ice shedding from a turbine rotor [4]

The area on the ground that is covered by the ice rainfall around the wind turbine is dependent, among other parameters, on the wind strength and direction, and on the mass and the size of the removed ice. Safety reasons for people and objects adjacent to the turbine, economical motivations related to the de-icing system operating cost, and structural unbalance dynamics are important issues related to those variables. Since the de-icing practice has the potential of controlling the size of the ice pieces that are periodically detached, different risk scenarios can be produced through the control of the de-icing strategy.

Literature on ice shedding from wind turbines appears to be rather lack: observations and measurements on ice fragments found on the ground around wind turbines were collected as part of the WECO project [3]. Frank and Seifert [37] performed an experimental campaign to investigate the aerodynamic forces on iced airfoils and ice fragments themselves. On the basis of these results, a first ice throw numerical model was presented by Morgan et al. [38]. The authors determine the safety distance for the chosen level of allowable risk for a three-bladed turbine having a diameter of 50 m, as reported in Fig. 2.44. The results were based on a rate of ice accretion averaging 75 kg/day during icing conditions, with the recommendation that allowable risk should be scaled pro rata under different assumptions.

This work was followed by another work from Seifert et al. [39], in which maps of ice hits probability on the turbine vicinity were proposed. The observed ice fragments from the WECO data base [3] and own additional data were collected by DEWI [39]. The results are given in Fig. 2.45. The size of the ice fragments is estimated here. Visual observations indicated that the ice fragments do not hit the ground as long slender parts but break off immediately after detaching from the blade into small fragments.

On the basis of the experience gathered in the WECO Program, some ice throw calculation allowed to define a graphic representing the possible hits on the ground. The trajectories, computed on the basis of the ice fragment mass and lift and drag, were combined with the topographical map of the site concerned. The typical ellipsoidal curves in the steps of wind speed depicted in Fig. 2.46 have been obtained, circumscribing the risk area on a map. Note that several maps will originate, one for each wind direction.

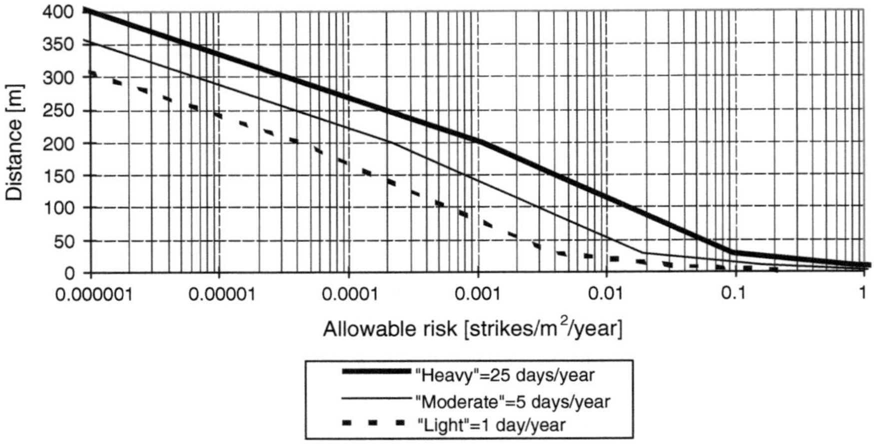

Fig. 2.44 Safety distance for different icing levels (50 m rotor) [38]

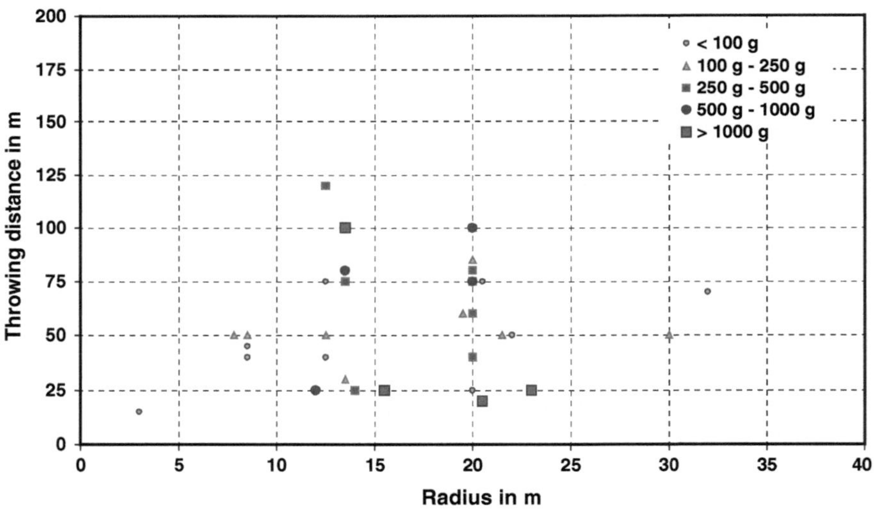

Fig. 2.45 Observed ice fragments from the WECO database and own additional data [39]

A simplified empirical equation has been introduced in WECO project to map a risk circle without detailed calculations to determine a safety circle around the turbine representing the overall risk area. The following empirical equation resulted:

$$4d = (D + H) \cdot 1.5$$

where d is maximum throwing distance, D is the rotor diameter and H is the turbine hub height. All dimensions are given in metres.

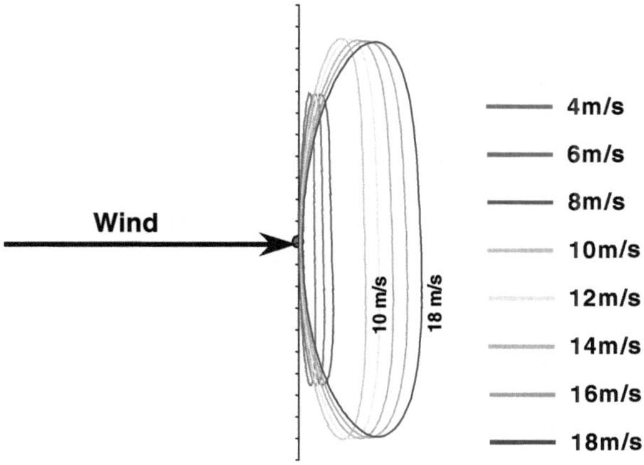

Fig. 2.46 Result of the ice throw calculation. The curves represent the worst case width per wind speed [39]

Further analysis of the ice fall from a wind turbine at standstill, leads to the graph of Fig. 2.47. The overall falling trajectories for different, wind speeds and rotor positions, and one ice fragment mass is shown.

The observations allowed to determine a simplified empirical equation valid for a still standing turbine:

$$D = V_{hub}\frac{D/2 + H}{15}$$

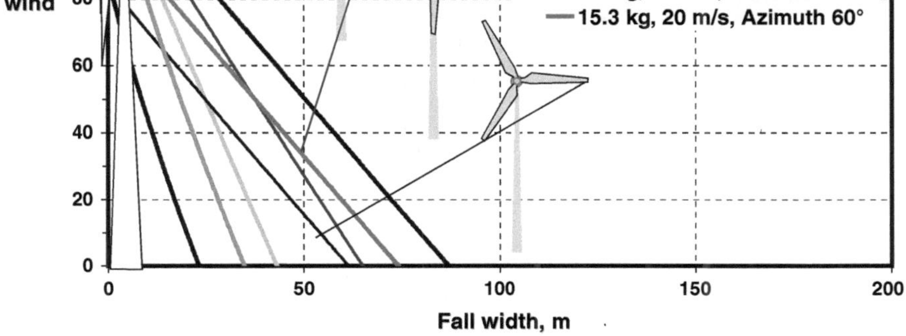

Fig. 2.47 Typical result of an ice fall width calculation for a turbine at standstill. Parameters: wind speed, rotor position and size of ice fragments [39]

where V_{hub} is wind speed at hub height (in m/s), d is the maximum falling distance (in m) and the other parameters are the same as defined before.

Maybe the only known study of ice throw associated wind turbine with integrated blade IPS was carried out in 2004, on a 600 kW Enercon E-40 (diameter 40 m, and hog height of 46 m) installed on the Gtsch mountain, in Switzerland, at 2,300 m a.s.l. since the turbine is located close to ski slopes, ice throw was considered as an important safety issue. Four winter seasons from 2005 and 2009 were matter of observation and the results published [40, 41]. Size and weight and distances from turbine of the collected fragments were ground-mapped and photographed to form a database, but no distinction between ice fall and ice throw was made in the study. One of the resulting maps is given in Fig. 2.48 [42] showing the distribution of the ice fragments around the wind turbine.

More than 250 fragments were recorded, even during summer time. The maximum distance was 92 m, the maximum weight 1.8 kg. The theoretical maximum distance according to Seifert empirical formula [39] of 135 m was not reached so far. The figure confirms the dependency of the ice throw on the prevailing wind conditions during icing events.

The following model performs a risk assessment relative to the ice shedding from wind turbines, integrated with the rotor blade de-icing system design. The investigation was performed applying a Monte Carlo method to the ballistic model

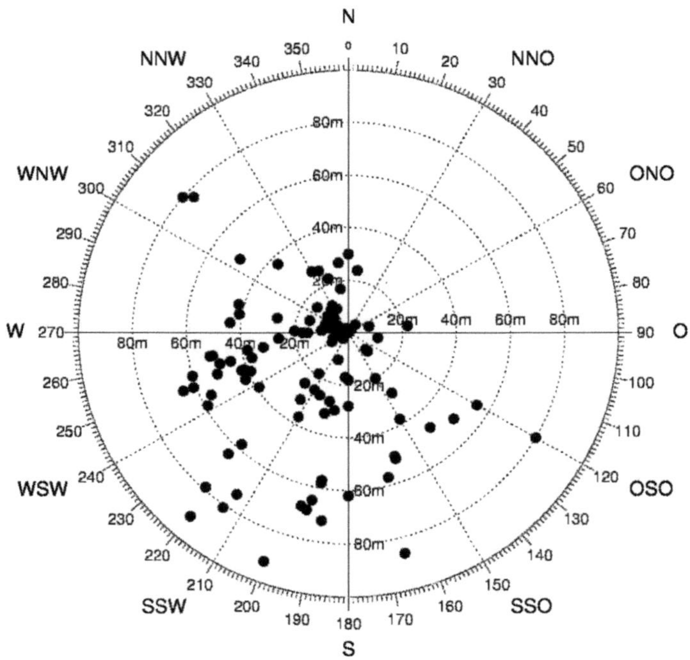

Fig. 2.48 Distribution of the ice fragments around the wind turbine [42]

of the ice pieces, which solves the ice pieces trajectory. Suitable density functions were set for each of the input parameters to the trajectory equation. The output fragments distribution on the ground was presented in terms of strikes probability and recurrence period, as a function of the input parameters: for this purpose, a parametric analysis was performed varying ice pieces mass and shape.

The trajectory of the ice throw can be evaluated by means of the second law of dynamics as shown in Eq. (2.12) written in vectorial notation as:

$$m \cdot \ddot{x} = m \cdot \bar{g} + \bar{D} + \bar{L} \tag{2.12}$$

The relationship can account for an ice piece of arbitrary mass and shape that is subject to mass and surface forces. Gravity, drag and lift forces on the ice body were considered.

Figure 2.49 shows the vectorial representation of the velocities and the forces. Drag and lift are computed as:

$$L = \frac{1}{2}\rho C_L A \bar{W}^2 \tag{2.13}$$

$$D = \frac{1}{2}\rho C_D A \bar{W}^2 \tag{2.14}$$

Being A the piece aerodynamic surface and w the relative velocity calculated according to Eq. (2.15)

$$\bar{W} = \bar{V} - \dot{x} \tag{2.15}$$

The differential Eq. (2.12) can be solved once initial position and velocity of the ice piece close to detach are specified. With reference to Fig. 2.50, the sets of Eqs. (2.17) and (2.18) establish the initial conditions relative to each of the three space components x, y and z:
for x_{in}:

$$\begin{aligned} i - component &: \ r \cos\theta \cos\gamma \\ j - component &: \ r \cos\theta \sin\gamma \\ k - component &: \ r \sin\theta + H_{tower} \end{aligned} \tag{2.16}$$

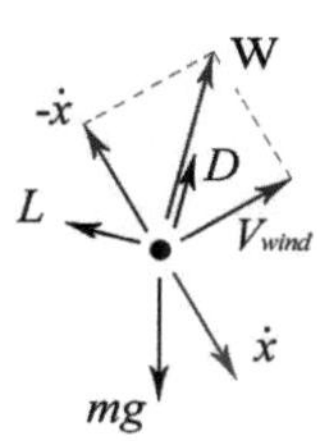

Fig. 2.49 Vectorial representation of the velocity and force fields

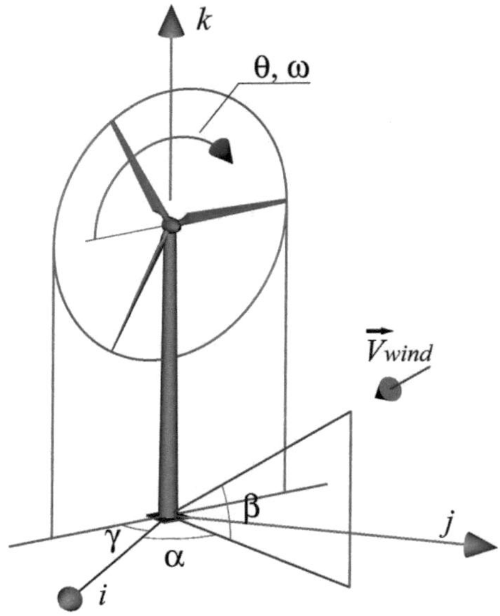

Fig. 2.50 Wind/turbine relative angles

for $\dot{x}_{in}$:

$$\begin{aligned} &i - component: \quad -r\ \sin\theta\ \cos\gamma\cdot\omega \\ &j - component: \quad -r\ \sin\theta\ \sin\gamma\cdot\omega \\ &k - component: \quad r\ \cos\theta\cdot\omega \end{aligned} \tag{2.17}$$

Equations (2.17) and (2.18) were stated assuming that:

- the rotor plane is always orthogonal to the wind speed direction (neither gusts nor transitory operation were considered): $\gamma = \alpha - \pi/2$;
- the blades rotation is clockwise, with reference to the wind approach direction;
- the rotor tilt has a negligible effect on the ice throw (and is therefore disregarded);
- the rotor hub centre lies on the tower longitudinal axis;
- the blade elasticity effect on the initial ice fragment acceleration is disregarded.

The input variables can be grouped into turbine, wind-and-site and ice fragments characteristics, as shown in Fig. 2.51.

During the wind turbine operation, these parameters can assume a large range of values due to variable wind speed and direction, and to the consequent adjustment of the turbine working conditions. Ice pieces with a variety of masses and shapes can shed from a generic position (r, θ) on the rotor, leading to different lift and drag coefficients for each ice piece. Accordingly, the fragments follow a wide range of trajectories; the most rational approach to address this huge range of variables combination is to use the Monte Carlo method. It allows to generate distributions of the input variables, allowing the computation of a number of trajectories. In this way, the landing position and velocity of each considered ice fragment was recorded for a

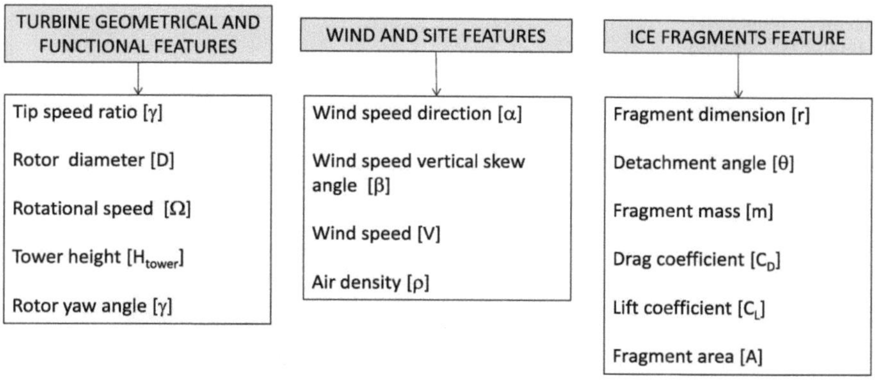

Fig. 2.51 Variables used for the simulation

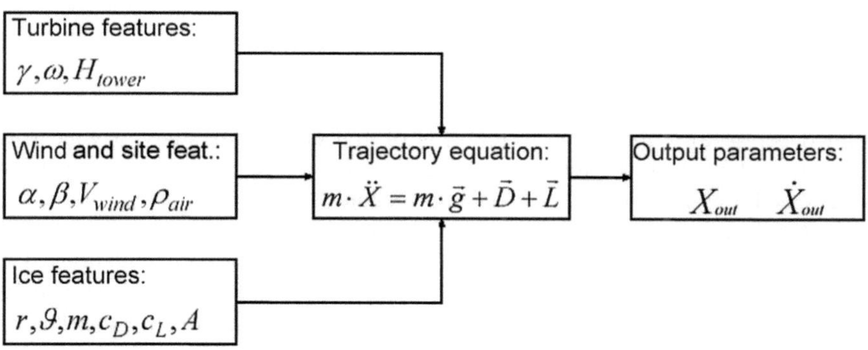

Fig. 2.52 Monte Carlo procedure for the trajectories computation

variety of possible combinations of the input parameters. Figure 2.52 shows a scheme of the computation process. Between 10^4 and 10^5 samples were used in each simulation in order to have a sufficiently large database and therefore obtaining repeatable results. In the following the distributions of the input variables are addressed.

2.9.1 Site Parameters

The area around the wind turbine is discretised in radial sectors of given angular amplitude; a wind rose diagram is then evaluated [43] that shows the temporal distribution of the longitudinal α and azimuthal β directions of wind speed for each sector (see Fig. 2.48). The reference time for the wind rose diagram evaluation can be seasonal or yearly based.

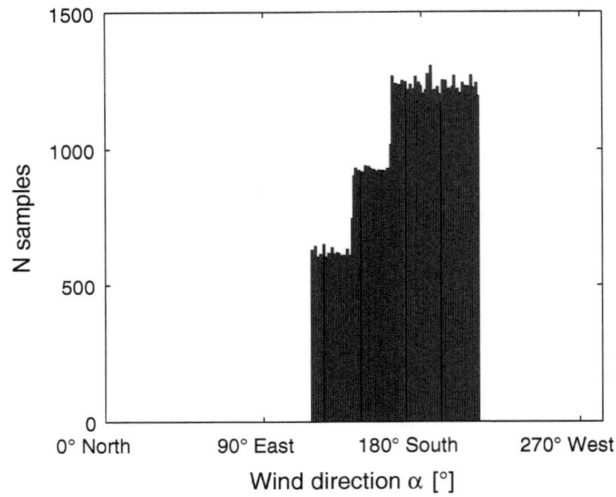

Fig. 2.53 Icing wind diagram; 10^5 samples

Following this routine, the wind direction (α and β) density functions can be defined per sectors. Since ice accretion is of concern in this study, a wind rose diagram evaluated on the total year icing time has to be considered: only wind directions during ice periods are taken into account.

A generic site was considered with an icing wind rose diagram spread over 120° and radial 30° sectors as shown in Fig. 2.53 was selected for the simulations. The wind direction density functions were created imposing rectangular density functions within each sector, in such a way that the maximum of the wind direction distribution function results equal to 1. Figure 2.53 shows the outcome of the extraction of 10^5 random numbers from the density function of wind direction α. Even though the trajectory model was developed to account for the wind azimuthal angle β, a flat terrain was considered for the simulations in the following; β equal to zero was therefore employed.

For each sector of the icing wind rose diagram, a Weibull wind speed distribution was imposed. The shape and scale parameters of the Weibull function in every sector were set in order to obtain, at the blade hub, average wind speeds between 6.5 and 7 m/s allover the 120° of the rose diagram. The maximum acceptable wind speed was fixed at 50 m/s, distinctive of the class I wind turbines (see IEC 61400 standards). The wind shear was modelled according to Eq. (2.18):

$$V(x, y, z) = k_t(x, y, z) V_{hub} \left(\frac{z}{V_{hub}} \right)^{\delta} + c_s \tag{2.18}$$

The power law value δ was chosen equal to 0.2 corresponding to a wind blowing on a flat terrain with some houses and hedgerows higher than 8 metres [43]. The k_t and c_S parameters account for local distortions of the wind speed profile (due to the tower shadow, additional shear wind and gusts); the simulations presented in the following are obtained ignoring the last two effects.

The blades rotational speed is directly inferred from the wind speed distributions, once the tip speed ratio TSR is stated for each turbine working condition. Since blades cyclic de-icing is considered in this work, it was assumed that the limited ice accretion allowed onto the blades has negligible effect on the wind turbine performance; thus TSR is the same during operation in wet as well as in dry conditions. For the present analysis it was considered the case of full-variable wind turbines with constant TSR in the range of wind velocities between zero and the rated speed. Constant rotor turning speed was set between the rated and the cut-off ($V_{cut-off} = 25\,m/s$) wind velocity.

2.9.2 Ice Fragments Mass

Available data about the mass of the ice fragments cast from the rotor blades are extremely incomplete. They are mainly due to the work of Seifert [38, 39], who records a limited number of ice pieces collected on the ground around wind turbines. Estimates of their mass indicate them to be in the range 0.1–1 kg. The mass evaluation of ice pieces shedding from the rotor when a cyclic de-icing system is adopted is absolutely arbitrary, since de-icing is a novel practice on wind turbines. Producing a mass density function of the shedding ice fragments is therefore a hard rehearsal. As a starting point, open to improvement, a Weibull-like density function was employed to describe the ice pieces mass distribution. A first simulation campaign was carried out setting the Weibull shape and scale parameters to values that allow obtaining a mean cast mass equal to the one of the Seifert data (0.36 kg), and to force 95 % of the computed samples in the range 0–1 kg (see Fig. 2.54). A parametric analysis was then carried out varying the mean of the mass density function. The maximum considered fragment mass was set to 25 kg, since this is estimated to be the maximum ice accretion mass on a MW-class wind turbine rotor blade, during an average icing day. The ice fragment thickness was varied between 1 and 9 cm, while tests were performed with ice pieces average mass of 0.18 and 0.36 kg.

2.9.3 Detachment Radius and Azimuthal Angle Distributions

The detachment radius density function was evaluated assuming that:

- the mass distribution of the cast ice fragments is unique and independent on the fragment separation blade spanwise position on the blades;
- the number of ice fragments is proportional to the mass of ice growing at a given spanwise position along the blades.

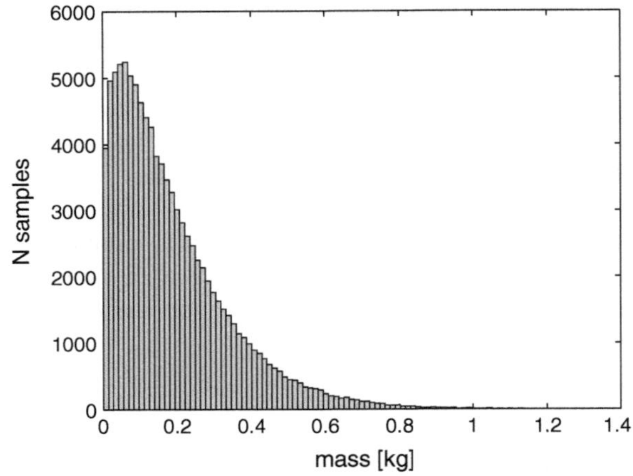

Fig. 2.54 Ice pieces mass density function; 10^5 samples

Experimental observations show that the mass of the ice growth onto the blades increases linearly with the distance from the hub. Moreover, depending on the blade size, part of the blade, towards the hub, may remain free of ice. Typically, if a MW-class wind turbine is considered, the inner 1/3, 1/2 of the blade span is not affected by ice formations. Therefore, the probability of a fragment to detach from the rotor at a given radius can be expressed through the next relationship:

$$f(r) = k_c \cdot (r - R_{min}) \tag{2.19}$$

The constant k_c can be computed assuming that f(r) is a density function, which related distribution function must have a maximum at $r = R_b$ equal to 1:

$$f(r) = \int_{R_{min}}^{R_b} k_c \cdot (r - R_{min}) dr = 1 \tag{2.20}$$

Once k_c is determined, the detachment radius density function of equation (2.19) is used to generate a set of input detachment radius values. Figure 2.55 shows the outcome of the extraction of random numbers from the density function relative to a 25 m long blade, with the assumption that the inner 1/3 of the blade is free of ice.

The probability of a fragment to be released from the rotor at a given azimuthal angle θ is proportional to the force produced on the ice piece. If aerodynamic forces produce about the same action on the ice accretion for every θ, the gravity and centrifugal forces control the θ ice detachment probability. Again, the mass distribution of the ice fragments is independent on the separation azimuthal location; therefore, the instantaneous acceleration can be considered, instead of the force, to calculate the detachment azimuthal angle density function:

$$f(\theta) = k_c \cdot |a| k_c [(\omega^2 r \cos\theta)^2 + (\omega^2 r \sin\theta - g)^2]^{\frac{1}{2}} \tag{2.21}$$

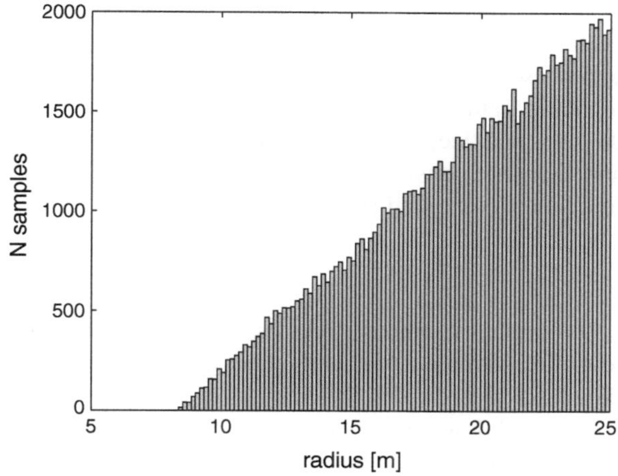

Fig. 2.55 Detachment spanwise position; 10^5 samples

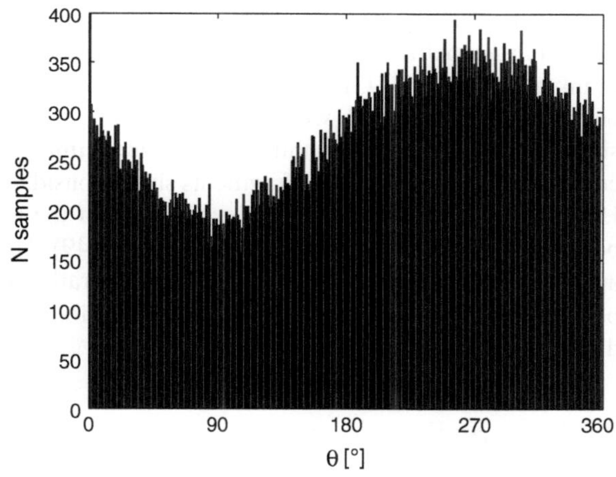

Fig. 2.56 Detachment azimuthal position; 10^5 samples

The unknown k_c might be computed as described above. The ω and r datasets previously generated are used in Eq. (2.21). Figure 2.56 shows that the ice fragments are more likely released when the blade is facing the ground ($\theta = 270°$); on the contrary, a minimum probability of ice removal is obtained when the blade is upward ($\theta = 90°$).

2.9.4 Drag and Lift Distributions

Once ρ_{air} is set, drag and lift distributions are generated through the definition of the density functions relative to the surface area of the ice fragments (A) and to the drag and lift coefficients (C_D, C_L). The density function of the surface area A can be evaluated from the mass density function of the ice fragments with the simplifying hypotheses that the ice pieces have all the same specific weight and the shape of regular parallelepiped. The thickness of the cast slabs is a control parameter of the de-icing practice; therefore it can be decided, to a certain extent, on the basis of safety and energetic considerations (see foe more detail Chap. 5). The surface area density function is computed as:

$$f(A) = \frac{f(m)}{\rho_{ice}\, t_h} \tag{2.22}$$

A parametric analysis was carried out varying the ice pieces thickness between few millimetres and typical thicknesses of ice fragments found on the ground (natural de-icing).

Lift and drag coefficients vary along the fragments trajectory; the ice pieces are in fact subject to a complex 3D spin, around their centre of gravity, that cannot be computed deterministically. For this reason, constant values of lift and drag coefficients are related to each cast ice fragment, distinctive of the average aerodynamic behaviour all along its trajectory. The evaluation of the density functions for the lift and drag coefficients was made on the basis of experimental data presented by Frank and Seifert [37]. For the investigated ice fragment shape, they found C_D to be in the range $0.5 \div 1.7$, while C_L to be in the range $-0.6 \div 0.6$. Rectangular density functions comprised within the mentioned limits were used for the drag and lift coefficients relative to all the fragments shape considered in the present simulations. However, while the drag lies in the direction of the relative velocity W, the lift is positioned in a plane orthogonal to it (see Fig. 2.49). Therefore, since the body twists and the average lift coefficient is null, the overall effect of the lift on the landing position is reasonably negligible. The contribution of the lift in Eq. (2.12) was thus disregarded.

2.9.5 Ice Strike Occurrence

Finally, the annual number of ice strikes N was evaluated on the basis of the data proposed by Seifert for a "moderate icing" site. The number of icing days per year n_{idy} (5 days/year) was multiplied by the average ice accretion mass per day on the rotor $m_{idy,avg}$ (75 kg/day), and divided by the average mass of the cast ice pieces $m_{ice,avg}$ (0.18 or 0.36 kg):

$$N = \frac{n_{idy}\, m_{idy,avg}}{m_{ice,avg}} \tag{2.23}$$

From this value, the strike probability per year and square metre P_y is computed from the ice fragments distribution on the ground. The strike probability is also evaluated in terms of recurrence period, as:

$$T = (P_y\, S_{cov})^{-1} \tag{2.24}$$

Table 2.7 Turbine geometrical and operating conditions

R (m)	H_{tower} (m)	Ω (RPM)	TSR (−)	ρ_{ice} (kg/m^3)	ρ_{air} (kg/m^3)
25	60	60	5.45	750	1.3

Here S_{cov} represents the surface on the ground covered by the object for which the risk analysis is carried out; in this study, the area covered by a person of about 1 m^2 was considered.

2.9.6 Ice Fragments on the Ground

Numerical simulations were carried out to predict the ice fragments distribution on the ground around a MW-size three-bladed wind turbine. In addition to the above mentioned input density functions, the variables in Table 2.7 were used for the simulations.

2.9.6.1 Standstill Operation

The ice fragments distribution on the ground, computed by considering the standstill condition, was taken as a reference; in fact, it is representative of the current recommended control strategy to stop the rotor in case of blades icing. Two configurations were investigated for the three-bladed turbine, the first with one rotor blade facing upward and the second with one blade facing downward. Ice fragments with average mass of 0.36 kg and 5 cm thick were considered for the simulations.

Figure 2.57 shows the product of the computation with 10^4 shedding events for the upward facing blade configuration; the solution for the downward facing blade configuration is slightly different, although it might be preferred since it could to some extent avoid ice from falling onto the nacelle. The majority of the ice fragments fall below the turbine blades, as they follow the wind direction. The turbine tower position is represented by the black dot. Due to the wind action, a few pieces fall outside the area swept on the ground by the rotor yawing in the prevailing wind direction (from west to south-east, see Fig. 2.53). Apart from an area very close to turbine tower, the strike probability P_y is anywhere lower than 2 strikes/m^2/year. The yellow contour presents a recurrence period of 10 year, meaning that a person or an object that lies on the contour has the average probability of being struck once in ten years. As can be noticed a distance higher than 25 m (0.5 D_r) from the turbine tower is satisfactory to reduce the probability of ice accidents below one in ten years. The above results are in agreement with those exposed by Seifert about the safety distance from wind turbines.

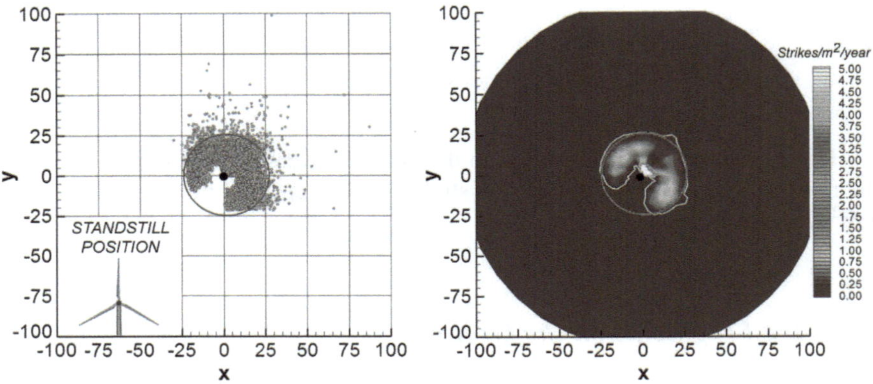

Fig. 2.57 Ice fragments distribution on the ground, and ice fragments strikes probability; standstill condition

2.9.6.2 Operation with Ice Debount/De-Icing

Artificial ice debound and throw during turbine operation were then analysed as a consequence of the de-icing, but results are of general validity; in case of de-icing the allowable ice accretion thickness on the blade surface before its removal depends on the penalties accepted on the blades aerodynamics, the dynamic loads (see Chap. 3), the de-icing energy operating cost (see Chap. 5), and finally on the strike risk. The comparative analysis of these parameters establishes the de-icing system control strategy; different risk scenarios result as a consequence. Simulations were carried out changing the ice fragments thickness for each considered average mass.

Figure 2.58 shows the strike probability in strikes/m^2/year for ice fragments with 0.36 kg average mass, being 1 to 5 cm thick, respectively. Once again, the yellow

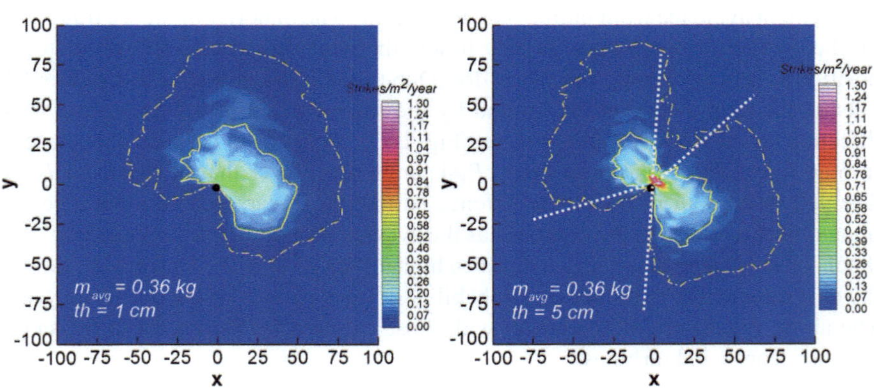

Fig. 2.58 Strike probability; 1 cm thick and 5 cm thick ice fragments

contours represent the recurrence period: 10 years recurrence period is represented with the continuous lines, while the dash-dotted lines are for 50-years period. The comparison of the two figures shows the different ice pieces distribution on the ground. The strike probability is anywhere lower than 0.6 strikes/m^2/year for 1 cm thick fragments, while it is higher than 1.3 strikes/m^2/year for 5 cm thick fragments in the area very close to the turbine. Since 10^4 ice shedding events were considered for both simulations, in the first case the fragments have a more scattered distribution on the ground than in the latter case. This is due to the marked wind transport onto the thinner and on average larger ice pieces (see Eq. 2.22). In both simulations the 10 years recurrence period enclosure is slightly higher than the one in Fig. 2.57 (with still rotor) due to the initial imposed acceleration and again to the wind action on the thinner pieces; the 10 years contours have a maximum extension of about 30–40 m (0.6–0.8 D_r), and the 50 years areas have an extension of about 80–90 m (1.6–1.8 D_r).

The maximum throwing distance for each computed case was compared to the one calculated by means of the relationship proposed in [39]. For the cases presented, a maximum distance of about 200 m was computed that has to be compared with the solution of the empirical equation in Seifert [39], which gives a value of 165 m. For the considered wind-and-site and ice pieces features the solutions according to the present model result are more conservative.

The comparison of the recurrence period enclosures shows that the contours areas for the 1 cm thick fragments are slightly larger than the ones relative to the 5 cm thick pieces. However, relevant difference is detected regarding the shape of the areas: the ones relative to the 1 cm thick fragments are extended in the prevailing wind direction (west south-west). The ones relative to the 5 cm thick fragments are extended in the prevailing initial direction at the time of their removal from the blades; shapes with two lobes are in fact obtained in the latter case.

Simulations were carried out supposing that the debound produces ice pieces with different average weight and same thickness give the results of Fig. 2.59.

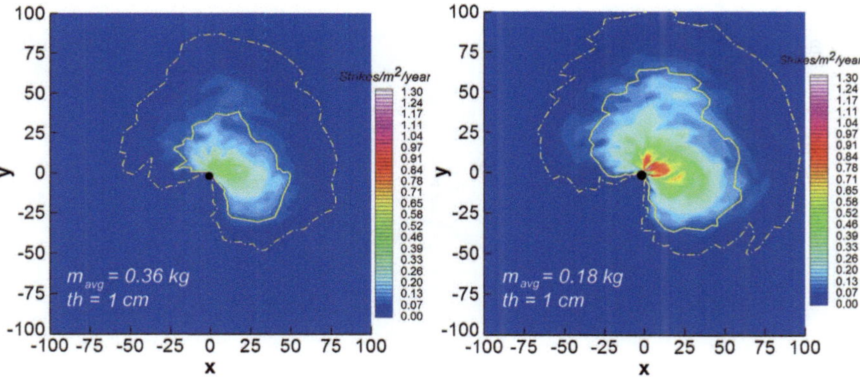

Fig. 2.59 Strike probability comparison; 1 cm thick ice fragments and different average weights

The case with a lower average mass (0.18 kg) compared to the case of Fig. 2.58 indicates that the number of cast ice fragments per year and square metre is higher (see Eq. 2.23); the maximum strike probability was evaluated of about 1 strike/m^2/year. The recurrence period contours are also wider when lighter pieces are cast form the blades; in fact, the effect of the wind transport becomes predominant on the bodies inertia.

The simulation shows that the wind turbine operation under icing conditions determines higher probability of being struck at a given distance from the turbine, if compared with the probability that results from stopping the rotor. This is due to the initial acceleration on the fragments that are about to detach, and to the significant wind transport onto light and thin ice pieces that are produced by means of the de-icing practice. However, even in the worst cases, the contours corresponding to a recurrence period of 50 years restrict an area that is not exceeding the safety distances of usual applications. In the same manner, minimum distances among wind turbines in wind farms are not influenced.

The ice fragment thickness has a minor effect on the assessment of the recurrence period contours; however, the removal of light and thin pieces at sites with high average wind velocities (where the wind transport effect becomes predominant) produces wider areas covered by ice fall.

The model has been used to analyse how these results can be transferred to larger wind turbines. For this purpose, the type of turbines shown in Fig. 2.60 have been considered. Differences in diameter, hub height, rotational speed, tip chord, number

	R [m]	H_{tower} [m]	Ω [RPM]	TSR [-]	Chord tip [m]	M_{blades} [kg/m^3]
a	25	60	25	5.45	~0.9	75
b	14,5	40	50	5.45	~0.45	73
c	25	60	25	5.45	~1.35	43

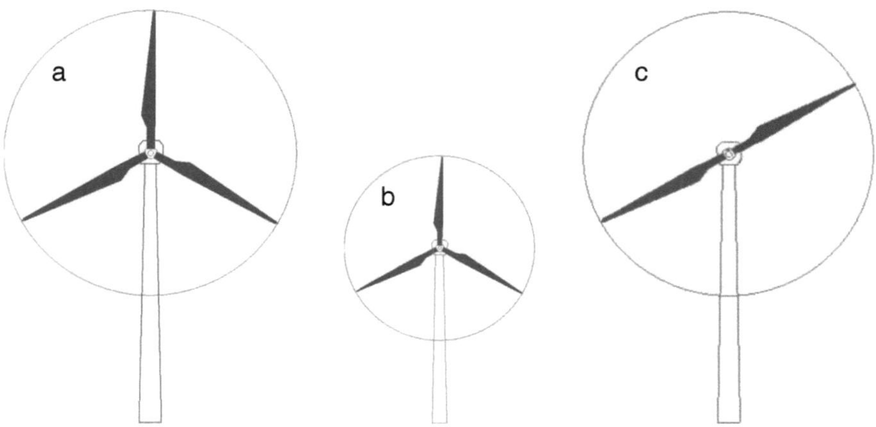

Fig. 2.60 Details of the WT used in the simulations of ice strike probability

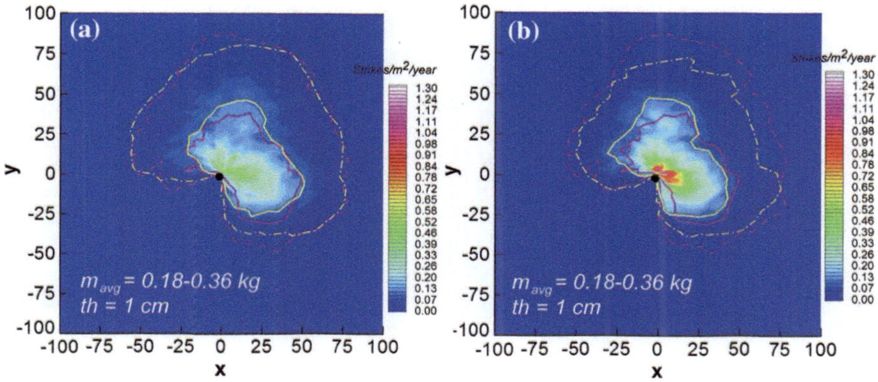

Fig. 2.61 Three-bladed turbines of different diameter, ice strike probability contours of 0.18 and 0.36 kg average mass ice blocks

of blades and hence accreted mass on the blades are indicated in the top table of the figure. Turbine b, despite of the smaller blade length accumulate a comparable mass of that of turbine a, because of the unfavourable icing conditions caused by the relatively thin blade thickness.

Three bladed turbines of different diameter generate the situation shown in Fig. 2.61. The ice strike probability on ground for the 50 m diameter turbine (Fig. 2.61a, on the left), and for the 25 m (Fig. 2.61b, on the right) are indicated. The yellow contours are used to present the results relative to 0.18 kg average mass, while the purple contours refer to 0.36 kg heavier blocks average mass. Again 10 years recurrence period is represented with the continuous lines, while the dash-dotted lines stay for a 50 years period. Downscaling indicate that the area of falling is more or less the same, with an higher probability of strikes for the smaller turbine in the near area surrounding the tower. The lower tower and diameter size are compensated by the higher rotational speed of the smaller turbine, thus giving a comparable result.

This would suggest that similar tip speed velocity designs lead to the same effects in term of risk probability, providing the blade size is not producing too different ice fragment masses. Therefore, the strike risk should be not so different also for multimegawatt size turbines. In fact, the relative velocity is given by:

$$W^2 = (\Omega\ R)^2 + V^2$$

or

$$W^2 = V^2(TSR^2 + 1)$$

Therefore, the lift and drag forces (see Eqs. 2.13 and 2.14) scale with TSR^2.

Finally Fig. 2.62 compares two turbines having the same rotor solidity but different number of blades. The area subjected to strikes is a little more confined, compared to

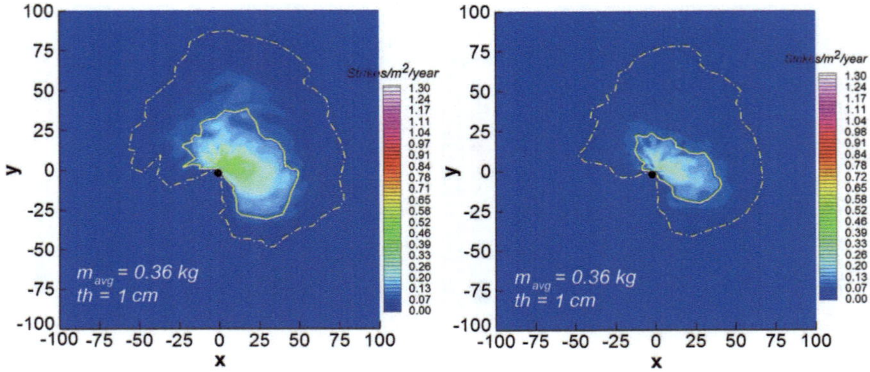

Fig. 2.62 Ice strike probability contours for 0.36 kg average mass ice blocks for three and two-bladed turbines of the same diameter and different solidity

DE-ICING PRACTICE FEATURE	Frequency of detachment		Mass of fragments		Number of fragments	
ISSUES	high	low	high	low	high	low
Rotor aerodynamics	•			•		
Rotor loads	•			•		
IPS energy savings		•	•			•
Risk for people (safety distance)	←	•	•	→	←	•
Risk for people (risk of strikes)		•	•			•
Risk for people (risk of injuries)	•			•	•	

Fig. 2.63 Guiding strategies for analysis of ice throw risk

the three-bladed turbine, Therefore, the overall amount of ice fragments is smaller. In fact, the two-bladed turbine, with longer chord blades, collects a smaller amount of water and hence ice (see table of Fig. 2.60).

The following Fig. 2.63 summarises up the consideration of this paragraph and gives a guiding strategies for analysis of ice throw risk during winter seasons.

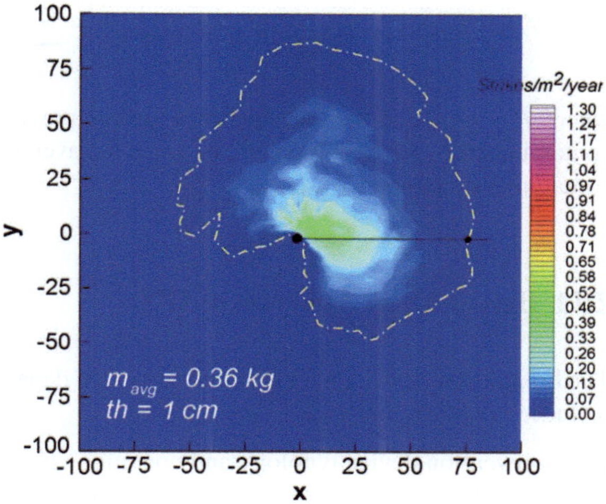

Fig. 2.64 Representation of the walking path for the WT maintenance during ice period

Worked Example on Estimation of Risk for Operators

The safety problem has to be addressed for the operators, who need to approach the tower closely. A risk assessment can be performed assuming a possible intervention scenario. If one suppose that the 50 years recurrence period contour sets the safety distance where the operator has to leave his maintenance car, then he needs to walk towards and then away from the turbine. Assuming that he is walking on the most dangerous rectilinear path (in order to make a conservative analysis), as shown in Fig. 2.64, the average number of strikes/h on the path can be evaluated as:

$$P_h = \frac{L_{path}\ w_{path}\ \bar{P}_{y,path}}{t_i}$$

If the average number of strikes/m^2/year on the path $\bar{P}_{y,path}$ is set equal to 0.2, and the path width w_{path} is set 1 m, the number of icing days is 5 (5 days/year = 120 h/year), one obtains 0.13 strikes/h. This number represents the number of time a person is struck by an ice fragment for every hour of permanence on the path during an icing day.

2.10 Economic Risks of Icing

During the icing period, the energy yield losses due to ice accretions on blades have to be taking into account in the annual energy yield and in the economic feasibility. The annual energy yield, in this case, can be computed as:

$$AEP = P_{clean\ blade}\left(8760 - t_{i.\ blade,i}\right) + \sum_{i}^{n} P_{i.\ blade,i}\ t_{i.\ blade,i} \tag{2.25}$$

where $P_{clean\ blade}$ is the average wind turbine power output when the rotor is not contaminated by ice accretion, $P_{i.\ blade,i}$ is the average power output during the ice contamination period ($t_{i.\ blade,i}$).

2.11 Break-Even Analysis of IPS

The break-even cost of an ice prevention system depends on many parameters. They include:

- ice prevention device typology and operation efficiency,
- site specific parameters: the probability or the time of icing, the wind resource, the air temperature and the adopted operating strategy in icing conditions,
- turbine specific parameters: the effect of the icing on the turbine power curve and production.

Is the costs of the ice prevention system (installation, operation and maintenance) are less than the costs of the energy losses suffered because to icing, the IPS will improve the project economics. The latter represents the IPS benefits that are the revenues due to the extra AEP during the icing periods. At the break-even point, where the minimum benefits necessary to cover all expenditures are achieved, the following condition is obtained:

$$\text{Annual IPS costs} = \text{Annual IPS benefits} \tag{2.26}$$

This condition is translated into the following:

$$\frac{I_{IPS}}{a} + O\&M_{IPS} = c_{el}\ AEP_{loss} \tag{2.27}$$

where I_{IPS} is the cost of the de-ice/anti-ice device, a is the annuity factor, and c_{el} is electrical energy cost priced from the net. I_{IPS}/a is therefore representing the IPS capital to be paid annually. The $O\&M_{IPS}$ costs include both maintenance (M_{IPS}) and electric consumption (E_{IPS}) during the whole year, that is:

$$O\&M_{IPS} = M_{IPS} + c'_{el}\ E_{IPS} \tag{2.28}$$

where c'_{el} is the cost of electric energy (generally it can differ from c_{el}).

The IPS electric consumption depends on icing period duration, weather conditions, wind turbine characteristics (rotor tip speed, heating surface of blades and number of blades) and type of anti-ice/de-ice device. The $O\&M$ can be written as:

$$O\&M_{IPS} = M_{IPS} + c'_{el} \sum_{i}^{n} P_{ICE,ave,i}\, t_{i.\,blade,i} = M_{IPS} + \frac{c'_{el} \sum_{i}^{n} \dot{Q}_{IPS,ave,i}\, t_{i.\,blade,i}}{\eta_{IPS}} \quad (2.29)$$

where $\dot{Q}_{IPS}$ is the power required to maintain the external surface of blade at temperatures above 0 °C (for instance +2 °C), while $\dot{Q}_{IPS}$ is the installed IPS power and η_{IPS} is its overall efficiency (discussed in Chap. 5). The break-even condition is thus given by:

$$\frac{I_{IPS}}{a} + M_{IPS} + \frac{c'_{el} \sum_{i}^{n} \dot{Q}_{IPS,ave,i}\, t_{i.\,blade,i}}{\eta_{IPS}} = c_{el}\, AEP_{loss} \quad (2.30)$$

To link the IPS cost and the IPS electric consumption to the WT overall cost, Eq. (2.30) is better expressed as:

$$\frac{b\, I_{WT}(D)}{a} + M_{IPS} + \frac{c'_{el} \sum_{i}^{n} \dot{Q}_{IPS,ave,i}(D)\, t_{i.\,blade,i}}{\eta_{IPS}} = c_{el}\, AEP_{loss} \quad (2.31)$$

Here both I_{WT} and $Q_{IPS,i}$ can be conveniently expressed as functions of the WT diameter D through suitable parametric relationships.

Equation (2.31) was used to carry out a break-even simulation for a 1,300 kW three-bladed wind turbine, located at 1,900 m a.s.l. in a mountainous site. The turbine is equipped with an IPS which heats the blade surface up to 2 °C when icing conditions are detected. Only the outer half part of each blade is heated. The IPS characteristics and costs are illustrated in Table 2.8. Both electric and hot air based IPS, each with its own overall efficiency, have been used.

Table 2.8 Site weather conditions and IPS data used for break-even simulation of Figs. 2.65, 2.66 and 2.67

Meteorologic parameters	
Mean temperature	−4.89 °C
Temperature Gauss parameters	m = 2, s = 7
Wind scale Weibull parameter	k = 1.5
Mean LWC	0.8 g/m^3
LWC Weibull parameters	k = 2, C = 0.9 m/s
Turbine parameters	
WT rated power	1,300 kW
Electric IPS	$\varepsilon_{IPS} = 0.9$, Cost = 8 % of WT cost
Hot air IPS	$\varepsilon_{IPS} = 0.3$ for standard blade (open channel)
	Investment cost = 2 % of WT cost
	$\varepsilon_{IPS} = 0.6$ for modified blade (closed channel)
	Investment cost = 3 % of WT cost
Price of electric energy c'_{el}	0.08 Euros/kWh
Annuity factor	14.28

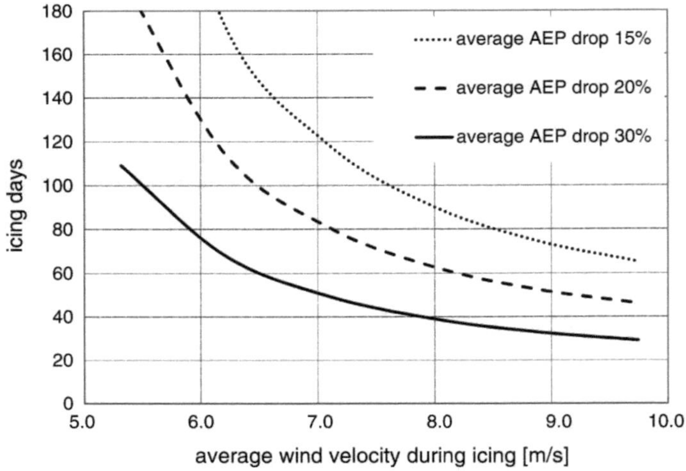

Fig. 2.65 Minimum iced days to achieve IPS economics with electric IPS, wind resource (expressed as average wind speed) and the annual energy production during icing period (expressed as percentage of the AEP in absence of ice accretion)

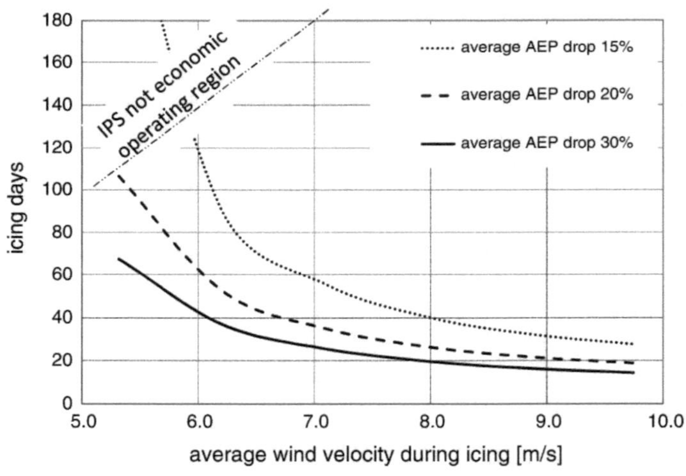

Fig. 2.66 Minimum iced days to achieve IPS economics with hot air IPS (open channel arrangement), wind resource (expressed as average wind speed) and the annual energy production during icing period (expressed as percentage of the AEP in absence of ice accretion)

Figures 2.65, 2.66 and 2.67 shows the results of simulations carried out by this simple model. The minimum icing days period to achieve economics of IPS, vs the wind resource (expresses as average wind speed at the hub height of 50 m) are displayed as function of different expected annual energy yields drop during the icing

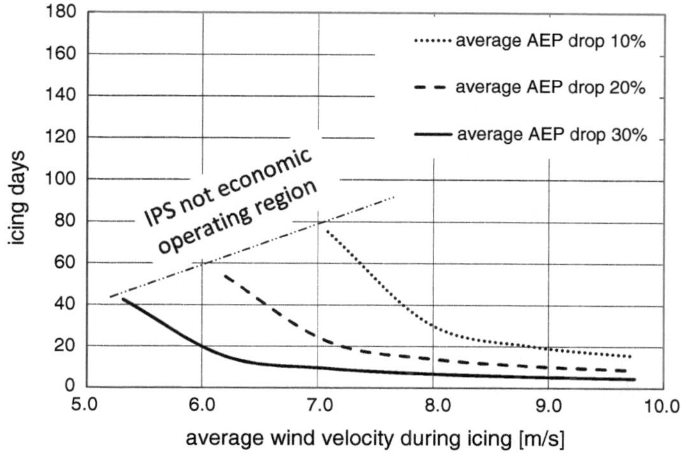

Fig. 2.67 Minimum iced days to achieve IPS economics with hot air IPS (closed channel arrangement), wind resource (expressed as average wind speed) and the annual energy production during icing period (expressed as percentage of the AEP in absence of ice accretion)

period if IPS would not be installed. The annual yield drop is expressed as percentage of the AEP in absence of ice accretion.

The electric anti-icing system needs higher investment cost compared to hot air systems (see Chap. 5), while its higher thermal efficiency (supposed at 0.9) allows compressing the annual operating costs.

The break-even analysis illustrated in Figs. 2.65, 2.66 and 2.67 indicates that the critical duration of icing depends on wind resource and on IPS investment costs. Light icing conditions suggest that the IPS is not generally convenient for low wind speed sites. In harsher environments, leading to more relevant AEP losses, hot air IPS break-even is achieved for less icing days at the same site compared to electric one. IPS installations at windy sites tend to be more economically viable, since the high energy yield is justifying the additional investment and operating costs of the equipment.

This profitability is higher for the simplest and cheapest technical solution of the hot air IPS, but such system becomes unprofitable for low, medium winds because the very high energy consumption due to the low thermal efficiency of the IPS [44]. These technology aspects are treated in greater detail in Chap. 5.

References

1. Laakso T, Holttinen H, Ronsten G, Horbaty R, Lacroix A, Peltola E, Tammelin B (2003) State-of-the-art of wind energy in cold climates. http://arcticwind.vtt.fi
2. Seifert H (2003) Technical requirements for rotor blades operating in cold climate. In: Proceedings of Boreas VI, DEWI Deutsches Windenergie-Institut GmbH, p 5
3. Tammelin B, Cavaliere M, Holtinnen H, Morgan C, Seifert H (2000) Wind energy in cold climate–final report WECO (JOR3-CT95-0014). Finnish Meteorological Institute, Helsinki. ISBN 951-679-518-6
4. Maissan JF (2001) Wind power development in sub-arctic conditions with severe rime icing. TSYE Corporation, Circumpolar climate change summit and exposition
5. WSP environment and energy Sweden (2014) http://www.wspgroup.com, last visit October 2014
6. Battisti L, Fedrizzi R, Dell'Anna S, Rialti M (2005) Ice risk assessment for wind turbine rotors equipped with de-icing systems. In: Proceedings of the VII BOREAS conference, Saariselka, 7–8 March 2005
7. Battisti L, Fedrizzi R, Dal Savio S, Giovannelli A (2005) Influence of the and size of wind turbines on anti-icing thermal power requirement. In: Proceedings of EUROMECH 2005 wind energy colloquium, Oldenburg, 4–7 October 2005
8. Seifert H, Richert F (1997) Aerodynamics of iced airfoils and their influence on loads and power production. In: Proceedings of European wind energy conference, Dublin, pp 458–463
9. Seifert H, Richert F (1998) A recipe to estimate aerodynamics and loads on an iced rotor blades. In: Proceedings of the IV BOREAS conference, Enontekio, Hetta
10. Germanischer Lloyd Industrial Services GmbH, Business Segment Wind Energy (2010) Guideline for the certification of wind turbines
11. NEW ICETOOLS (2004) Wind turbines in icing environment: improvement of tools for siting. Certification and operation. EU Commission Project. Contract No: NNE5-2001-259
12. International Standard (2001) ISO 12494–atmospheric icing of structures, 1st edn
13. Makkonen L, Autti M (1991) The effects of icing on wind turbines. In: Proceedings of wind energy: technology and implementation (EWEC), pp 575–580
14. Papadakis M, Yeong HW, Wei H, Wong SC, Vargas M, Potapczuk M (2005) Experimental investigation of ice accretion effects on a swept wing. U.S. Department of Transportation-Federal Aviation Administration, Final report, DOT/FAA/AR-05/39, August 2005
15. Makkonen L (1984) Atmospheric icing on sea structures. U.S. Army Cold Regions Research & Engineering Laboratory. CRREL Monograph 84–2
16. Battisti L (2003) Relevance of ice prevention systems for wind energy converters. IV Italian-German Colloquium for Science, Trento, 16–18 January 2003
17. Cober SG, Isaaq GA, Strapp JW (2001) Characterization of aircraft icing environments that include supercooled large droplets. J Appl Meteorol, Am Meteorol Soc 40:1984–2002
18. FAA advisory circular (2003) Aircraft icing protection. AC No. 20–73A
19. Battisti L, Dal Savio S, Dell'Anna S, Brighenti A (2005) Evaluation of anti-icing energy and power requirement for wind turbine rotors in cold climates. In: Proceedings of the VII BOREAS conference, Saariselka, 7–8 March 2005
20. Fikke S et al (2006) COST-727, Atmospheric Icing on structures. Measurements and data collection on icing: State of the art, publication of meteoswiss 75:110
21. Baring Gould I, Cattin R, Durstewitz M, Hulkkonen M, Krenn A, Laakso T, Lacroix A, Peltola E, Ronsten G, Tallhaug L, Wallenius T (2011) IEA Wind recommended practice 13: Wind energy projects in cold climates. IEA Technical report
22. Cattin R (2012) Icing of wind turbines. Elforsk Technical report, January 2012
23. Botta G, Cavaliere M, Holttinen H (1998) Ice accretion at Acqua Spruzza and its effects on wind turbine operation and loss of energy production. In: Proceedings of the IV BOREAS Conference, Hetta, 31 March-2 April 1998
24. Albers A (2011) Summary of a technical validation of ENERCON's rotor blade de-icing system, Deutsche wind guard consulting Gmbh, pp 11035–V2

25. Homola MC et al (2006) Ice sensors for wind turbines. Cold reg sci technol 46(2):125–131
26. Tammelin B, Cavaliere M, Holttinen H, Morgan C, Säntti K, Seifert H (2000) Wind energy production in cold climate. Finish Meteorological Institute Publication 41:2000
27. Tammelin B et al (2003) Improvements of severe weather measurements and sensors EUMETNET SWS II PROJECT. Final report, Finish Meteorological Institute, Helsinki
28. Pedersen TF, Paulsen US (1999) Classification of operational characteristics of commercial cup-anemometers. Risoe National Laboratory/Wind Energy and Atmospheric Physics Department, pp 45–49
29. Fortin G, Perron J, Ilinca A (2005) Behaviour and modeling of cup anemometers under icing conditions. In: Proceedings of XI IWAIS, Montreal, 16th June 2005
30. Tammelin B, Peltola A, Hyvönen R, Säntti K (1996) Icing effects on wind measurements and wind energy potentials prediction. In: Proceedings of an international conference BOREAS III. Finish Meteorological Institute
31. Dobesch H, Zach S, Viet Tran H (2003) A new map of icing potentials in Europe–problems and results. In: Proceedings of an international conference BOREAS VI. Finish Meteorological Institute, Pyhätunturi
32. Harstveit K (2002) Using routine meteorological data from airfields to produce a map of ice risk zones in Norway. In: Proceedings of 10th international workshop on Atmos. Icing of structures, Brno
33. Ryerson CC (2000) Remote sensing of in-flight conditions: operational, meteorological, and technological considerations. NASA Langley Research Center, NASA/CR-2000-209938
34. WEMSAR (1999) Wind energy mapping using synthetic aperture radar. ERK-C7-1999-00017
35. James GE (1998) Random number generation and Monte Carlo methods. Springer, New York
36. Botura G, Fisher K (2003) Development of ice protection system for wind turbine applications. In: Proceedings of the VI BOREAS conference, Pyhätunturi, 9–11 April 2003
37. Frank R, Seifert H (1997) Ice im Kanal, DEWI Magazine, vol. 10. pp 4–13
38. Morgan C, Bossanyi E, Seifert H (1998) Assessment of safety risks arising from wind turbine icing. In: Proceedings of the IV BOREAS conference, Enontekio, Hetta, Finland
39. Seifert H, Westerhellweg A, Kröning J (2003) Risk analysis of ice throw from wind turbines. In: Proceedings of the VI BOREAS conference, Pyh, Finland, 9–11 April 2003
40. Cattin R et al (2012) Ice throw reloaded studies at Guetsch and St. Brais. In: International wind energy conference winterwind, Skelleftea, Sweden, 7–9 February 2012
41. Cattin R et al (2007) Wind turbine ice throw studies in the Swiss Alps, EWEC 2007 conference. Milan, Italy, 7–10 May 2007
42. Cattin R et al (2009) Four years of monitoring a wind turbine under icing conditions. In: 13th international workshop on atmospheric icing of structures (IWAIS), Andermatt, Switzerland, 8–11 September 2009
43. Battisti L (2012) Gli Impianti Motori Eolici, Ed. Lorenzo Battisti, ISBN 978-88-907585-0-8
44. Brighenti A (2006) Wind turbine installations in cold climates. PhD Thesis, University of Trento, Italy

Chapter 3
Aerodynamic Performances of Ice Contaminated Rotors

Abstract This chapter is dedicated to the study of the aerodynamic performances of ice contaminated airfoils. Generalities on aerodynamics of contaminated profile are explained, looking to the effect of roughness on lift and drag, with the help of an exhaustive review and comment of the results currently available. Successively, the type of ice contamination is analysed and classified according to the major effect on boundary layer and aerodynamics. To close the substantial gap in systematic studies in the field, which lacks a general assessment of the problem, a more consistent classification of ice shapes is proposed. The effects analysed were quantitatively used to evaluate the drop of the power and thrust curve of a wind turbine operating in light icing conditions. With regard to loads, a deeper analysis of the effect of ice has been accomplished by a modification of the Flex-5 aeroelastic code. In order to solve the basic lack of realistic ice shapes and masses on the blades usually employed, the concepts of arbitrary contamination levels and ice frequency levels introduced in Chap. 2 have been used to create realistic damage scenarios of the turbine in actual icing environments.

3.1 Generalities on the Flow Condition Past the Profile

From a macroscopic point of view, the interaction between a body and a fluid in relative motion generates two regions with different characteristics of flow, one where the effects of friction are negligible, and another where these effects are significant. The latter region regards the portion of the fluid in contact with the body and is called *boundary layer*. The extension of this region, i.e. its thickness, is variable from a fraction of a centimetre to a few centimetres or even metres (for example in the case of large vessels in motion in water) and changes along a streamlined body. In the boundary layer the fluid velocity varies from zero at the body surface up to a finite value equal to the undisturbed speed at its outer edge. The motion in the boundary layer may be laminar or turbulent. Laminar flow is orderly and uniform and the pattern of the stream lines accompanies the profile shape. From a phenomenological point of view it resembles that of the outer flow devoid of friction. The laminar layer can destabilise and change to turbulent. This process is called *transition*. The turbulent motion is instead characterised by incoherent structures. These features are illustrated in Fig. 3.1.

L. Battisti, *Wind Turbines in Cold Climates*, Green Energy and Technology,
DOI 10.1007/978-3-319-05191-8_3

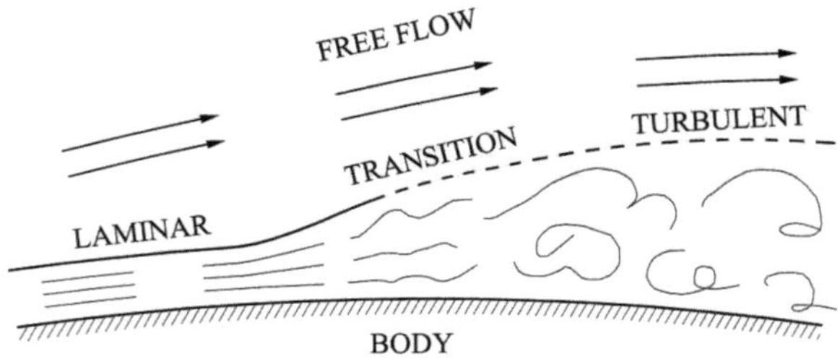

Fig. 3.1 General picture of the flow into the boundary layer [1]

The limit that divides the boundary layer from the motion without friction is defined as the thickness of the boundary layer. This is conventionally assumed where the velocity of the flow within the boundary layer reaches 99 % of the value of the undisturbed velocity outside of it [1]. The part of the drag due to friction is generated almost exclusively within the region of the boundary layer and its intensity depends on the characteristics of the boundary layer itself, the Reynolds number and the local roughness. Relatively low frictional forces occur within the laminar boundary layer, while in the turbulent boundary layer friction forces are much more intense. For this reason there is a tendency towards the realisation of laminar conditions over the body as a method to reduce the resistance. The aerodynamic profiles of old conception have a laminar boundary layer along the approximately first 20 % of the surface, while the most recent low-resistance profiles can present a laminar boundary layer for more than 70 % of the surface.

In both flow regions, laminar and turbulent, the frictional forces are inversely proportional to the Reynolds number. Indeed as the Reynolds number decreases the drag coefficient grows.

Both types of boundary layers are influenced by the pressure gradient in the flow direction. When the pressure decreases in the flow direction the fluid in contact with the body moves to favourable pressure gradient, while when the pressure increases in the direction of flow it experiences an adverse pressure. Referring to the principle of conservation of energy for an elementary stream tube enclosing a line current, the flow accelerates in a favourable pressure gradient and decelerates in the adverse one. Writing the principle of conservation between two sections A and B of an elementary stream tube (conditions are assumed homogeneous on the section) one obtains:

$$p_A - p_B = -\frac{1}{2}\rho\left(W_A^2 - W_B^2\right) \tag{3.1}$$

If the pressure increases between sections A and B ($p_A < p_B$), then the speed is reduced ($W_A > W_B$) and vice versa. Because of the frictional forces, the kinetic energy of the flow will reduce and in the presence of adverse gradients the situation of Fig. 3.2 occurs, where the variation of the speed within the boundary layer in subsequent stations along the direction of motion, both for a favourable gradient and an unfavourable pressure field respectively is shown.

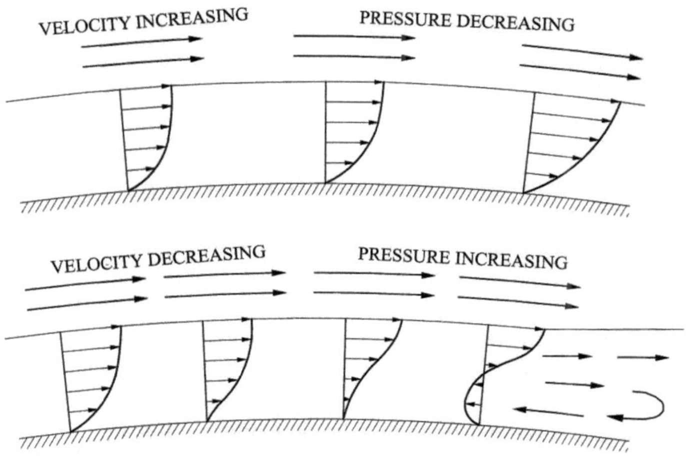

Fig. 3.2 Variation of the velocity within the boundary layer in successive stations along the direction of motion, for a favourable pressure gradient and an unfavourable pressure gradient, respectively [1]

The unfavourable pressure gradient, acting against the motion, can force the flow near the surface of the body to reverse its direction and to move backward in the direction of decreasing pressure.

In most situations, as the reversal of flow in the boundary layer is established, the flow separates from the body trailing edge (TE) forming a large turbulent wake as shown in Fig. 3.3. This phenomenon of separation takes the name of *TE stall* and may extend on a part only of the pressure and/or depression surface of the profile or over both. The pressure variations in the flow accompanying the separation cause a reduction in lift and an increase in the resistance, as evidenced by Figs. 3.6 or 3.7.

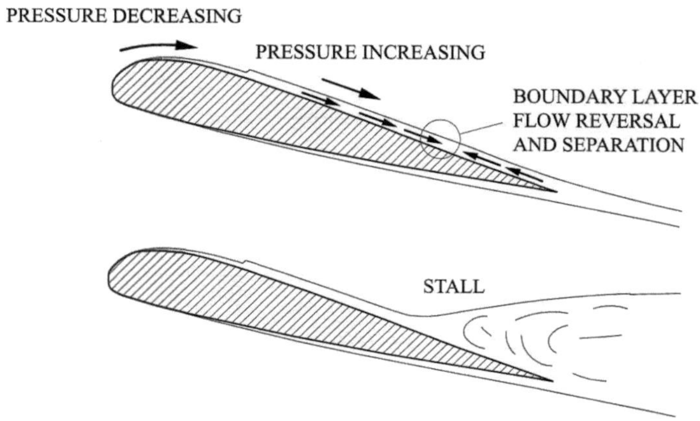

Fig. 3.3 Representation of TE stall onset [1]

The boundary layer is susceptible to destabilisation and transition to turbulent state. The stability of the laminar boundary layer is influenced both by the pressure gradient and the Reynolds number. Low Reynolds numbers and favourable pressure gradient increase the stability, while higher numbers and adverse gradients destabilise it. When subjected to an adverse pressure gradient, the laminar boundary layer may react in three ways: it can separate and produce stall, it can separate and reattach a little downstream as a turbulent boundary layer, or may definitely destabilise and become turbulent. These phenomena are illustrated in Fig. 3.4. The tendency to assume one of such features is a complex combination of factors that depends on the conditions of the boundary layer, the intensity of the pressure gradient, on the conditions of external turbulence, on the Reynolds number and on external disturbances of various nature (i.e., vibration, turbulence). In general terms, at reduced Reynolds numbers the laminar boundary layer tends to separate, at intermediate values it tends to separate and to reattach as a turbulent boundary layer, while at high Reynolds numbers it tends to destabilise and to become turbulent.

A profile generates lift through a pressure difference between the pressure and suction surfaces of the profile. This behaviour implies an acceleration of the flow on the suction surface and a deceleration on the pressure surface. In Fig. 3.5, a typical

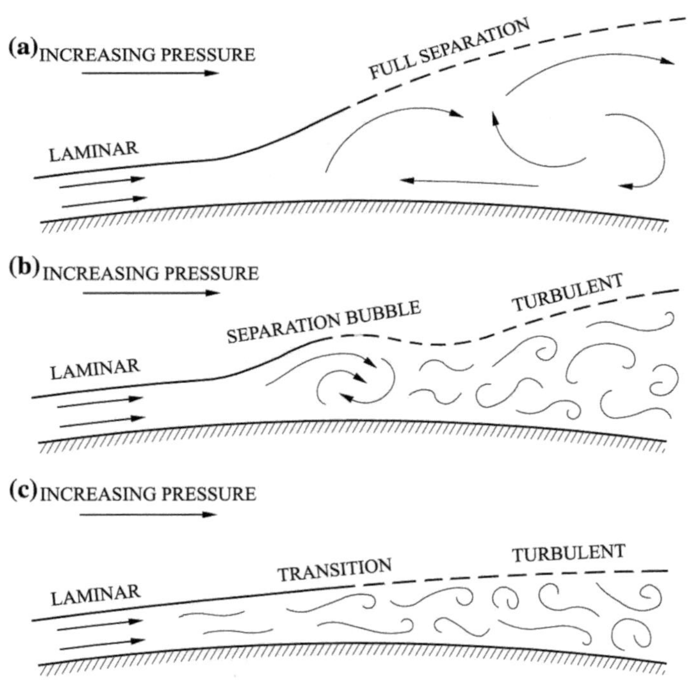

Fig. 3.4 Possible scenarios of boundary layer evolution in adverse pressure gradients [1]. **a** Low Reynolds numbers: full separation and stall. **b** Medium Reynolds numbers: separation and reattachment as turbulent. **c** High Reynolds numbers: transition to turbulent

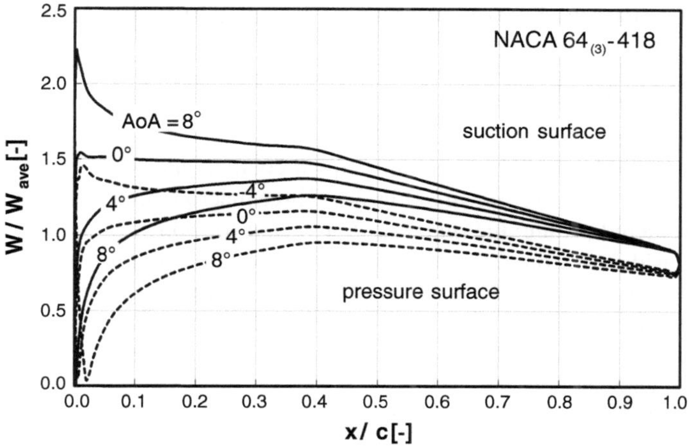

Fig. 3.5 Typical pattern of speed along the profile of the pressure and suction surfaces for four different AoA for the profile NACA $64_{(3)} - 418$

pattern of the normalized relative velocity along the profile for both pressure and suction surfaces at four different AoA is given.

Generally, the relative velocity is normalized to the velocity of the outer flow, the said velocity being the undisturbed upstream one. The performance of the profile (lift, drag, aerodynamic moment) is represented as a function of the relative value of the speed along the profile. With increasing AoA, the flow velocity on the suction surface increases, accelerating the flow very rapidly in the area of the leading edge (LE). At the same time, the velocity on the pressure surface decreases. The trend indicated in Fig. 3.5 shows that the adverse pressure gradients on the back of the profile occur. The intensity of the latter rises with the increase of AoA and increases on the pressure surface as AoA drops. According to the reaction of the boundary layer to the adverse pressure gradients, the separation and the stall are promoted on the suction surface at increasing AoA and grow on the pressure surface with decreasing AoA. Accordingly, since the laminar boundary layer is relatively weak and cannot withstand too high adverse pressure gradients, the transition to turbulent flow establishes before reaching areas with too high adverse gradients, otherwise the boundary layer separation takes place and the stall onsets. In summary, the profile acts as a device that efficiently produces lift only until the boundary layer is able to manage the pressure distributions that are established during operations.

The reduction in lift and the increase in resistance are caused by an increased separation area of the boundary layer when the angle of attack increases (for both positive and negative AoA). Separation and stalling can occur either on the suction surface when the AoA increases, and on the pressure surface as the AoA decreases. This occurs because the coefficient of resistance increases both in the positive and negative AoA field.

3.2 Generality of Aerodynamics of Wind Blade Profiles

3.2.1 Symmetric Airfoils

2D experimental data of lift, drag and moment coefficients for a typical symmetric profile over 360° AoA are shown in Fig. 3.6. Three macro regions can be recognized for the considered profile: the *attached flow region* (or *potential flow*) ranging (for the case considered) from 0° to about 8° AoA, the *separate flow region* extending from about 8° to 172°, and the *reattached flow region* from about 172° to 180° AoA.

The separated flow region can be further subdivided into two sub-regions: the first dominated from an incipient flow separation of the TE and important dynamic instabilities, with stall onset.

As the AoA increases, the separated flow extends towards the LE and the behaviour resembles that of the flat plate with separation phenomena originating from LE and TE, and generation of *von Kármán* street vortices.

In the second sub-region, as the flow faces the TE, the separation bubble contracts progressively and allows the flow to reattach on the downwind side while the flow remains attached on the upwind side with a mirror separation region emanating from the LE. After passing 180° AoA the behaviour mirrors as in Fig. 3.6 shows.

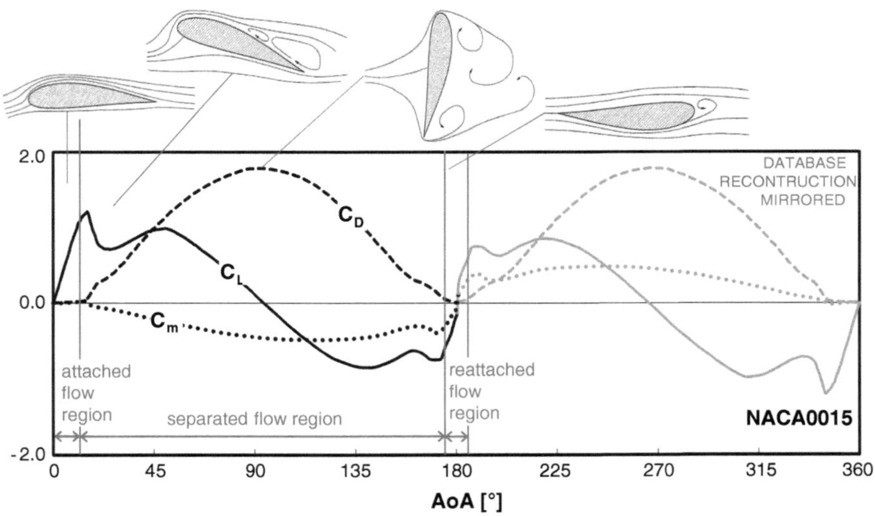

Fig. 3.6 Example of lift C_L, drag C_D and pitching moment C_m curves [2] for symmetric profiles (NACA0015 airfoil [3])

3.2.2 Asymmetric Airfoils

2D experimental data of lift, drag and moment coefficients for a typical wind turbine blade profile, the DU 97-W-300 [4], airfoil are given in Fig. 3.7 over 360° AoA. Compared to Fig. 3.6 we note that due to the camber, the mirroring effect above 180° is not reproduced. In general, the presence of a finite camber shifts the C_L at 0° AoA, but according to the thin profile theory, the slope is not affected (the latter is rather influenced by the profile thickness).

Higher lifts to drag performance are also obtained.

We can still recognize the presence of the three mentioned macro regions (see Sect. 3.2.1) in the first 180° path, which are replayed from 180° to 360° AoA, although the lack of symmetry compared to symmetric airfoil due to the camber effects are quite evident [2].

Experimental data are normally highly scattered in the separated region due to the hysteresis in the flow behaviour. Figure 3.8 shows a typical C_L, C_D e C_m data for increasing AoA (measurements obtained in the wind tunnel) and the corresponding decreasing AoA, generating a separated flow condition with the typical hysteresis loop.

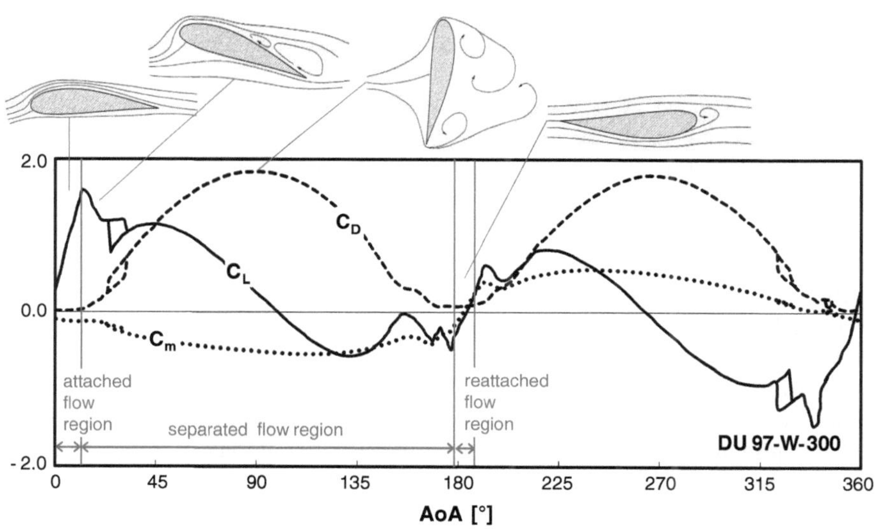

Fig. 3.7 Example of lift C_L, drag C_D and pitching moment C_m curves [2] for asymmetric profiles (DU 97-W-300 airfoil [4])

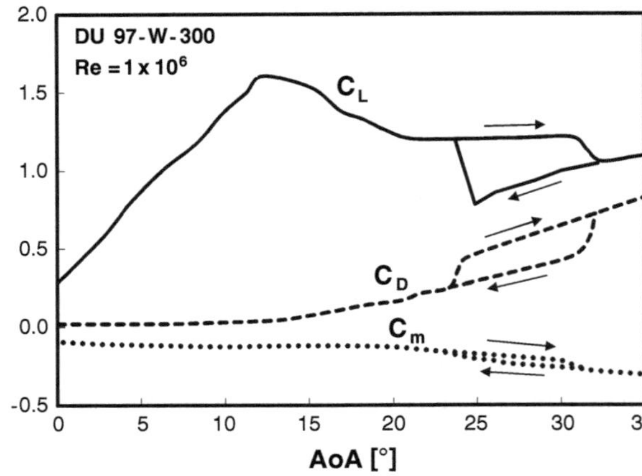

Fig. 3.8 Example of hysteresis on C_L, C_D e C_m curves [5]

3.3 Generalities on Aerodynamics of Contaminated Profiles

3.3.1 Effect of Surface Fouling and Deterioration

Surface fouling, erosion or general deterioration lead to an alteration of the clean (original) profile surface.

These phenomena share the common effect to increase the surface roughness, and as a consequence the behaviour of the boundary layer leads to a variation in the aerodynamic lift, drag and moment. It can be said that the major efforts made in recent decades in the development of innovative profiles is focused on creating surfaces that are only little affected by surface degradation effects. As the wind turbine processes millions of cubic metres of air per year and although the volume fraction of contaminants carried by the wind is extremely low in certain areas, the overall effect is significant. In addition, the turbine operates in environments where maintenance and repair are complex.

To investigate the effect of the roughness on the profiles aerodynamic performances, a common approach is to test artificially altered surfaces in wind tunnels.

Each research institute generally uses its own definitions of roughness. For instance the standard NACA roughness is obtained through typical size of grains distributed in a uniform manner from the leading edge downstream on the pressure surface and the suction surface (usually to 7.5 % of the chord length).

Subsequently, NASA has introduced staggered distributed rough tapes (within the limits of extension of 7.5 % of the chord).

Adhesive tapes are used alternately arranged in a zigzag pattern with the aim of promoting the transition.

One of the most known and wide experimental tests on the effects of light roughness on the leading-edge area on the 2D aerodynamic performance and control characteristics of single-element lifting surfaces was published by Abbott and von

Doenhoff [6] in 1945. The researchers tested over and collected data from previous NACA experiments on many different airfoils at low speed (freestream Mach number ≤ 0.2). Of particular importance was the collection of data at a chord-based Reynolds number of 6.0×10^{-4} and a chord relative roughness of about 4.6×10^{-4}, which would correspond to about 0.5 mm roughness height on an airfoil chord of 1 m. The roughness was applied using carborundum grains, and extended from the LE to 8 % chord on both the upper and lower surfaces. Although dated, the measures are still a reference for the field and allow discussing some general effects of the surface roughness on the profile geometric features. The first outcome was that the aerodynamic response of a particular airfoil to leading-edge roughness contamination is strongly influenced by the specific stall behaviour on each airfoil. Lynch et al. [7], in their exemplar work on the effects of ice accretions on aircraft aerodynamics, organised these data trying to nucleate important relationships.

Figure 3.9 focusses on lift. Reductions and increase of the maximum lift can be observed, spread on a very wide range of lift ranges. Artificial roughness improved or little depleted the maximum lift level attainable by profiles having a clean maximum lift coefficient of about 1.0 or less, or a thickness ratio of 8 %, i.e. profiles having a pronounced thin-airfoil-type leading-edge separation (abrupt stall behaviour). For thickness ratios of 9 % or higher, or for geometries with a clean maximum lift coefficient above about 1.1, a much more pronounced reduction in the maximum lift from about 15 % to nearly 40 % is observed. This picture indicates that a more complex correlation would hold between maximum clean lift capability and airfoil thickness ratio.

Figure 3.10 reports the data assembly for parasitic drag by warning that when drag is of concern, more complex effects have to be taken into consideration, when data are analysed, among them the transition movements due to the roughness (at the considered test Reynolds number, and the fact that in considering 2-D data, only changes to the basic 2-D parasite drag of the lifting surface are involved). Increases in minimum parasite drag level vary anywhere from about 50 % up to nearly 200 %.

It has to be emphasised that such high Reynolds numbers (6×10^{-4}) occur for very large wind turbines usually at blade mid radius, while middle/large wind turbine blades exhibit Reynolds numbers in the range ($3 - 5 \times 10^{-4}$). Extrapolation of data below or above the presented experimental data is dangerous because, as shown by Lynch by a comprehensive analysis of available tests, the results do tend to indicate that a greater non-dimensional roughness height is required at low Reynolds numbers in order to simulate the (incremental) penalty that would be incurred at higher Reynolds numbers. The problem is to understand how leading-edge non-dimensional roughness (k/c) can be scaled with the reduction of the maximum lift expected at lower Reynolds numbers. Figure 3.11 is a collection of Reynolds number ranging between 1 to 4×10^{-4} (original references for the data used can be found in [7]. The airfoils have thickness ratios of 9 % and above, and a clean maximum lift coefficient of at least 1.0. On the same graph the Brumby curve [8, 9] is superimposed, originally developed to indicate the effects on the maximum lift coefficient of light leading-edge roughnesses.

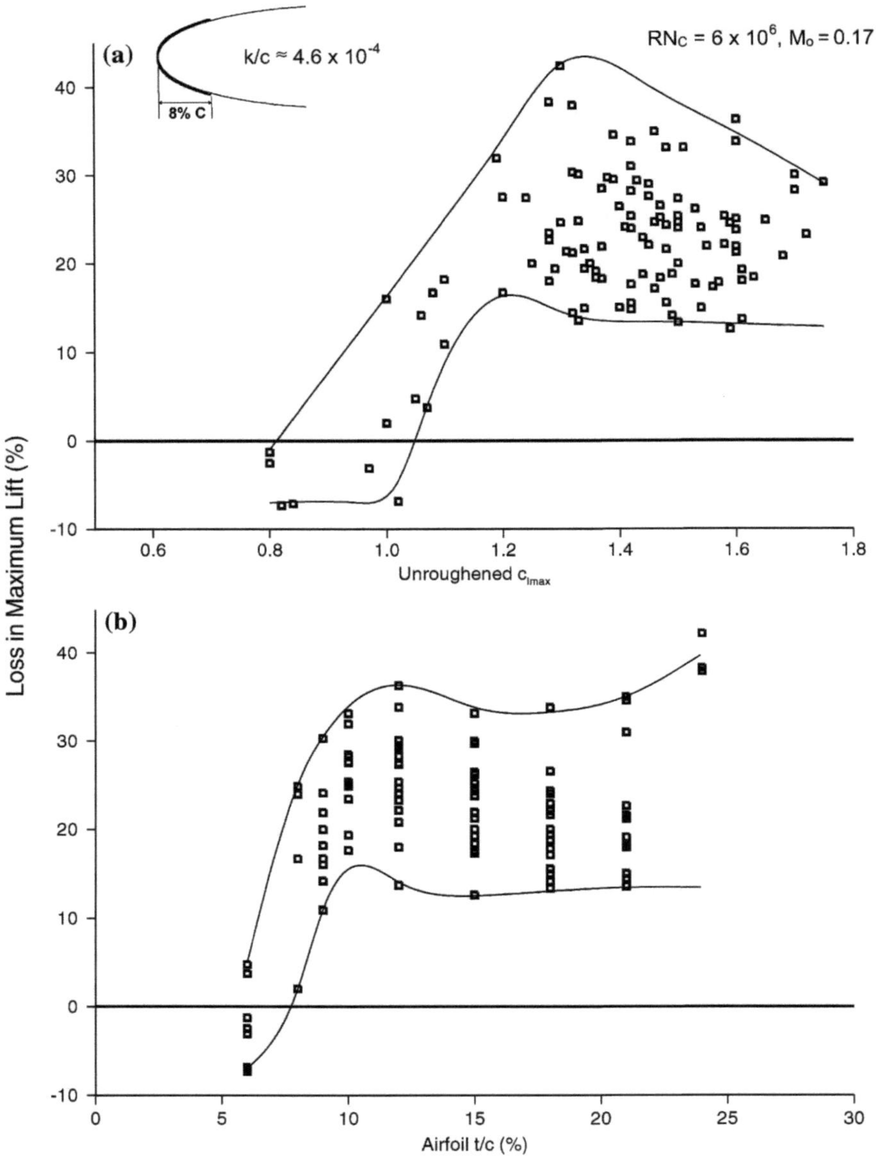

Fig. 3.9 Maximum lift penalty caused by leading-edge roughness [7]

By considering typical wind turbine blades tests, Fig. 3.12 shows the lift and drag coefficients of a NACA 63-425 measured in clean conditions and dirt contamination ($Re = 1 \times 10^6$), simulating the latter by means of a zigzag ribbon with thickness of 0.35 mm at a distance from the leading edge equal to 5 % of the length of the

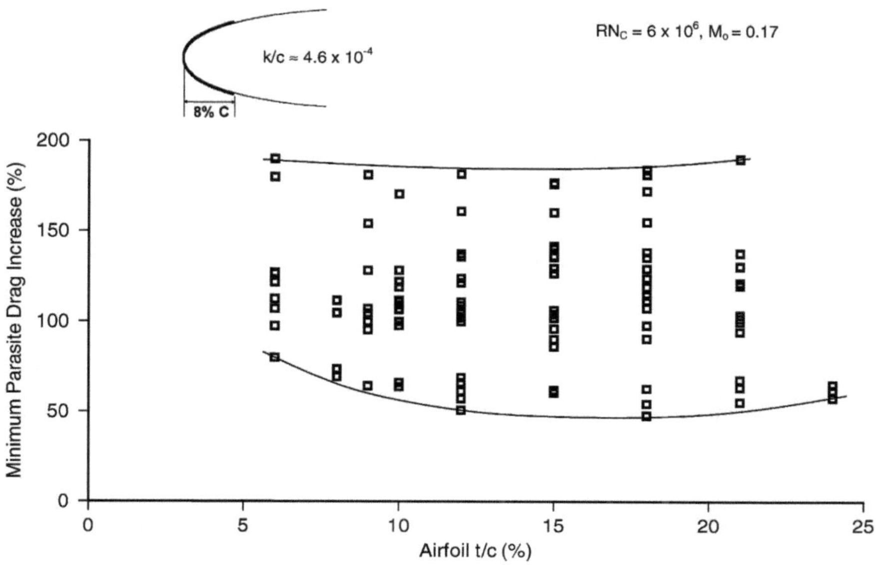

Fig. 3.10 Minimum parasite drag increase due to leading-edge roughness [7]

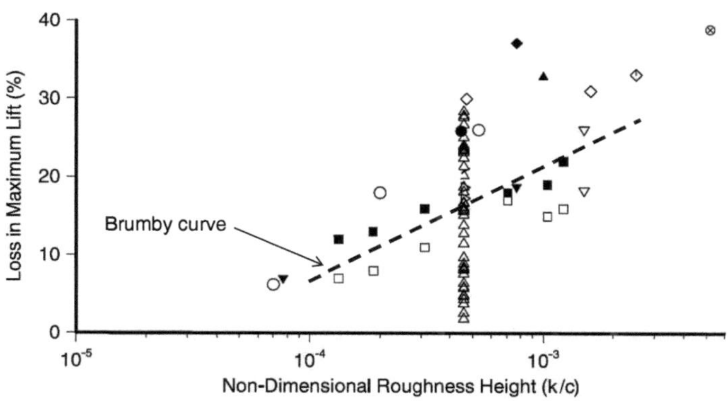

Fig. 3.11 Roughness size effects on maximum lift penalty at low Reynolds numbers and Brumby curve, adapted from [7]

chord [10]. Artificial roughness introduces a premature boundary layer transition from laminar to turbulent and an increase in its thickness with a considerable reduction of lift. Another example of the effect of the performance degradation is indicated in Fig. 3.13 for the profile FX-W-270S, where the effects of the application of a promoter of transition (*tripping wire*) on both pressure and suction surface at a distance from the edge of attack of 3 % [10] are reported. Due to anticipated stall the profile lift drops by about 30 %.

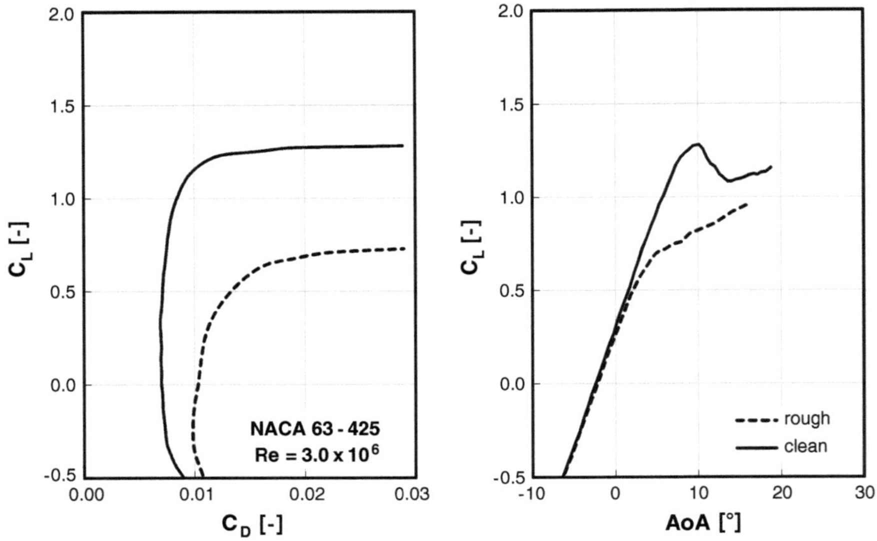

Fig. 3.12 Coefficients of lift and drag of a NACA 63-425 measured in a clean condition and under conditions of fouling, simulating fouling through the affixing of a tape in a zigzag pattern with thickness 0.35 mm at a distance from the leading edge equal to 5 % of the length of the chord [10]

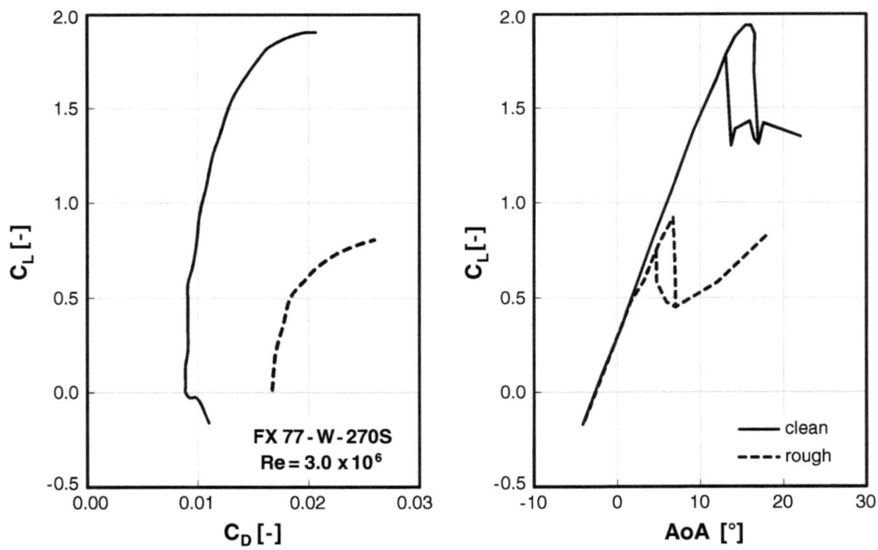

Fig. 3.13 Example of the effect of degradation of performance for the profile FX-W-270S, through the application of a *tripping wire* on both surfaces under pressure and depression at a distance from the edge of attack of 3 % [10]

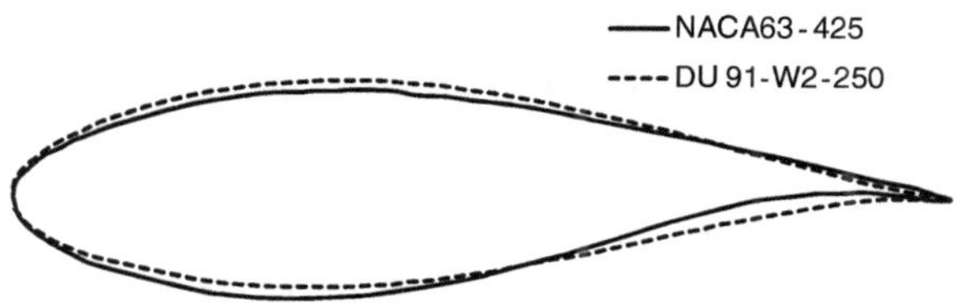

Fig. 3.14 Comparison of the NACA 63-425 and profile DU 91-W2-250, developed at the University of Delft [10] for wind turbine blades

The fouling determines a thickening of the boundary layer and shifts the onset of the transition in the vicinity of the leading edge. This induces a premature separation, creating a high resistance and a reduction of lift.

In the profile design stage one can counteract the performance degradation due to roughness by ensuring that on the clean surface the transition is already close to the leading edge when it reaches the maximum lift coefficient. This obviously does not eliminate the effects of increased drag in the condition of fouling, but in this case the difference between the clean and dirty configuration is much less because the pressure distribution is designed in such a way that the turbulent boundary layer can better face the pressure gradient against moving towards the trailing edge.

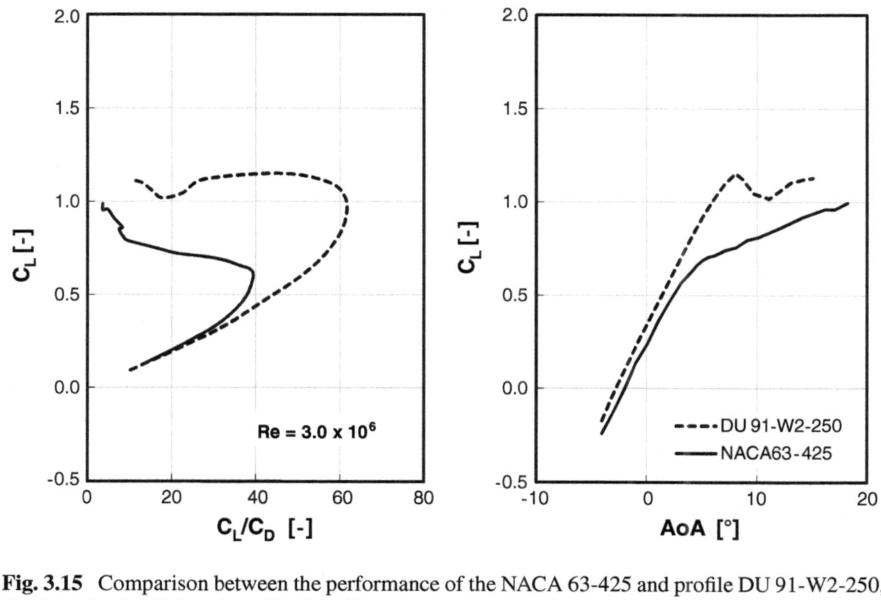

Fig. 3.15 Comparison between the performance of the NACA 63-425 and profile DU 91-W2-250, developed at the University of Delft [10] under simulated fouling (tape is applied in a zigzag pattern with a thickness of 0.35 mm on the upper surface up to 5 % of the chord)

A typical example of a profile designed with this approach is the DU 91-W2-250, developed at the University of Delft [10]. In Fig. 3.14 one can appreciate the difference in contour compared to the typical NACA 63-425.

This profile shows a lower thickness of the upper surface to reduce the speeds and adverse pressure gradients. This approach reduces the maximum lift on the suction surface, but is compensated, however, by the effect obtained on the opposite pressure surface (called *aft loading*). While in the clean configuration both profiles have approximately the same performance, when the tape is applied in a zigzag pattern with a thickness of 0.35 mm on the upper surface up to 5 % of the chord, the performance differs significantly in favour of the profile DU 91-W2-250 as shown in Fig. 3.15.

3.4 Effect of Ice Contamination on Aerodynamics

There are two areas of investigation in airfoils icing. The first is the analysis of the aerodynamic effects of a given ice shape on a selected type of airfoil behaviour (*icing effects* analysis). The second concerns the capability to capture (numerically or experimentally) the ice shape under specified atmospheric conditions (*ice accretion* simulation). This chapter deals with the first class of problems, requiring models of ice accretion as treated in Chap. 4.

Phenomenological effects of ice accretion on aerodynamic profiles are known for decades, initially reported from the direct experience of pilots flying in icing conditions, and condensed in empirical rules for safe flying, and after scientifically analysed with more and more sophisticated tools in the years. It was immediately recognised that ice on wings and propeller blades alters the profile aerodynamics determining decreased (or sometime increased) lift, increased drag and moment instability. From the mass forces point of view, the generation of additional gravitational loads causes a shift in the blade barycentre and an increase in fatigue loads. Historically speaking, in the pre-computer aided era, the scientific problem was experimentally treated in icing wind tunnels with environmental controlled conditions. In the following years with the introduction of computers and the increase in computing power, CFD simulations have joined the experimental tests, nowadays even with fully 3D field analysis. The level of accuracy can be very high in some cases. The standards in aircraft icing states that if a numerical code can reproduce the same aerodynamic effect on the aircraft as evidenced in flight, this tool can be a useful aid, and can even be sufficient for anti-ice systems design and potentially for certification purposes.

Both experimental and numeric analyses are advisable, but experimental tests are of paramount difficulty and cost compared to numerical tests. In Fig. 3.16, the development times required for tests runs and the technical difficulties of the different approaches are qualitatively compared.

Although ice accretion analysis on aircraft has been initiated some 70 years ago (among the wide literature available see for all [11–14]), only a few observations and tests are up to now available for wind turbines [15–19], and the attempt to take

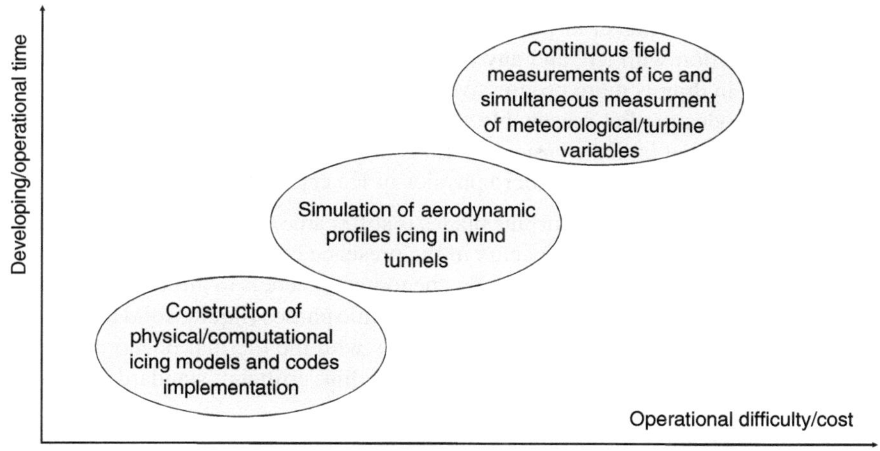

Fig. 3.16 Times required and difficulties of the different approaches to study icing effects

advantage from the results obtained in the aircraft field inevitably collides with the following problems:

- among the few correlations recognised as valid in the aeronautical field, only a few data are usable for wind turbines because of the limited angles of attack ranges that have been investigated;
- specific wind turbine tailored profiles are entering in use. There are not at the moment methods allowing to predict the aerodynamic behaviour of a new profile on the basis of the behaviour of similar existing ones;
- the test conditions involve different physical parameter ranges (LWD, MVD, temperatures, velocities), hardly related to the parameters recorded on wind sites (i.e. relative humidity or clouds coverage, icing duration);
- it is complex to reproduce tests in climatic tunnels for wind turbines rotors, due to the rotor dimensions (and associated scaling problems, rotation shall be modelled together with Reynolds, Mach and Weber numbers), environmental variables, and duration of events.

Nevertheless, some general consideration can be drawn from the survey of this past and still running vast experience:

- the performance degradation caused by glaze or rime ice accretion differs significantly due to the differences in ice form and texture observable;
- ice masses increase at approximately constant rates with increasing time in a rime icing condition and progressively increasing rates in glaze icing conditions;
- experimental droplet impingement rates on the airfoil section agree with theoretical calculations for angles of attack less of that at which separation occurs, beyond which evident disagreements are observed;
- the aerodynamic penalties are greatly influenced by the airfoil shape (thickness of the airfoil, pointed leading edge, flow separation from the suction side at high

angles of attack). Thicker airfoils with blunt leading edge exhibit few or no cases of drag reduction with ice, and any correlation among ice shape, angle of attack and change in drag is more progressive;
- numerical codes predict reasonably the shape of ice accretion in rime (running dry) condition, but indicate large deviations in glaze (running wet) condition, due to the difficulty to model the micro physics of ice crystals' growth at the surface.

Experimental data on wind turbine blades is still scarce, and it is evident that more severe effects of contamination occurs in the presence of ice compared to common environmental fouling. The aerodynamic phenomenon here is in this case more complex because it involves two or three thermodynamic phases (liquid, solid and vapour water phases), it is more massive and combines with the thermal phenomena connected to icing progress that makes it difficult to link artificial, standard roughness of existing tests to the different ice shapes.

Natural observations of the ice accretion process during natural icing conditions, as well as during controlled experiments, indicate that the initial ice accumulation on the LE of aerodynamic surfaces is assimilable to uniformly distributed roughness in the form of small hemispheres, with a typical height equal to the thickness of the ice buildup. This "first phase" roughness can be helpfully simulated by existing tests with artificial roughness. As will be demonstrated later, even small non-dimensional roughness (k/c), of the order of 20×10^{-4} on the leading-edge region will cause significant aerodynamic performance and control degradations.

Ice contamination, in addition to the depletion of the blades and the overall WT performance, affects safe operation causing WT stop for maximum vibration onset due to uneven ice mass accumulation on blades. Usually, the large ice accretions visible in aircraft practise are not recorded in WT during operation, since WT stops rotating at relatively lower mass deposits. Instead, large and wider ice shapes accrete during standstill or parking conditions since either IPS (if present) does not operate, or very large parts of the blade become wetted and get iced, since unprotected (see Fig. 2.5). Icing exposure durations can be much longer (some hours) during standstill or parking, compared to standard icing flight envelopes, as WT cannot, of course, escape in clouds operations, this strategy being possible through a change in flight schedule for aircraft.

The most rational way to analyse the behaviour of the turbine rotor is to reconstruct the aerodynamic behaviour of the single blade profile and to produce new, contaminated aerofoil data. This approach can be performed either numerically by CFD analysis and/or experimentally in icing wind tunnels. The method should be possibly validated by the following procedure:

- taking field measurements of an instrumented turbine during icing;
- comparing computer data with measurements and
- tuning the semi-empirical model in order to better capture the in-field behaviour.

Figure 3.17 shows a typical experimental-based procedure [20] resembling the aircraft practise. This work is cited for its historical relevance, being the first one carried out on the topic. It was aimed to reproduce the modification of the coefficients of lift, drag and moment caused by different degrees of contamination by ice

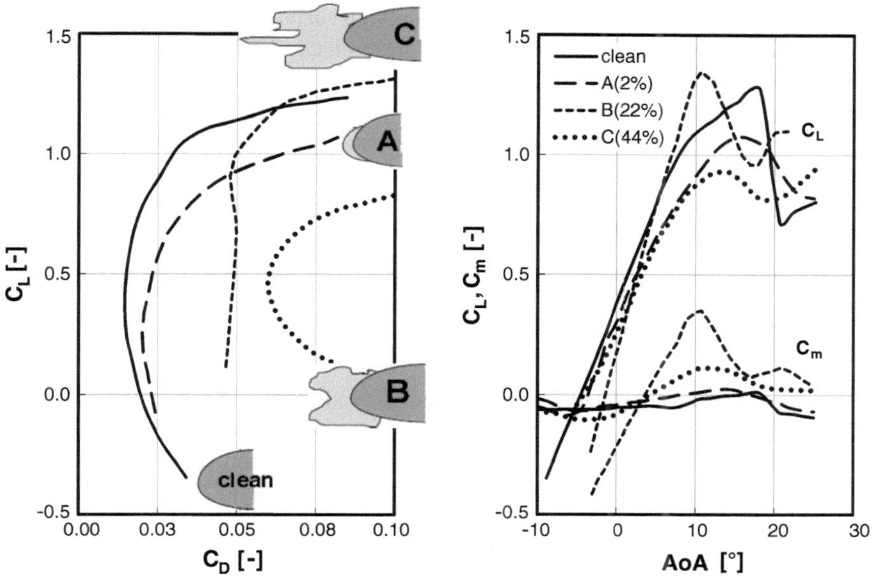

Fig. 3.17 Example of alteration of the lift, drag and moment coefficient at growing icing for NACA 4415, modified from [20]. The different accretions increment the chord length of 2, 22 and 44 %

Fig. 3.18 Example of reconstruction originated artificial shapes from [20]

on a NACA 4415, a popular profile in use in wind turbine design. Artificial accretions of ice (replicating the real ice shapes) were placed on the leading edge of the profile protruded upstream by 2, 22 and 44 % of the original length of the chord. The shapes were glued on the LE and the modified profiles tested in the wind tunnel (see Fig. 3.18). This very important and quite pioneering work was very instructive in an early stage of investigation of the problem, but suffered from a lack of generality and could hardly be extended to quantify, for instance, the expected power drop of a WT operating in icing environment.

3.5 Numerical Simulations

Relatively numerous aeronautical ice accretion-prediction codes developed to simplify certification of aircraft for flight in icing conditions are available: the LEWICE code from NASA (USA [21, 22]), the CANICE code from cole Polytechnique (Canada [23–25]), the CAPTA code from ONERA (France [26]), the MULTICE code from CIRA (Italy [27]) and the 2DFOIL-ICE code from the University of Twente (The Netherlands [28]), and FENSAP-ICE (Newmerical Technologies International [29]). These codes adopt similar methodologies for heat modules and they can predict water catch rates, water droplet impingement limits, ice accretion shapes and overall thermal ice protection system requirements.

The use of these codes for wind turbines in icing conditions analysis is not straightforward. Important differences are related to climatic conditions at which they are calibrated, the operating angles of attack of the profiles and the effect of rotation.

Analysis of iced wind turbine rotors are not so numerous, and even less numerous are the studies including from in-field feedback and code calibration based on experimental data. The inspection of the few number of recent experimental and CFD-based works available [30–39], lead to the following considerations:

1. when one tends to extrapolate aircraft data to wind turbine field, it should be borne in mind that wind turbines cannot operate with the medium-high levels of contamination (lobster or scalloped shapes) reported in many aeronautical reports, because also light ice weight on blades causes a suggestive drop in performances and WT stop is quickly achieved. For instance, for a typical tip section, by setting the hypothesis of a uniformly distributed impingement area extending by 5 % in the chordwise direction on the suction surface and 20 % on the pressure one, it ends that by supposing 2 mm glaze thickness, one obtains about 0.5 kg of ice per metre of blade. Therefore, well before reaching a limit weight causing unbalance, the WT will probably experience a large performance drop which prevents operation with higher levels of contamination, as will be demonstrated in the following paragraphs. One can therefore state that dispersed ice roughness is the first cause of blade performance depletion and WT stop during icing conditions;
2. a substantial absence of systematic studies in the field, which lacks a general assessment of the problem. Literature studies have shown that ice accretions can have dramatically different effects on pressure distributions and boundary layer

development and hence on lift and drag values. But there is scarce evidence of efforts to link different ice shapes to comparably similar performance drop.

For this reason a survey of the huge experimental work undertaken in the aeronautic field is instructive to gather indications on the performance degradation mechanism.

3.6 Experimental Tests in Aeronautical Field

As mentioned, several experimental tests have been carried out in the aeronautical field since the first recognition of potential icing hazard in early 1930. A further boost in the research was imparted after the ATR-72 accident in 1995. From that era, the experimental research in icing moved from profile-specific tests, i.e. characterisation of a single profile behaviour in rime and glaze icing growth conditions, to experiments aimed to understand the general effect of some ice shapes on the profile performance, i.e. more pointed towards classifying the effect of ice shapes on boundary layer alteration and to deduce general laws on icing effects.

This approach, throughout a better understanding of the aerodynamics of iced profiles, has led to the development of tests on new profiles other than the classic NACA 0012, as NACA 23012, NLF 0414, and NACA 6-series, and of new other airfoils more suited to contrast the adverse effect of icing. The essential literature is listed in the references of this chapter at the end of the book.

3.6.1 Ice Geometry Identification

Ice accretion on surfaces is in general a typical 3-D problem, originating from stochastic ambient conditions. Nevertheless, observations indicate that given ranges of ambient conditions tend statistically to create recognisable 2D shapes. These shapes can be replicated in icing wind tunnels, and according to the resulting geometrical features, detailed aerodynamic tests on the altered shapes of the profiles can be made. This methodology has some limitation, because for instance for rotating frames, as helicopter and turbine rotors, where relevant inflow asymmetry occurs, the ice accretion cannot be reduced to a 2D problem.

A comprehensive presentation of experimental methods used to perform such tests is given in the research reports prepared by the Department of Aerospace Engineering and the NASA John H. Glenn Research Center at Lewis Field (see for all [40]). The ice shapes tested have been prepared by castings of ice accretions obtained from icing tests at the NASA Glenn Icing Research Tunnel (IRT) and the simulated ice shapes were replicated with the LEWICE 2.0 ice accretion code, commonly used in the industry, and developed by NASA Glenn Research Center. Adjustments are necessary to account for the 3D flow effects, to improve the construction of the ice shape, the droplet trajectory, the heat transfer and ice growth calculations modules. Table 3.1 lists the test conditions for two cases of an iced GLC-305 airfoil, while in Figs. 3.19 and 3.20 the comparison of ice accretion output and LEWICE simulation for three

Table 3.1 Icing conditions for IRT ice shape casting, reproduced from [40]

Icing condition	Description	AoA (°)	V (m/s)	T^0 (°C)	LWC (g/m^3)	MVD (μm)	Spray time (min)
Glaze	Complete scallop condition (ID: Ice1 or IRT-CS10)	4	111.8	−3.9	0.68	20.0	10.0
Rime	Scaled condition from 2D tests (ID: Ice3 or IRT-SC5)	6	90.0	−11.3	0.51	15.5	5.0

different wing sections are shown. Test conditions included Reynolds number of 1.8×10^6 based on the wing mean aerodynamic chord.

The investigation, only partially reported here, shows that the general rime ice growth feature is captured from the CFD code, despite the apparent differences, in terms of global mass and ice distribution. It can be seen that important discrepancies still exist to replicate the shape of glaze conditions. In Fig. 3.21 the C_L and C_D coefficients for the two situation listed in Table 3.1 are given.

From the inspection of Fig. 3.21 it is evident the scalloped glaze shape accretion tends to be more penalising compared to the less pronounced rime shape, in terms of both C_L and C_D. The percent deficits of C_L and C_D with respect to clean date are given in Fig. 3.22. Reductions up to 35 % results for C_L in glaze conditions, while this parameter remains relatively invariant for rime. The C_L records the maximum deviation, up to 1,000 times at low AoA and 100 times for rime. These figures cannot however be taken as absolute, but only as a trend, because in-depth investigation is needed, taking into consideration the ice roughness, and distribution on the surface, as shown in the following paragraphs.

The correspondence with LEWICE simulations is given in Fig. 3.23. The clean C_L and C_D data have been tuned from the author's report to match with the experimental ones. The LEWICE output showed that in general the trends in aerodynamic performance degradation of the wing profiles with the simulated ice shapes were similar to those obtained within the IRT with casted ice shapes. However, LEWICE ice shapes tend to underestimate aerodynamic performance losses compared to those obtained from ice castings.

Also from these high qualified results, ambiguities still exist for classification of the ice structure, especially in the identification of the dimensions that are critical for an accurate prevision of the aerodynamic behaviour. This represents a problem of concern for comparison of CFD simulations and experimental tests.

Deviations derive partly from the inherent stocastic nature of the phenomena (although limited by the controlled ambient variables of the wind tunnel), partly from 3D phenomena (well analysed in [40]) and partly from some deficiencies in modeling of micro-physics o ice growth and water beds motion on the surface. In real contests, additional events intervene to alter further the icing mechanism, as vibrations, cohesive detachment, and drift of ambient conditions in time, therefore more differences are expected.

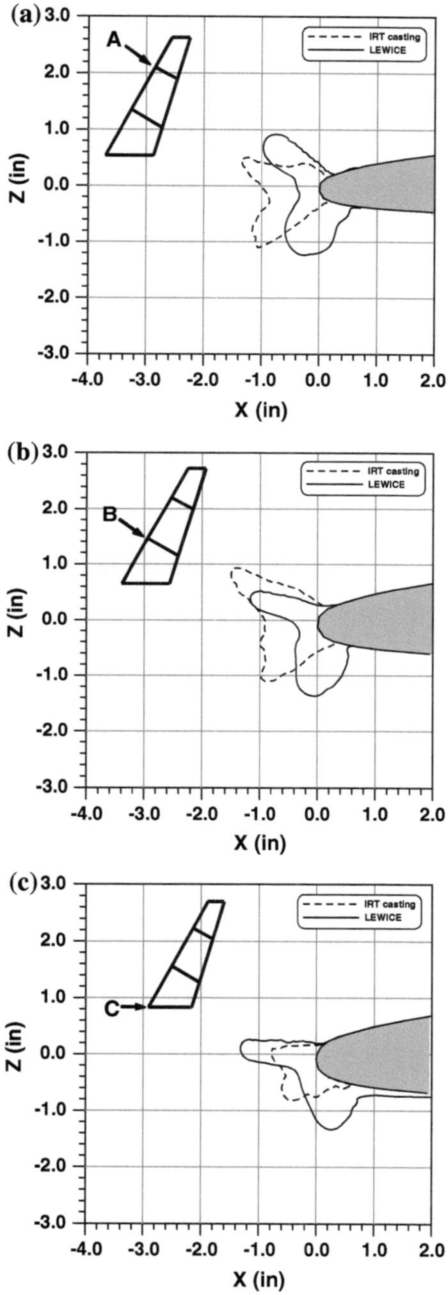

Fig. 3.19 Comparison of glaze ice accretion tests and LEWICE simulation for three different wing sections [40]. **a** Profiles at station A (50 inches from root). **b** Profiles at station B (25 inches from root). **c** Profiles at station C (wing root)

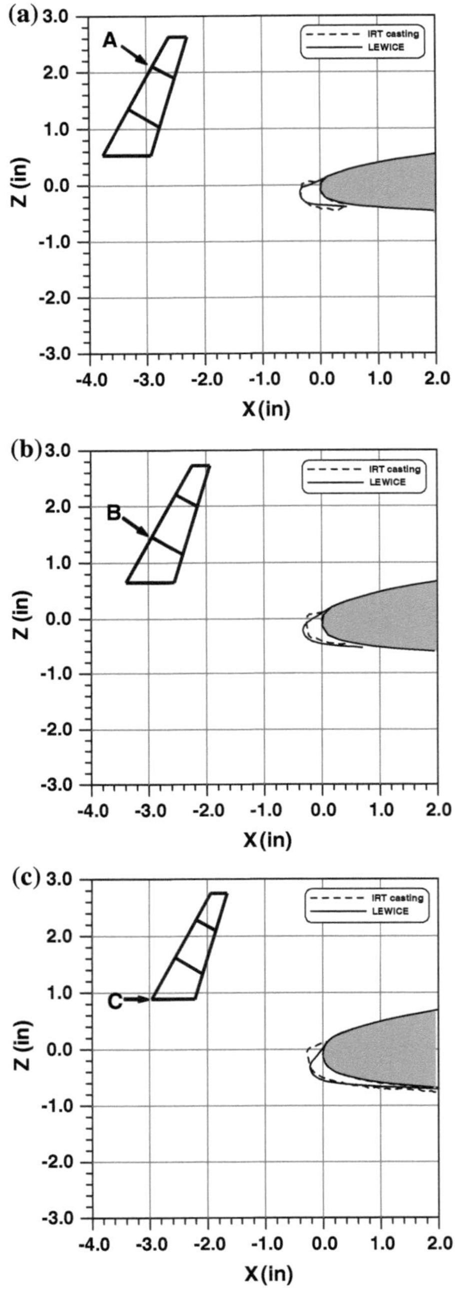

Fig. 3.20 Comparison of rime ice accretion tests and LEWICE simulation for three different wing sections [40]. **a** Profiles at station A (50 inches from root). **b** Profiles at station B (25 inches from root). **c** Profiles at station C (wing root)

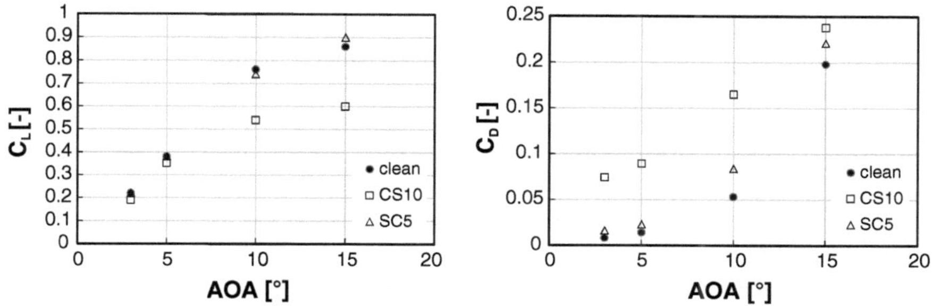

Fig. 3.21 C_L and C_D coefficients for the two situations listed in Table 3.1 (rearranged from [40])

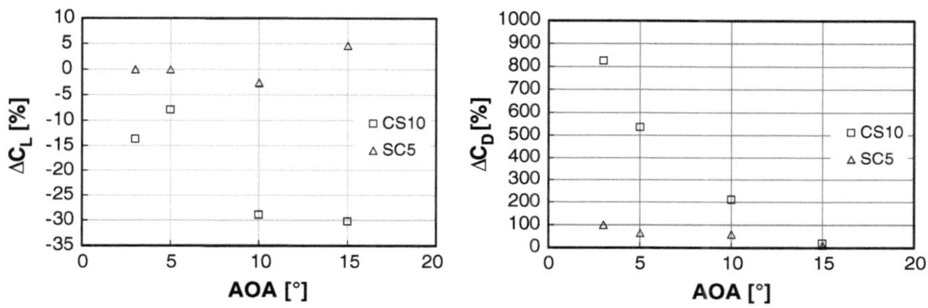

Fig. 3.22 C_L and C_D coefficients for the two situations listed in Table 3.1 in terms of percent deviation from the clean data (rearranged from [40])

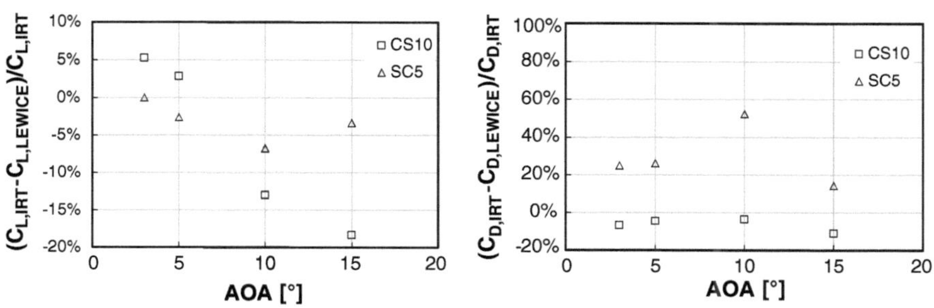

Fig. 3.23 Percent deviation of experimental C_L and C_D coefficients compared to LEWICE simulations for the two situations listed in Table 3.1 (rearranged from [40])

Under conditions of incipient icing the boundary layer is disturbed by raindrops or water particles that impact on it. This fact determines a shift of the transition point towards the blade leading edge area, which can also possibly coincide with the point of stagnation. This transition process is promoted by the complex local flow field and is called bypass transition, since it bypasses the classic Tollmien-Schlicting mechanism. At icing onset crystals are formed on the surface, usually induced by dirt or irregularities. They promote boundary layer turbulent mixing and this in turn enhances the heat exchange. The result is an acceleration of the phenomenon of ice formation and growth of protuberances altering the original shape of the profile. Depending on the thermo-fluid-dynamic conditions on the surface, growth can occur in *wet* or *dry* conditions (depending on whether there is runoff or not of water on the profile), characterised by very different forms and different ice consistencies. The analysis of the behaviour of the boundary layer is very complex, since in addition to the roughness of the irregular surface that generate tens of points of stagnation from which many boundary layers develop and interact, there is the additional problem of the runoff water on the surface. This problem is common when inter-cycle ice accretion or residual ice occurs due for instance to cyclic heating/icing steps induced by a de-icing system. Here various shapes can originate. As example, the characteristics of residual and intercycle ice accretions for a NACA 23020 airfoil along with the aerodynamic performance penalties of intercycle ice accretions have been anlysed by Broeren and Bragg [41]. The experimental ice shapes in Fig. 3.24 were generated according to the test conditions listed in Table 3.2. The nominal height (ignoring the larger ridgelike features) of ice shape 290 was $k/c = 0.0056$.

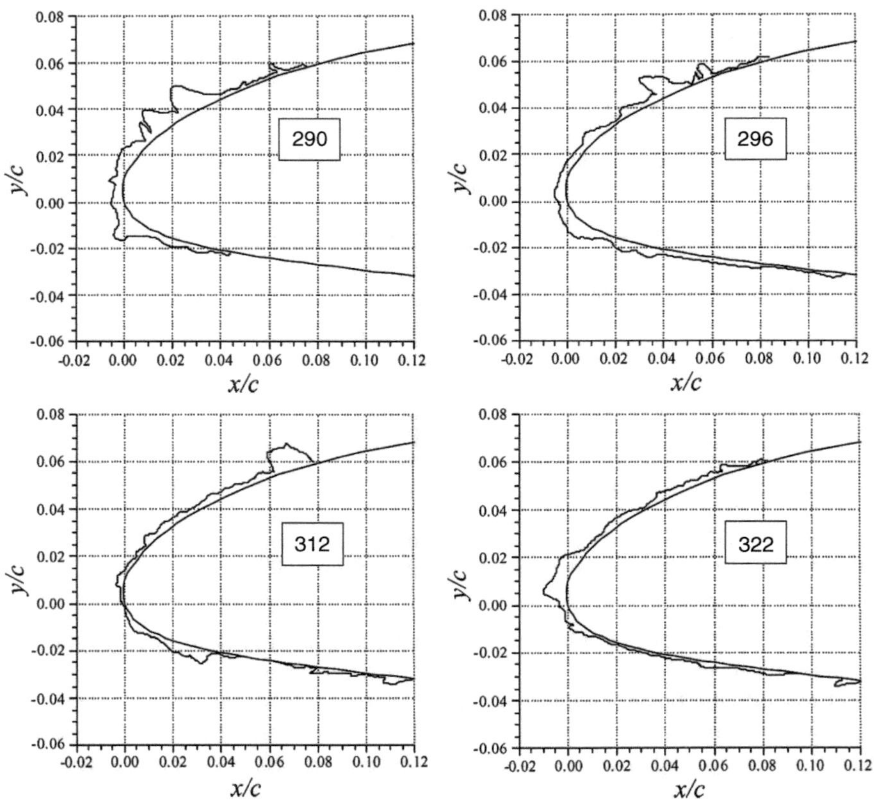

Fig. 3.24 Experimental ice shape 290 (*up* and *left*), 296 (*up* and *right*), 312 (*down* and *left*) and 322 (*down* and *right*) from [41]

Table 3.2 Ambient data leading to selected ice structures, reproduced from [41]

Ice shape name	AoA (°)	V (m/s)	Re_c (−)	MVD (−)	LWC (g/m^3)	T (°C)	Spray time (min)
290	0	89.5	6.5×10^6	20	0.45	−10.0	12
296	0	89.5	6.5×10^6	20	0.65	−6.1	12
312	0	89.5	6.5×10^6	40	0.25	−6.1	12
332	0	89.5	6.5×10^6	40	0.40	−20.0	3

All geometries, with the exception of shape 322 brought to a similar performance degradation, despite the evident differences in the geometry. As shown in Fig. 3.25, the maximum lift coefficient showed a reduction of about 60 % from the clean value of 1.8 with a remarkable degradation for AoA greater than 4°. The stalling angle was reduced from 17.5 to about 8.5°. The fast heating/icing cycle of shape 322 resulted in a slightly lower maximum lift penalty of about 50 %. Also significant is the variation in the airfoil pitching moment that becomes highly AoA dependent.

The results of lift to drag reduction are given in Fig. 3.26. The dramatic drop in aerodynamic efficiency ranges from about 15 to about 80 %.

From the examples presented it is apparent that it is not possible to establish some general quantitative conclusions on the effect of the icing conditions on the coefficients of lift, drag and moment. In fact shapes 290, 296 and 312 show almost the same aerodynamic efficiency drop despite the different ice accretion and test conditions under which they are obtained. It can be concluded only that always an alteration of the coefficient of lift, drag and moment is expected.

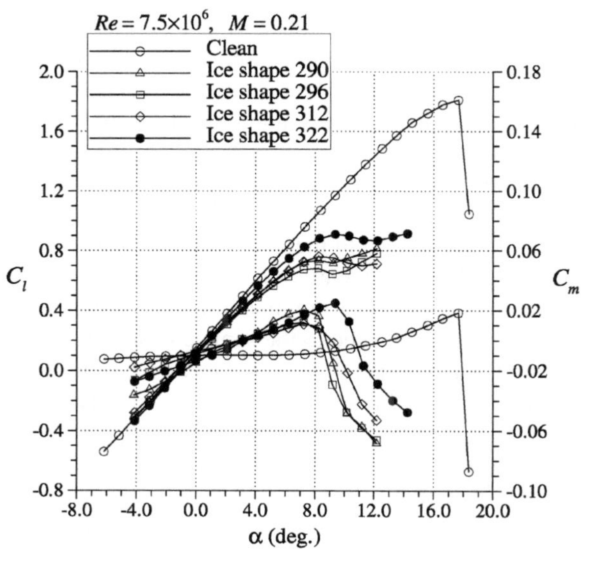

Fig. 3.25 Lift and moment versus AoA (i.e. α in the original graphic) of clean and contaminated profile [41]

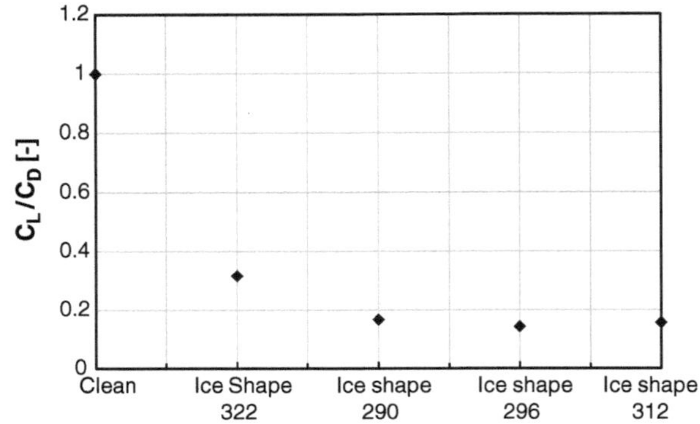

Fig. 3.26 Results of lift to drag reduction due to ice-induced roughness. Elaboration of data from [41]

Accepted evaluation criteria to establish to which extent the results of ice prediction codes adequately match the ice shapes generated in experimental test under the same boundary conditions are not yet apparent. It is nevertheless evident that the geometric comparisons alone are not sufficient to determine similarity. Analysis based on visual appearance (o digitally imagined) only indicates that different ice shapes can produce significantly different aerodynamic effects, but also the contrary holds, since small details of the geometry can make large changes.

3.6.2 Replication of Ice Real Geometry

Although the application of sandpaper to the airfoil leading edge is a common procedure to replicate ice roughness, and can be indicative of the initial state of ice accretion, it has to be emphasised that a critical problem for correct artificial reproduction of the ice roughness is the accuracy of the subscale simulations against the full-scale data. This problem is somehow related to the above discussion on the relationship between Reynolds number and relative roughness scaling. It was found (see for instance [42]) that geometric scaling of the uniform ice roughness height tended not to capture the $C_{L,max}$ and C_D when the roughness was applied at high concentration. Employing more dispersed roughness concentration would increase the fidelity of results. It is worth to mention that until now no accurate method exists for measuring roughness concentration on an ice accretion. Therefore, a crucial operation in ice subscaling derives from modeling both roughness height and concentration.

Ashenden [43] come to the same conclusion using artificial ice shapes in a low-Reynolds wind tunnel. Lower impact on lift is due to ridges growing on the pressure surface although the drag increases consistently. As shown in the case of dispersed roughness, ice ridges are expected to anticipate the stall angle. Tests carried out on 23012 airfoils with anticipated ridge at 10 % chord indicate anticipation of 9° for $56\times10^{-4}\,k/c$ and 13° for $139\times10^{-4}\,k/c$. The drop penalties obtained from these tests gave a profile drag increase by a factor of 4 and 10 respectively. Usually, the sandpaper roughness is not high enough to simulate the normal heights of ridgelike features and protruding icing elements observed in natural environment and icing tunnel. Its use is therefore limited to simulations with uniform, low dispersed roughness, typical of the first minutes of icing process.

3.7 Type of Ice and Boundary Layer

Due to the "natural" varieties of ice shapes and their 3D observed native characteristics, not only in open field but also in controlled environment, the only consistent way to qualify ice contamination can be made on the basis of the general effect on boundary layer and aerodynamics of different ice shapes (possibly 2D).

This consideration is based on the fact that the boundary layer structure depends on how pressure gradient develops along the profile, which in turn depends on profile thickness distribution along the camber line, its curvature and the pointing flow conditions. In clean conditions, families of profiles designed to have similar pressure coefficients will exhibit similar performances. This is quite evident for instance if one refers to stall behaviour. Trailing-edge stall is induced by anticipation of the turbulent boundary-layer separation point from the trailing edge with increasing angle of attack. Leading-edge stall is more abrupt with sudden flow separation near the leading edge, generally with no subsequent reattachment. The abrupt separation is usually promoted from the bursting of a small laminar separation bubble located just after the leading edge that results in a sharp decrease in lift. Thin-airfoil stall experiences flow separation at the leading edge, with reattachment (laminar separation bubble) at a point that moves progressively downstream with increasing angle of attack. It will be demonstrated that although in some cases ice can change stall types of profile, for some ice shapes (streamlined and roughness) where the clean, original shape is almost maintained the original aerodynamic behaviour tends to be preserved. It can be said therefore that profiles with a similar c_p distribution will be similarly affected by icing as the shape is maintained.

To demonstrate this statement, the following discussion presents the behaviour of a very common aircraft wing profile, the NACA 23012.

Instead of focusing on classification of ice shapes, (length and angles of horns, etc.) a much more profitable approach is to identify ice accretion having relevantly similar and recognisable effects on aerodynamics.

A very fundamental work on iced profile classification has been undertaken from the University of Illinois at Urbana, and NASA John H. Glenn Research Center at Lewis Field of Cleveland [44, 45]. Bragg et al. recognised four categories of ice accretion inducing different aerodynamic behaviour:

1. *disperse roughness*,
2. *horn*,
3. *streamwise*,
4. *spanwise ridge*.

These ice accretions are documented in Figs. 3.38, 3.39 and 3.40, taken from [46]. These ice shapes have been reproduced by moulds and the aerodynamic performance of NACA 23012 measured in a pressurized test tunnel to decouple Reynolds and Mach number effects.

3.7.1 Dispersed Roughness

Dispersed roughness is defined as the type of ice accretion that affects the boundary-layer transition process by extraction of momentum from the boundary layer. This leads to premature trailing-edge separation and increases drag and reduces lift. The distinctive nature of this aerodynamic disturbance is to produce native 3D separation. This kind of phenomena represents often the first stage of the icing process, which can evolve further to a variety of shapes. The roughness height is usually greater than the local boundary-layer thickness, also at the very early growth stage. *Roughness* is characterised by three distinctive parameters:

1. height; Shin [47] measured bead heights (k) from 0.28 to 0.79 mm, much thicker than the expected local boundary layer (δ);
2. superficial density; usually with the term *Roughness* one refers to distributed roughness on the surface. Instead single roughness elements act as isolated bodies, and aerodynamically behave as flow obstacles, originating localised 2D separation that have the characteristic separation length of the order of the single element size;
3. surface location.

The phenomena of transition is described by the achievement of a critical roughness-based Reynolds number Re_k, defined with the help of Fig. 3.27

$$Re_k = \frac{\rho \, v_k \, k}{\mu} \tag{3.2}$$

The most important feature of the dispersed roughness is the superficial concentration or density. This parameter is of utmost relevance when one tries to replicate by artificial contamination in wind tunnels the effect ice grains on airfoils surface. About the basic work of Abbott and von Doenhoff [6], details are given in the preceding paragraph on Generalities on aerodynamics of contaminated profiles. Kaups

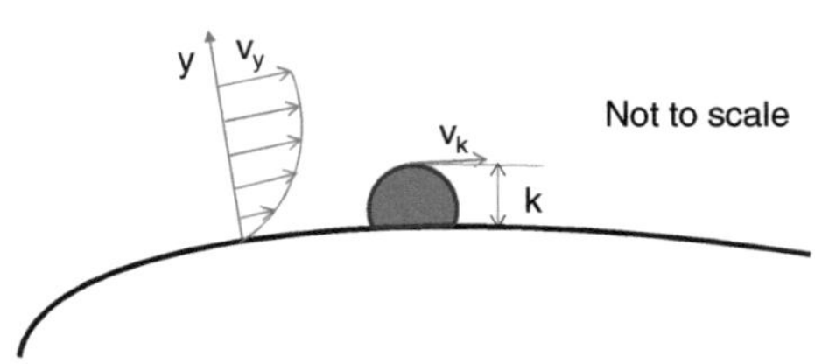

Fig. 3.27 Definition of roughness height (k) and local velocity (v_k)

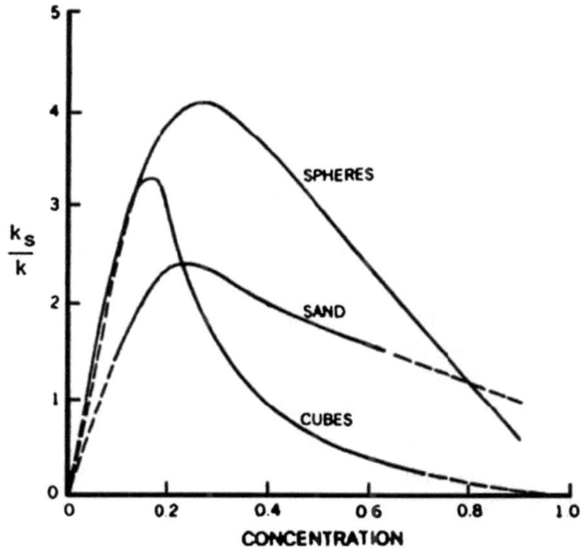

Fig. 3.28 Equivalent sandgrain roughness as a function of concentration and shape [48]

in the late 1960s [48] related an arbitrary roughness to the uniform sandgrain roughness leading to the same aerodynamic effect. He defined the concentration as the mean value of the area covered by the roughness elements. The curves in Fig. 3.28 give the ratio of equivalent sand grain roughness (k_s/k) versus roughness elements concentration and shape.

Regardless of the kind of artificial elements used to create the artificial roughness (spheres, sand or cubes), the same trend emerges: the increasing roughness density causes initially the equivalent effect to increase the roughness. However, as the density increases further the roughness effect declines due to the interaction of the elements (the downstream elements are in the wake of the upstream ones) and the equivalent effect is that of a smaller roughness.

The equivalent sand grain roughness (k_s/k) can be related to the performance drop. A highly cited study of the effect of roughness on $C_{L,max}$ and stall reduction is that of Brumby. He published a very popular graph [49], where he shows the effect on reduction in airfoil maximum lift of size, chordwise extent and location of roughness, primarily dedicated to NACA airfoils, for a wide range of roughness types and locations for a variety of Reynolds numbers. His studies prove that the maximum lift coefficient decreases as the roughness moves from the trailing edge towards the airfoil leading edge.

According to all authors cited, the leading edge location seemed to be the most sensitive, and this is reasonable because in this location the highest depression occurs in the flow, and a destabilising effect on the boundary layer can be triggered by roughness. It is also reasonable to expect that this contingency will depend not only on roughness size but also on airfoil type. This statement seems to be proved by more recent tests at Illinois University and NASA Lewis Centre (see Lee and Bragg [50]).

More severe lift drop occurs at increasing roughness size. Loss in angle of attack margin to stall decreases almost linearly with roughness.

Small leading edge roughness of the order of $k/c \geq 20 \times 10^{-4}$ causes significant degradation of aerodynamic performances. This means that if we take two typical chord lengths of MW size WT, a mid span (about 3 m) and tip span (about 1 m) section, k should be less than about 5–6 mm for the mid section and about 2 mm for the tip one to avoid unwanted aerodynamics effects. Small outboard sections are therefore less tolerant to ice formation compared to the outboard ones. A typical *optimum adverse* problem derives for the involved variables: the small dimensions of the tip sections collect more water compared to the inboard ones, and the cooling thermal flows that are also more intense in the outboard locations.

It follows that ice starts accreting always on the tip sections, and since they actually are responsible for most of the aerodynamic torque, a quite rapid decay of the performance occurs with a marginal level of ice contamination on the surface.

A data collection of maximum C_L reduction is produced in Fig. 3.29 obtained by reporting the ratio of the actual maximum lift to the clean one as function of the k/c parameter.

3.7.2 Horn Ice

Horn ice is a far more massive accretion compared to dispersed roughness that exhibits at least one protrusion orientated at a significant angle to the oncoming flow as shown in Fig. 3.30. This results in a large laminar separation bubble aft of the ice accretion with major effects on the aerodynamics. Surface velocity and pressure measurements, PIV [51] and CFD simulation [41] have put in evidence that the flow field shows a strong unsteady feature. However, looking at the flow time-averaged characteristics, the bubble locates at the airfoil leading-edge region

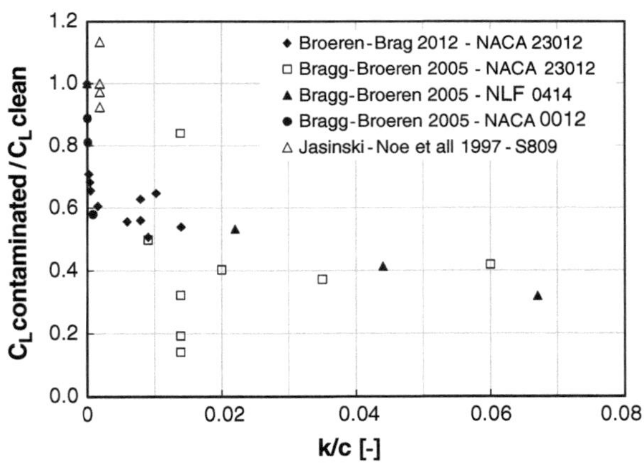

Fig. 3.29 Experimental data of C_L reduction due to surface roughness from different reference sources

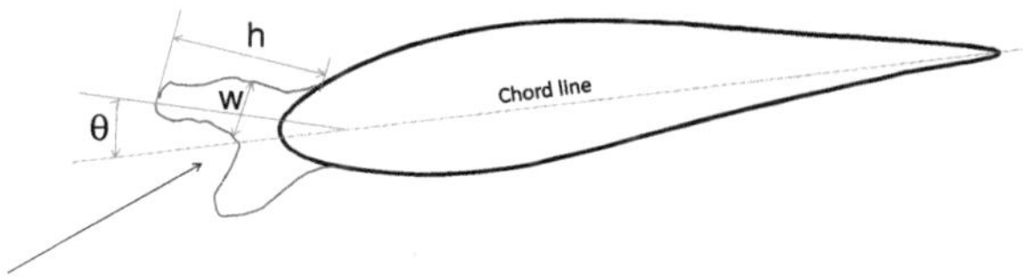

Fig. 3.30 Geometry of a horn ice shape

with the consequence that the stagnation point is shifted on the ice accretion. The phenomenon occurs usually on both airfoil surfaces, but with different intensity. On clean airfoils, this bubble is generated when the laminar boundary layer encounters an adverse pressure gradient sufficient to cause the flow separation. The larger the flow angle, more far and extended will be the region of separated streamlines. The obstacle of the horn causes the flow to separate from its edge forming a dead zone after with possibly reversal of flow at the surface. If the turbulent, rolling shear layer formed outside of the bubble conveys enough high energy from the external flow field, a pressure recovery becomes possible and the bubble reattaches somewhere just after.

As the unsteady feature of this flow is large in scale, the effects on the forces and moments are also large, with force fluctuation and unsteady behaviour on stall characteristics. This effect emphasises at high angles of attack since the reattachment point moves progressively to the trailing edge. Kim and Bragg [52] tested a typical horn shape by varying height, horn-tip radius and airfoil surface location on an NLF 0414 profile. On an about 445 mm chord profile, the roughness conditions of $k/c = 0.022$, 0.044, and 0.067, respectively, were tested with sharp, 25, and 50 % radii ($r/w = 0.00, 0.25, 0.50$, where w is the horn base width as shown in Fig. 3.30), i.e. the tip of the horn formed a wedge when the shape was sharp, and then was rounded to give the other shapes. As can be noted in Fig. 3.31, the separation bubble significantly reduces the maximum lift, and the stalling angle of attack. Little effect of the horn tip radius profile of the ice shape is expected on maximum lift coefficient and stall angle.

Further findings shows that, moving the horn location downstream caused both $C_{L,max}$ and stall angle to decrease (see Fig. 3.32. This information conjugated with the preceding ones, leads to the conclusion that the aerodynamic performance is relatively insensitive to the specific feature of the horn (roundness) and instead depends on the horn height and location. These parameters control the separation region. Since the separation point always originates from the horn tip, the surface roughness plays only a minor role in aerodynamic performance.

3.7.3 Streamwise Ice

Streamwise ice is a shape replicating the leading edge area and thus does not cause the large separated-flow regions of the horn ice, as shown in Fig. 3.33. This accretion

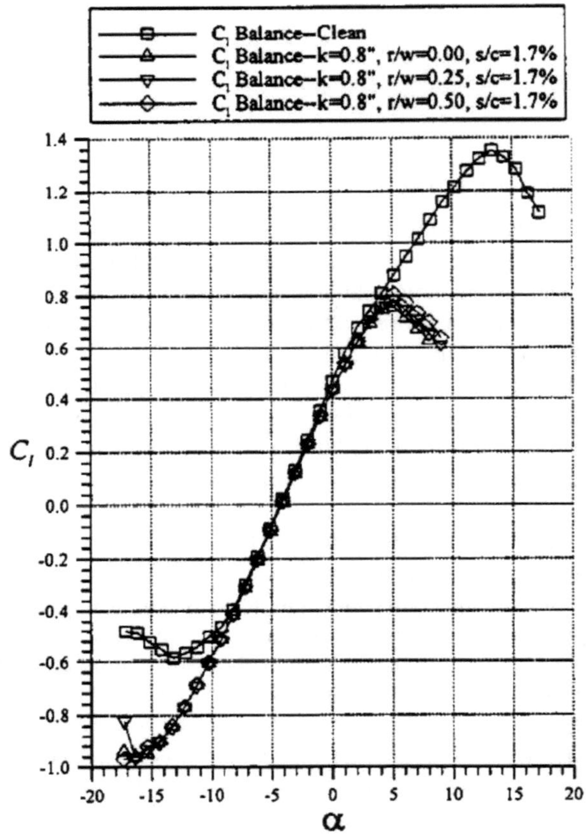

Fig. 3.31 Effect of horn tip radius on C_L at $Re = 1.8 \times 10^6$ and $M = 0.18$, for $k/c = 0.044$ horn at $s/c = 1.7\,\%$ on NLF 0414 airfoil [52]

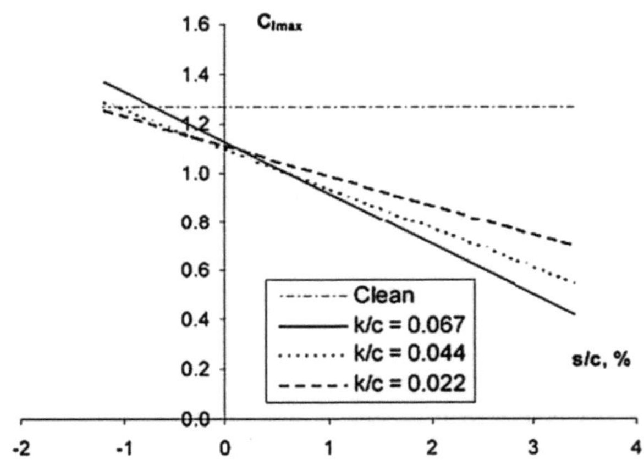

Fig. 3.32 Effect on C_L of horn location at $Re = 1.8 \times 10^6$ and $M = 0.18$, for $k/c = 0.044$ horn for NLF 0414 airfoil [52]

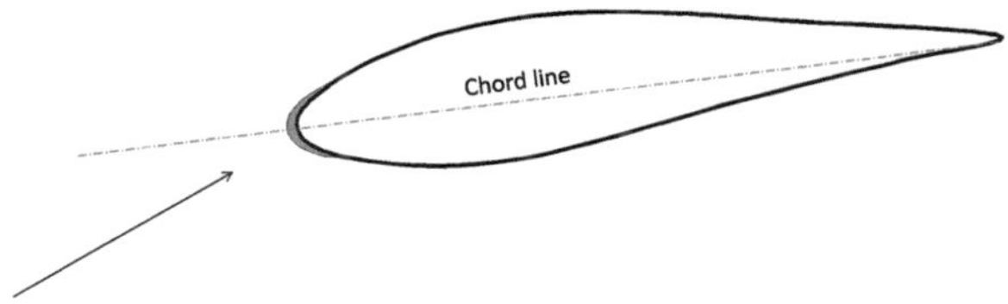

Fig. 3.33 Example of streamwise ice

often forms as a result of rime icing conditions, which occur at cold temperatures when the incoming droplets freeze on the surface at impingement. Moderate separation at ice shape scale at the ice/airfoil junction can here occur. Addition of roughness to the ice shape causes an additional increase in drag with no appreciable change in maximum lift (see the results shown for shape SC5 in Figs. 3.21 and 3.22). In some cases the extra chord generated by the ice cup is expected to compensate the light loss of lift coefficient.

3.7.4 Spanwise Ridge Ice

Spanwise ridge ice is characterised by a large separation bubble aft of the ridge formation, but differs from horn ice in that ridges are located downstream of the leading edge as shown in Fig. 3.34.

The airfoil surface located upstream of the ice ridge is comparatively smooth or can be even clean when, for instance, an anti-icing system is operating. This keeps the leading region uncontaminated, so that the stagnation point is not located on the ice accretion, and a regular boundary layer develops upstream. This leads to an additional separation bubble upstream of the ridge, thus making the three-dimensional characteristics of the ridge more important than in the horn ice case. Already in 1940, Johnson [53–55] showed in a low Reynolds number wind tunnel that lift losses and drag for suction side ridge edges were larger compared to ordinary full ice accretion on leading edge.

The protuberances caused by ridge accretion depend on ridge shape and location on the surface. Lynch et al., in a very detailed work, showed the effects of [7] small $k/c \leq 25 \times 10^{-4}$ and large $k/c \geq 50 \times 10^{-4}$ protuberance location on maximum lift as illustrated in Figs. 3.35 and 3.36 respectively. The common feature of the

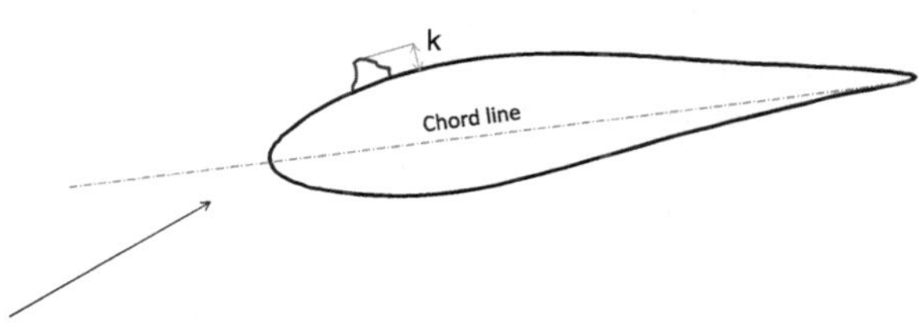

Fig. 3.34 Example of spanwise ridge

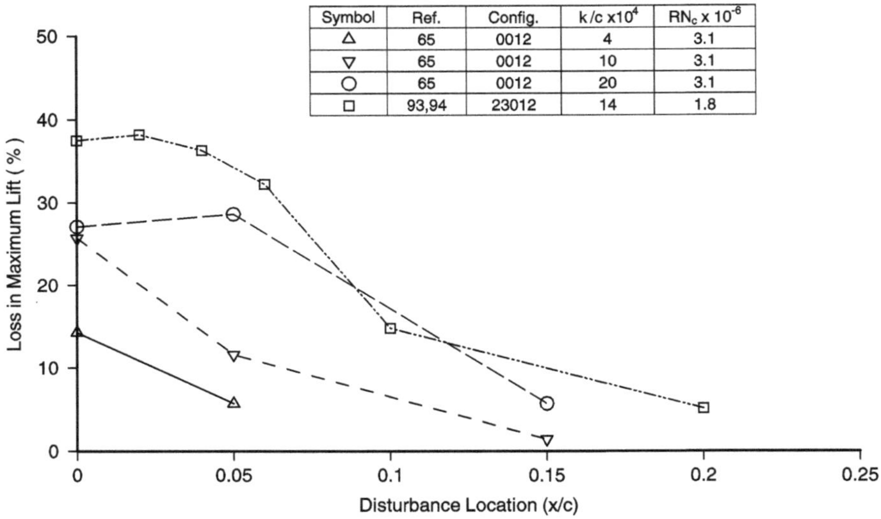

Fig. 3.35 Effect of small protuberances on lift loss [7]

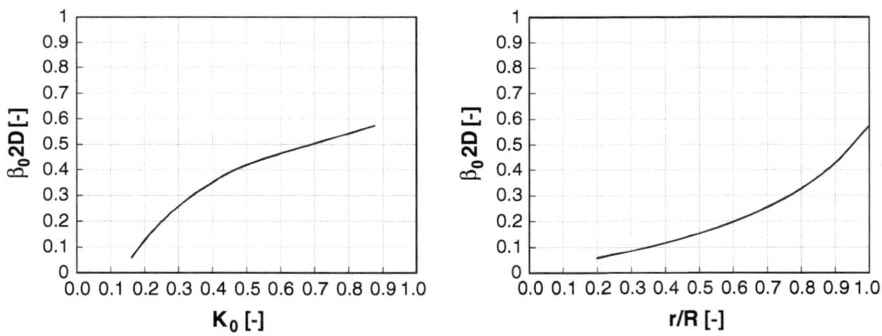

Fig. 3.36 Effect of large protuberances on lift loss [7]

diagrams is that the most critical location appears to be the leading edge area and the increase of C_L penalty with the protuberance size. Small protuberance location aft the 15 and 20 % of the chord do not seem to affect performances significantly. Also, the protuberance shape feature is important. Figure 3.37 shows clearly that the larger protuberances do not always cause the more severe penalty. Similar size protuberances can cause different penalty levels.

The latter considerations lead to the conclusions on the effect of the specific location of ice on the surface. If one looks at a typical W/W_{ave} distribution on the blade, as shown as instance in Fig. 3.5, it follows that most of the lifting capacity of the profile is produced on the near-leading edge suction surface area, where the suction peak is positioned. This region is aerodynamically very delicate, because

Symbol	Ref.	Simulation Geometry	k/c x10^4	RN_c x 10^{-6}	Clean Max. c_l
△	79	T 7 x 14	115	2.0	1.28
▽	79	T 6.5	107	2.0	1.28
□	79	S 6.5	107	2.0	1.28
▷	79	T 2.5 x 8.2	41	2.0	1.28
●	95	¼-Round	139	1.8	1.33

Airfoil: Mod 63A212 (Twin Otter Tail)

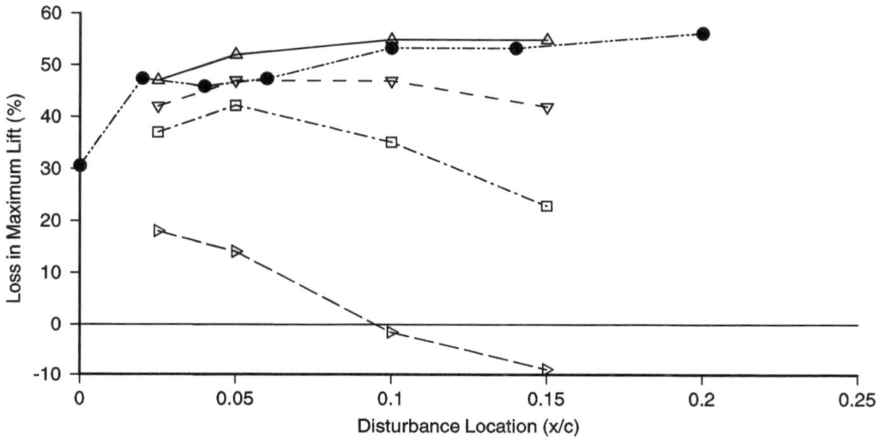

Fig. 3.37 Effect of protuberances shape on lift loss [7]

it is dominated by a severe adverse pressure gradient just after the peak, where the boundary layer loses momentum and becomes more prone to separate with a resulting loss of lift and increase of drag. The roughness changes the boundary layer characteristics in a way that separation is promoted earlier. The situation is not so dramatic on the pressure side, so as we can say that for a proper operation it is essential to keep clear from ice the upper surface aft the leading edge point. Since the angle of attack can change during operation, the impingement area will achieve a larger compass, and larger will be the wetted area to be protected against ice accretion. This is particularly important to correctly set the maximum upper (suction side) impingement limit.

An exemplary work was carried out by Bragg and the research group of the Illinois University [46] on aircraft NASA 23012 airfoil, and a wide selection of this work is reported in the following because it is helpful from the simulation and prediction perspective in wind energy. On the basis of the conclusions drawn, some consideration can be extended to comment on WT operation in icing conditions. Tests in icing wind tunnel were carried out with the ambient data given in Table 3.3, generating the basic ice structures classified in the same table.

The horn shape in Fig. 3.38 (left) shows a typical upper-surface horn occurring in glaze-type accretion. The streamwise shape in Fig. 3.38 (right) almost reproduces the leading-edge radius, with a smooth zone on the nose followed by downstream rime feather roughness.

Table 3.3 Ambient data leading to selected ice structures, reproduced from [41]

Test name	Type	Airspeed (m/s)	AoA (°)	MVD (μm)	LWC (g/m^3)	T (°C)	Spray time (min)
EG1164	Horn	78.2	5.0	20	0.85	−6.2	11.3
EG1162	Streamwise (streamwise 1)	67.0	2.0	30	0.55	−25.3	10.0
EG1126	Glaze roughness (roughness 1)	89.4	2.0	20	0.50	−7.4	2.0
EG1159	Spanwise ridge	67.0	1.5	20	0.81	−9.6	15.0
EG1125	Rime roughness (streamwise 2)	89.4	2.0	15	0.30	−20.7	20.0
EG1134	Fine rime roughness (roughness 2)	89.4	2.0	40	0.55	−20.7	2.0

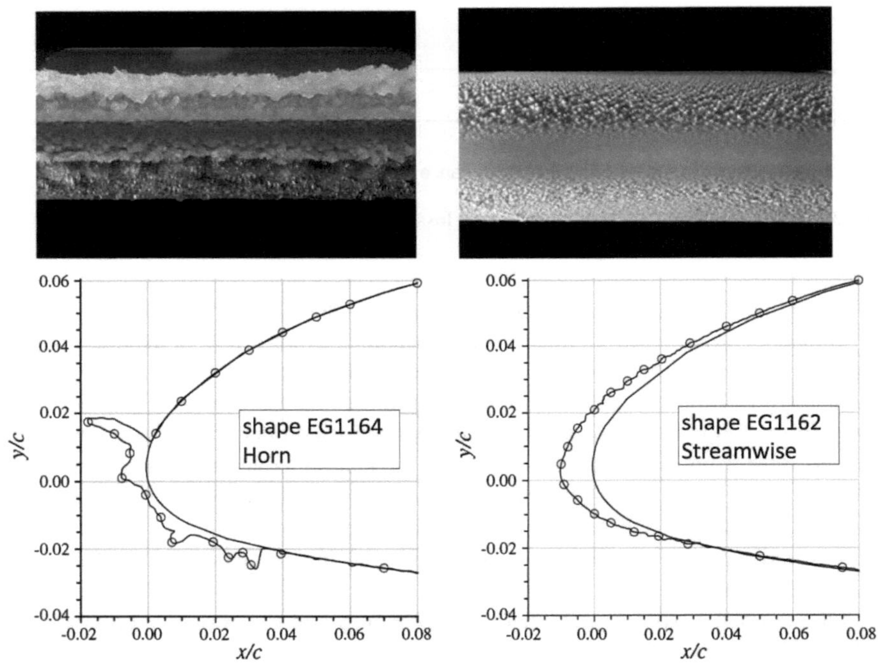

Fig. 3.38 Picture and mould reproducing shapes of horns (*left*) and streamwise ice (*right*) [46]

The glaze roughness in Fig. 3.39 (left) has a smooth zone in the stagnation region followed by large roughness downstream. The spanwise-ridge shape in Fig. 3.39 (right) was formed by the consequence of heating the leading edge. The heat input and icing conditions were adjusted to accrete the upper and lower surface ridges shown in Fig. 3.40 (left).

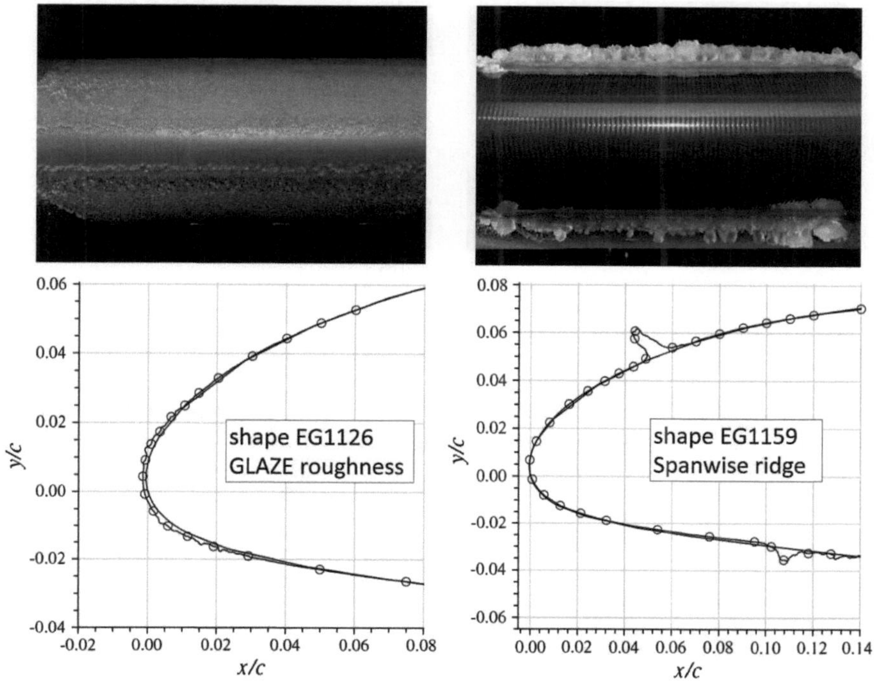

Fig. 3.39 Picture and mould reproducing shapes of glaze roughness (*left*) and spanwise ridge (*right*) [46]

The streamwise shape in Fig. 3.40 (right) has a more pointed geometry at the leading edge and is less conformal to the surface than the other streamwise shape of Fig. 3.39 (left). The roughness case in Fig. 3.40 (left) was formed in cold conditions, resulting in very fine rime feathers that were very different in size and distribution from the glaze roughness case in Fig. 3.39 (right).

The resulting aerodynamic performances, measured in tunnel tests are depicted in Fig. 3.41. In general, ice has detrimental performance effects as function of the differences in the ice-accretion geometries and roughness levels. Ice roughness occurs during the initial stages of the ice accretion process before a significant ice shape develops. Streamwise and roughness shapes have a similar effect on lift, drag and pitching moment. A drop as large as about 37–39 % in $C_{L,max}$ (from 1.85 to 1.16) is observed and an anticipation of the stall angle of about 34–35 % or 6.2° (from 18.1° to 11.6°) occurs resulting from boundary-layer separation on the aft portions of the airfoil.

Horn and spanwise-ridge shapes produce rather different effects. Their size and location cause large upper-surface separation bubbles that significantly alter the flow field. The resulting maximum lift coefficient drops down to a deep 0.86 at α_{stall} of 8.8°, which leads to a 54 % reduction in $C_{L,max}$ from the clean airfoil at these

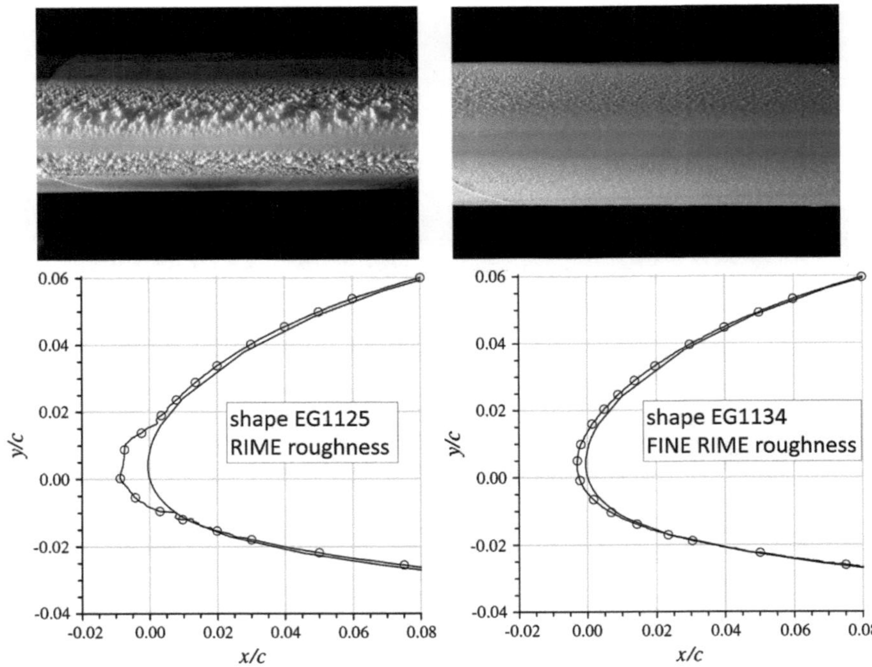

Fig. 3.40 Picture and mould reproducing shapes of low temperature formed rime roughness (*left*) and fine spanwise ridge (*right*) [46]

Reynolds and Mach number conditions. The effect of the spanwise-ridge shape is as expected even more severe, $C_{L,max} = 0.52$ at α_{stall} of 5.6°. This is a situation often encountered in electrically heated ice prevention system (IPS induced icing). Usually, only leading edge area is protected and Fig. 3.42 shows clearly that the actuation of the anti-icing system leads to melting and clearing off ice from the leading edge region, but the melted ice runs back towards the aft part causing large "ridges" accretion. This situation is schematically illustrated in the blade sections of Fig. 3.39 (right).

The highest iced-airfoil $C_{L,max}$ drop occurred with the roughness 1 shape. The drag penalty is also the largest for this shape for angles of attack greater than 9°. By contrast, the data for the roughness 2 reproduces the results of simulation of artificial 80-grit sandpaper on the 36 in or $k/c = 0.00023$, and shows the smallest maximum lift penalty measured with the $C_{L,max}$ of 1.28 being about 10 % higher than for the other ice simulations. The difference between the two surface roughnesses is due to the fact that the rime roughness 2 was accreted with a larger MVD drop size, resulting in a larger surface extent of ice roughness (the downstream iced area covering a larger surface wrap).

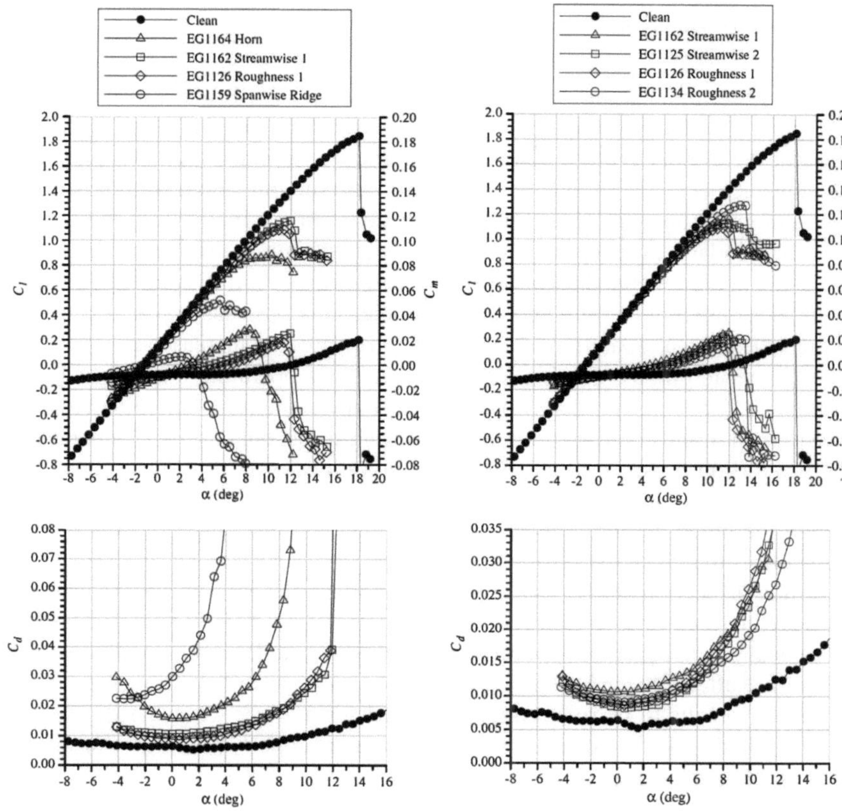

Fig. 3.41 Comparison of performance effects of the simulated ice configurations on the NACA 23012 airfoil at Re 15 to 9 × 10^6 and M = 0.20 (*left*) and comparison of performance effects of the roughness and streamwise ice simulations on the NACA 23012 airfoil at Re 15 to 9 × 10^6 and M = 0.20 (*right*) [46]

Fig. 3.42 Example of IPS-induced ice ridges accretion aft the electrically heated LE

3.7.5 Stall Behaviour

The topic of stall anticipation is particularly important for WT blades, since stall limit tends typically to change going from root to tip due to changing peripheral speed, and because of the thinner profiles employed towards the tip blade sections. Therefore tip sections, let us say with thickness to chord ratio ranging about from 9 to 12 %, exhibit a more abrupt stall onset. This stall mechanism originates from the leading edge with boundary layer separation. Transition from laminar to turbulent boundary layer occurs in the separated shear layer and reattachment of the turbulent boundary layer happens quickly entrapping a small bubble. This boundary layer feature can promote full, sudden separation and stall when triggered by increased roughness.

This discussion points out the fact that tip sections operating normally close to maximum C_L (about 0.8 $C_{L,max}$) do not have usually enough stall margin to tolerate any ice growth. Light contamination will cause an earlier stall onset accompanied by a significant drop in performance, and blade vibrations (see for instance Figs. 3.25 and 3.41). Drag is also dramatically enhanced by laminar to turbulent transition. Although the moderate thrust increase can be tolerated, the tangential force drops with the effect to reduce the aerodynamic torque.

A further inspection of Fig. 3.41 introduces another relevant comment which concerns the reduction of stall angle in contaminated operation. This effect that in aircraft leads to heavy consequences in aircraft manoeuvering under icing conditions, will have a severe effect also in WT power control. In stall controlled turbines, the maximum power results to be reduced and obtained for a lower wind speed, while in pitch controlled WT, the controller will operate not properly leading to unwanted oscillation of rotational speed and power.

3.7.6 Instationary Aerodynamic, 3D and Rotational Effects

Instationary measurements were performed for the iced and non-iced cross section in order to investigate the influence of icing on the dynamic stall behaviour. The results exhibited that no significant change in the behaviour of iced and non-iced $C_L(\alpha)$-curves arise. A shift in the zero-lift angle of attack of the iced airfoil compared to the clean one resulted as well as a lower amplitude of the maximum and minimum lift coefficients in the stall region for the iced section. In particular, these results suggested that an influence on the dynamic behaviour in the post stall region could be expected.

Aerodynamic 3D effects are pretty important and cannot in principle be neglected in WT. In particular, complex flow patterns develop in root blade regions due to the presence of local high solidity and proximity effect of large bodies as nacelle and spinner. Rotation will also cause an important migration of the water due to centrifugal force to the outer section, thus causing accumulation of water and ice in the tip sections and more severe effect of erosion and detachment of the ice shapes, making difficult the prediction of the final ice shapes.

3.8 Icing Effect on Power Production

To analyse the effect of blade contamination on the performance of the wind turbines, two rotors have been designed with the following profiles, having similar characteristics:

- an aeronautic profile traditionally used for wind turbine application: the NACA 63-425;
- a profile designed specifically for the wind turbine application: the DU 91-W2-250.

The clean and iced power and thrust curves of these wind turbines have been compared to analyse the effect of slight (roughness type) ice contamination on the profiles.

Since the performance of the profiles was only available for a Reynolds number of 3×10^6 for both clean and dirty conditions [10], the blades have been designed to have a Reynolds number of approximately this value at 75 % of the blade radius. In Fig. 3.15 the Lilienthal polars of the NACA 63-425 and DU 91-W2-250 profiles are given. It is noted that TU Delft profile is slightly more performing than the NACA in clean condition and has slightly higher maximum lift coefficient and maximum aerodynamic efficiency, which allows for a more slender blade layout. However, the differences in clean conditions are not as marked. Different is the situation for leading edge augmented roughness profile, where a dramatic reduction in performance of the NACA profile compared to Delft is observed, with regard especially to the maximum efficiency. The profile TU Delft, thanks to a careful design, based on the control of the laminar turbulent transition point and an additional load on the aft pressure side, limits more efficiency the losses in dirty conditions at leading edge. This is partially attained by lower thickness of the profile on the suction side which is proved to considerably limit the losses of the maximum lift coefficient.

Both rotors have been designed for a class III according to IEC 61400-1, (maximum average speed of 7.5 m/s), and optimized for an average speed of site of 6.75 m/s, intermediate between the maximum average speed of class IV (6 m/s) and of class III (7.5 m/s). The design speed of the rotor, according to the IEC 61400-1 and for normal design criteria, is given by $V_{\text{design}} = 1.4\, V_{\text{ave}} = 9.45$ m/s, which has been approximated to 9.5 m/s. The tip speed ratio used in the design was set to 7.5, slightly lower than that normally used in the megawatt class turbines, but more than suitable for a turbine of medium size such as is here intended to analyze. The rotor diameter was chosen to have a Reynolds number of about 3×10^6 at 75 % radius. It resulted in a rotor of 40 m in diameter.

Clearly, as the aerodynamic characteristics of the two profiles are different, two different design conditions were selected with regards to the maximum efficiency of the profiles, listed in Tables 3.4 and 3.5.

With regard to the TU Delft profile, the design point was not taken at maximum efficiency, but at a somewhat lower AoA, in order to maintain a margin from the maximum lift coefficient of about 20 %. The blade was profiled from 20 % of the radius outward, while between 20 and 10 % of the radius the blade has been joined to the hub with a circular section of 1 m diameter. The circular section is maintained

Table 3.4 Design condition for the NACA 63–425 profile equipped turbine

Profile	NACA 63–425
Design C_L (−)	1.06
Aerodynamic efficiency C_L/C_D (−)	118
Design AoA (°)	6.5
$C_{L,max}$ (−)	1.280
AoA $C_{L,max}$ (°)	10.0

Table 3.5 Design condition for the DU 91-W2-250 profile equipped turbine

Profile	DU 91-W2-250
Design C_L (−)	1.16
Aerodynamic efficiency C_L/C_D (−)	125
Design AoA (°)	6.2
$C_{L,max}$ (−)	1.370
AoA $C_{L,max}$ (°)	9.2

up to 6 % of the radius where usually the connection of the blade to the hub is placed. Optimum chord line and pitch angle have been determined by common relationships (see for instance Burton, etc. [56]).

Figure 3.43 shows, for NACA 63-425 profile, the comparison between the results of the ideal chord distribution c_{opt}, the linearized chord c_u and the optimum one

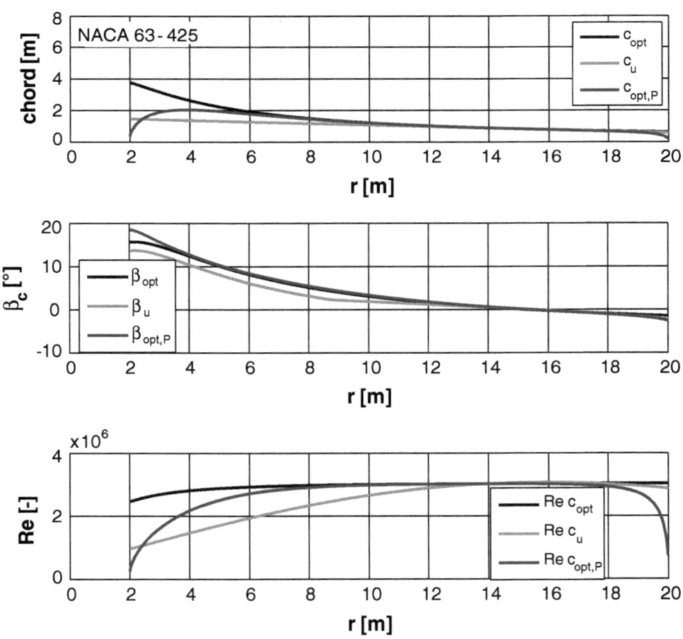

Fig. 3.43 NACA 63-425 blade geometric design

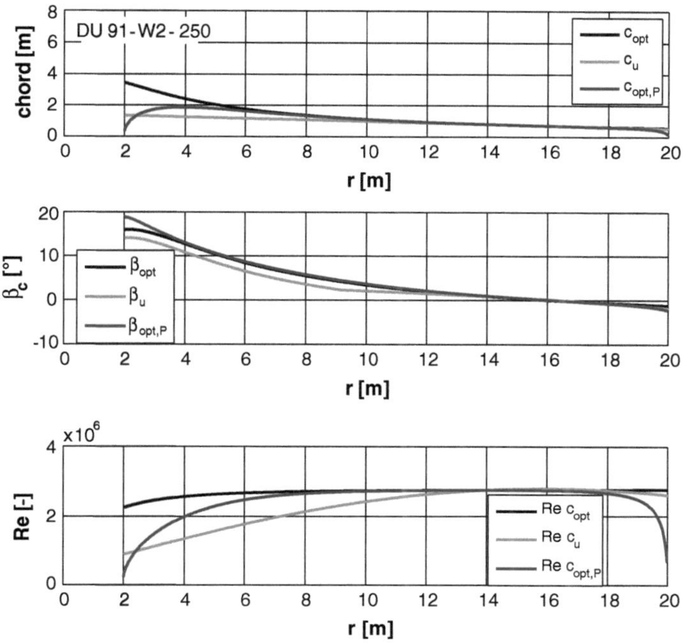

Fig. 3.44 DU 91-W2-250 blade geometric design

$c_{opt,P}$, by considering tip and hub losses according to the simplified formulation of Prandtl. The latter was then used for rotor design. Figure 3.44 shows the same output for DU 91-W2-250 profile.

These geometries were then included in the BEM code WT_Perf [57] to evaluate the performance of the two rotors under normal and light icing conditions. The profile database has been extended over 360 degrees with the method outlined in Battisti [2]. Because the behaviour of the profiles was not known up to overt stall conditions and there are no models (a part of the experimental data shown in the previous paragraphs for some profiles) to predict the change in stall angle for conditions of contamination, the performance of the rotor at stall has not been analysed.

This choice is however questionable, since nowadays almost the totality of the turbines of medium and large size works with pitch adjustment, for which the profile normally is operating in not stalled conditions. The results of the simulations have been performed for a speed of rotation of the rotor of 34 rpm, corresponding to the design condition of a wind velocity of 9.5 m/s and tip speed ratio of 7.5. In Figs. 3.43 and 3.44 twist β_c and Reynolds number Re as function of radius are shown.

Power coefficient, power, flapwise bending moment and thrust for the NACA 63-425 equipped rotor in clean conditions are shown in Fig. 3.45, while the same outputs are displayed in Fig. 3.46 for the DU 91-W2-250. In clean conditions both rotors give very comparable output. Performances of the same rotors operating in

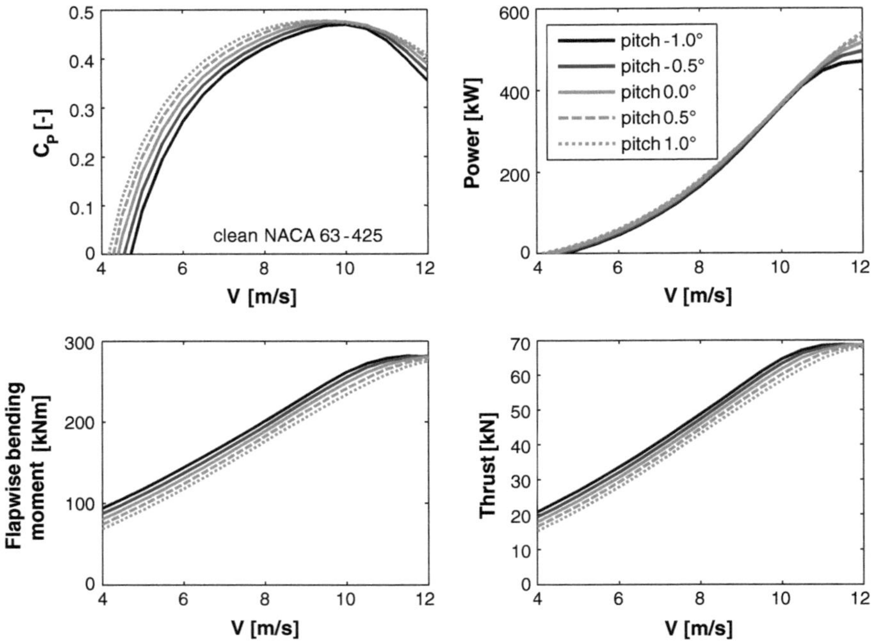

Fig. 3.45 C_P curve, power curve, flapwise bending moment and thrust for the NACA 63-425 equipped rotor in clean conditions

icing conditions are shown in Figs. 3.47 and 3.48, for the NACA 63-425 and the DU 91-W2-250 respectively. One can notice immediately that even light icing conditions cause disastrous effects on the behaviour of the NACA 63-425 equipped rotor. The power coefficient is strongly reduced and the power curve is greatly altered. The control setting becomes completely ineffective as consequence. Stopping of WT in this condition is advisable to avoid loss of control.

Contrarily, the DU 91-W2-250 shows a valuable bur gentle degradation, with a minor affect on the operation of the turbine, which keeps a quite predictable behaviour. To conclude, the example shows the ability of a profile designed for fouled condition to operate satisfactorily also in light icing conditions.

3.9 Influence of Ice on Turbine Aeroelastic Behaviour

The aerodynamics and loads of iced wind turbine rotors have been analysed in the context of the WECO programme [19, 58] and in [59]. In [19, 58], a set of airfoils reproducing natural icing was produced by recreating the ice fragments collected from the ground near a wind turbine operating under icing conditions. The aeroelastic behaviour of wind turbines operating during icing events was addressed in [60, 61] by comparing measured time series with numerical simulations. In [60], the iced rotor

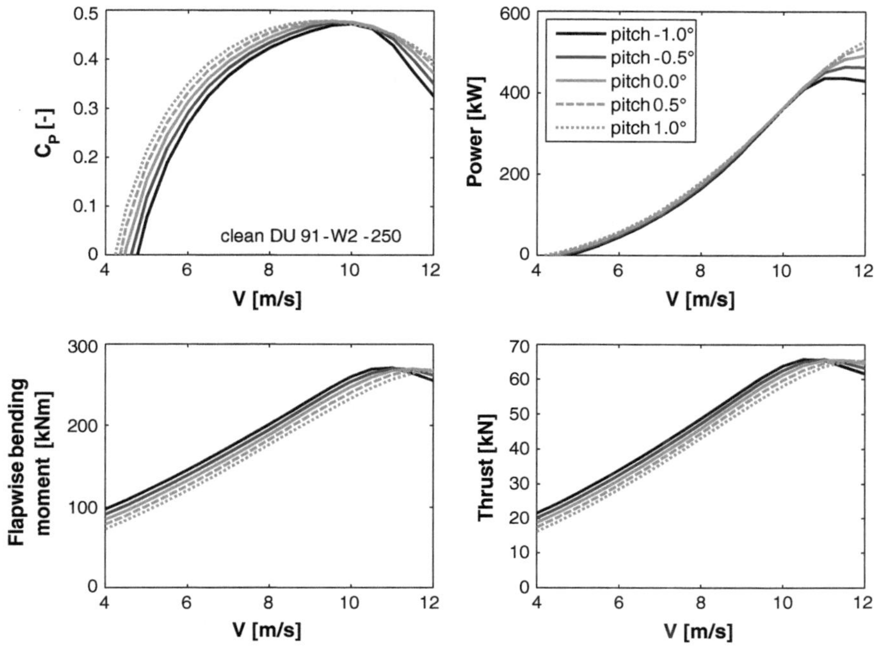

Fig. 3.46 C_P curve, power curve, flapwise bending moment and thrust for the DU 91-W2-250 equipped rotor in clean conditions

was modelled by changing the aerodynamic performance of the airfoils. A constant increase in the drag coefficient and a constant decrease in the slope of the linear part of the lift coefficient curve were applied. In [61], the iced rotor was modelled by adding a pitch difference and a mass unbalance to one of the three blades. A sensitivity analysis was conducted to achieve better matching between the frequency spectra of the measured time series and numerical simulations.

The aeroelastic analysis is possible on the strength of the typical tools available: accessible information concerning the characteristics of wind turbine rotors contaminated by ice and commercial aeroelastic codes. Thanks to Stig Øye, of the Fluid Mechanics Section at DTU, who developed a specific version of the FLEX5® code [62] to take the characteristics of the iced rotor into account, and the Centro Italiano Ricerche Aerospaziali (C.I.R.A.) for conducting a series of ice accretion simulations on the data of the Tjæreborg Turbine a series of specific numerical tests were carried out. The results have been extensively discussed with Prof. M. Hansen of DTU, who contributed to this section and my former Ph. D student G. Soraperra, who devoted a long work to customize the FLEX5® code [62].

The main characteristics of the FLEX5® aeroelastic code [62] are as follows:

1. it simulates the operation of horizontal-axis, fixed or variable-speed, pitch or stall-controlled wind turbines with 1-3 blades;

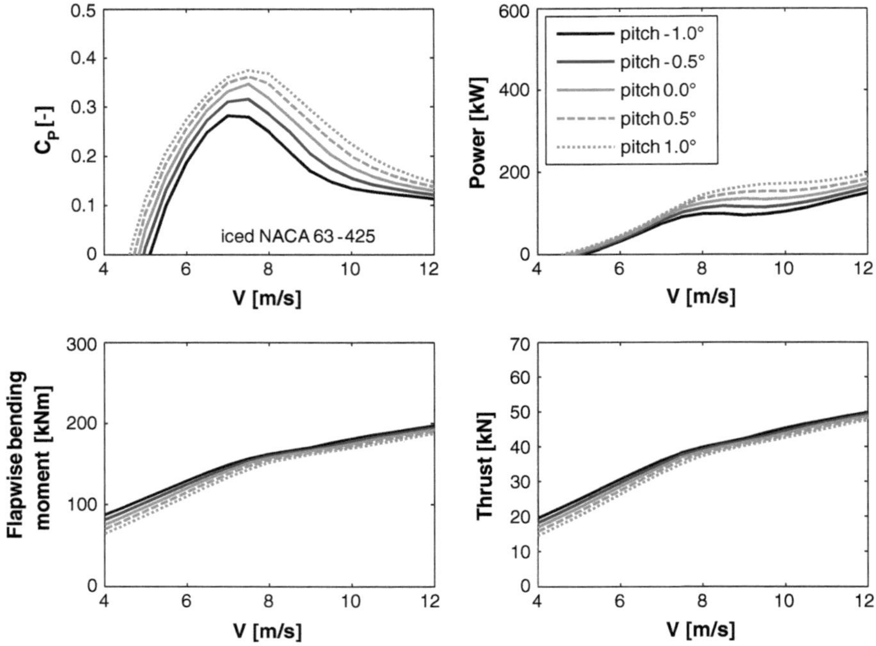

Fig. 3.47 C_P curve, power curve, flapwise bending moment and thrust for the NACA 63-425 equipped rotor in icing conditions

2. it runs in time domains, producing time series that are directly comparable with measurements;
3. it is based on a structural model with relatively few, but important degrees of freedom to describe the rigid body motion and elastic deformation of the turbine;
4. it simulates transient operation, e.g. start-ups and shutdowns due to pitching or breaking.

These characteristics are believed to come close to the optimum as concerns the trade-off between computational efficiency and accuracy. Further details are available in [27]. The code is widely used by wind turbine manufacturers and wind turbine installers to predict loads in the design stage, test the control systems and verify the calculations prescribed by the IEC 61400 series norms.

3.9.1 The Aeroelastic Model

The modelled wind turbine was based on an actual MW-class wind turbine built in the late 1980s for research purposes. The original wind turbine, named the Tjæreborg turbine after the location where it was installed, was a horizontal-axis, three-blade, upwind and pitch-regulated wind turbine. It was equipped with a 4-pole asynchronous

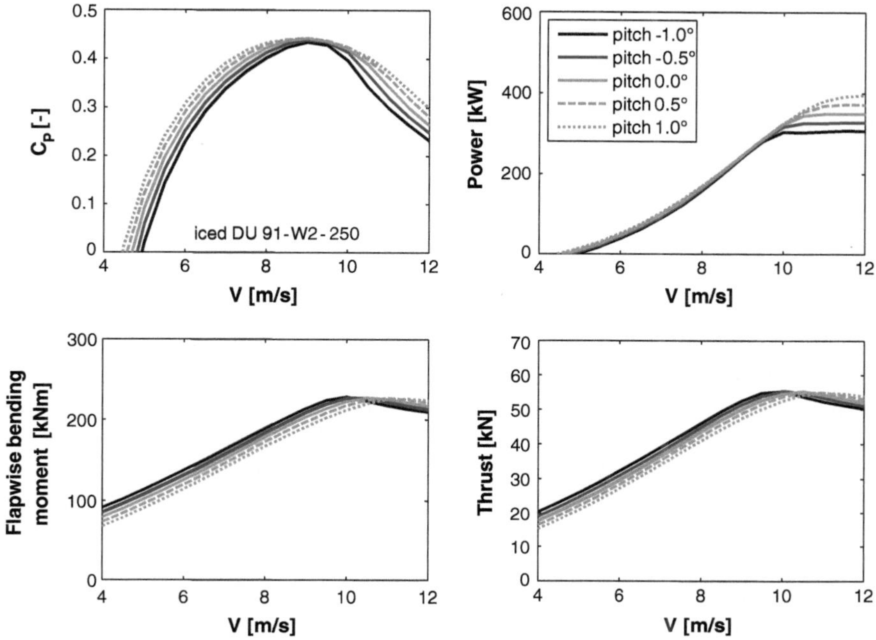

Fig. 3.48 C_P curve, power curve, flapwise bending moment and thrust for the DU 91-W2-250 equipped rotor in icing conditions

Table 3.6 Main characteristics of the original Tjæreborg turbine

Variable	Value
Rotor diameter (m)	61
Number of blades	3
Hub height (m)	0
Cone angle (°)	0
Tilt angle (°)	3
Synchronous rotor speed (rpm)	21.93
Rated power rotor speed (rpm)	22.36
Cut in wind speed (m/s)	5.0
Rated wind speed (m/s)	14.3
Cut-out wind speed (m/s)	25.0
Rated power (kW)	2000.0

generator and it had a reinforced concrete tower. Table 3.6 lists the main characteristics of the original Tjæreborg turbine. The airfoil sections of the rotor blades were the NACA 44XX family with a thickness-to-chord ratio ranging from 12 to 30 %. The same profile family was tested in [19, 58] at different icing levels.

Table 3.7 Light versus heavy configuration of the Tjæreborg turbine

Wind turbine	Lighter Tjæreborg	Original Tjæreborg
Blade mass (kg)	7,963	7,963
Rotor mass (kg)	42,500	42,500
Nacelle mass (kg)	80,000	154,000
Tower mass (kg)	86,200	550,000
First tower eigenfrequency (Hz)	0.60	0.81

A complete set of inputs for the aeroelastic model, FLEX5®, for the Tjæreborg turbine was available because the machine has been extensively analysed in the past. The main changes needed to simulate an iced rotor concerned the airfoil database, the introduction of a magnification factor for the blade mass distribution and the overall mass of the nacelle, and the features of the tower. The iced rotor model is presented in the next section. A lighter nacelle and a lighter steel tower were taken into account for most of the analysis. These changes are believed to make the aeroelastic behaviour of the Tjæreborg turbine more similar to that of modern MW-class wind turbines. The characteristics of the new, lighter Tjæreborg turbine with a steel tower are given in Table 3.7 and compared with those of the original. The blade mass was not changed, so neither was the weight of the hub. The first eigenfrequencies of the Tjæreborg turbine tower were calculated with FLEX5® and are shown in the last row.

In real applications, sensors (inertial sensors, etc.) might stop the turbine whenever vibrations become excessive. This possibility was not considered in the control system model developed for this study because this analysis focuses on evaluating the additional loads determined by generating power during icing events.

3.9.2 Iced Rotor Physical Model

The choice to start from a relatively poor level of information makes necessary to adopt arbitrary assumptions. This approach limits the accuracy of the estimations under a quantitative point of view, but makes the analysis applicable to a large number of cases. On the other hand, using high quality input information would make the results valid for a specific machine to a specific site.

The mass distribution for the iced rotor was either modelled analysing the results of numerical ice accretion simulation and following the Germanischer Lloyd standard.

A set of numerical simulations was done in collaboration with the Centro Italiano di Ricerche Aerospaziali (C.I.R.A.) using the MULTICE code [63] developed for aeronautical purposes and specifically adapted to wind turbine rotors. The meteorological input as liquid water content LWC, mean droplet diameter MVD, humidity, etc., are those of a typical mountainous site in potential icing conditions. The results of the simulation consisted of the production of the ice shapes for a set of spanwise blade sections. Figure 3.49 shows a typical ice accretion for the tip section (radius:

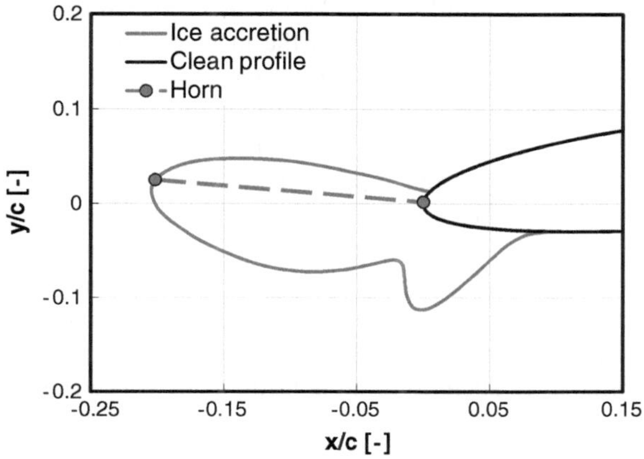

Fig. 3.49 Icing conditions for the rotor tip sections (radius: R = 30.5 m, profile: NACA 4412, liquid water content: LWC = 0.8 g/m^3, mean droplet diameter: MVD = 20 μm, absolute temperature T = 267.15 K, pressure: p = 90 kPa, humidity: Rh = 98 %, relative velocity: w = 71.92 m/s, chord size: c = 0.924 m, AOA = 5.07°, accretion time: 180 min)

r = 30.5 m, profile: NACA 4412, liquid water content: LWC = 0.8 g/m^3, median droplet diameter: MVD = 20 μm, absolute temperature T = 267.15 K, pressure: p = 90 kPa, humidity: Rh = 98 %, relative velocity: w = 71.92 m/s, chord size: c = 0.924 m, AoOA = 5.07°, accretion time: 180 min).

A point on the ice contour and a point on the profile contour have been selected for every ice shape to define the maximum ice thickness (horn length).

The main characteristics of the ice accretion for a set of radial stations along a rotor blade are listed in Table 3.8. The simulations were carried out for the same environmental condition stated before. The first row contains the radius r, the second row contains the extension of the iced area around the leading edge of the profile E, the third row contains ice mass per unit metre m_{ice} and the last row contains the maximum ice thickness t_{ice}. No ice formed for r< 12.46 m.

A data set consisting of an ice mass distribution m_{ice}(r), and the performances of the airfoils C_L, C_D and C_m as a function of the angle of attack, for a set of stations along the spanwise direction of the blade, is defined here as a Contamination Level (CL). The ice accretion simulations gave a first mass distribution that could be directly added to the blade mass distribution.

In alternative, the ice mass distributions m_{ice}(r) were calculated according to the Germanischer Lloyd standard for estimating the ice's distribution and the load it induces (Germanisher Lloyd GL Wind 2010—Chapter 4–4.2.4.2.2 Ice) [64]. The ice's mass distribution is assumed to grow linearly up to 50 % of the radius and to

Table 3.8 Ice accretion along the blade (profile: NACA 4412, LWC = 0.8 g/m^3, MVD = 20 μm, T = 267.15 K, p = 100 kPa, w = 71.92 m/s, c = 0.924 m, α = 5.07°, accretion time = 180 min)

r (m)	12.46	15.46	18.46	21.46	24.46	27.46	28.96	29.86	30.5
E (m)	0.114	0.194	0.203	0.104	0.359	0.163	0.175	0.162	0.120
m_{ice} (kg/m)	0.00	0.14	1.54	0.84	10.01	6.16	11.90	15.75	16.38
t_{ice} (m)	0.000	0.013	0.092	0.053	0.189	0.149	0.177	0.182	0.191

be constant from 50 % of the radius to the tip (trapezoidal ice mass distributions). Four values were chosen for the mass per unit length at half the rotor radius, m_E: 5.20 kg/m, 9.44 kg/m, 18.89 kg/m and 37.76 kg/m.

The mass distribution, achieved from the numeric simulations and the four trapezoidal ice mass distributions, define the five CLs used for the aeroelastic analysis of the Tjæreborg turbine operating during icing events.

The available version of FLEX5© code accepts as an input only a mass magnification factor to be multiplied to the clean blade mass distribution. For every CL, the magnification factor $K_{bending}$ that reproduces the blade root bending moment of the iced case was calculated. The performances of the iced airfoils where calculated by interpolating the results of experimental tests on a NACA 4415 airfoil with three artificial ice accretions at the leading edge. The ratio t_{ice}/c between the thickness of the ice layer and the chord size was used to characterise the airfoil performances database. The ice accretion simulations gave a first ice thickness distribution (see Table 3.8, last row). In alternative, the thickness of the ice layer for the CLs achieved from the Germanischer Lloyd standard was calculated as follows:

$$t_{ice}(r) = \frac{m_{ice}(r)}{\frac{1}{2}\rho_{ice} E(r)(k_{taper} + 1)}$$

where E(r) is the extension of the area covered by ice at a given radius, k_{taper} is an arbitrary parameter that takes into account for the shape of the ice cross-section and ρ_{ice} is the ice's density. E(r) and k_{taper} have in general to be assumed by observing the qualitative characteristics of typical ice accretion on wind turbine rotors, or by inferring them by ice accretion calculations using codes as MULTICE [27]. In this analysis, the mean value of the E(r) row in Table 3.8 was adopted together with $k_{taper} = 1/8$ for all the radial stations. This approach gave an ice thickness distribution that has the same trapezoidal trend of the ice mass distribution.

A linear interpolation between the experimental airfoil performance databases with respect to the actual value of t_{ice}/c was obtained for each station implemented along the spanwise direction of the blade. Three to five stations were implemented along the spanwise direction of the blade. The main data of the contamination levels are given in Table 3.9.

Table 3.9 Main data for the adopted contamination levels

Contamination level	CL-O	CL-S	CL-1	CL-2	CL-3
m_E (kg/m)	–	5.20	9.44	18.89	37.76
INT ($m_{tot} \cdot r \cdot dr$) (10^5·kgm)	71.7	71.7	73.5	77.6	85.7
$K_{bending}$ (–)	1.032	1.032	1.058	1.116	1.232
$t_{ice,max}$ (m)	0.191	0.075	0.135	0.271	0.541
$t_{ice,max}/c$ (–)	20.31 %	7.93 %	14.40 %	28.80 %	57.60 %

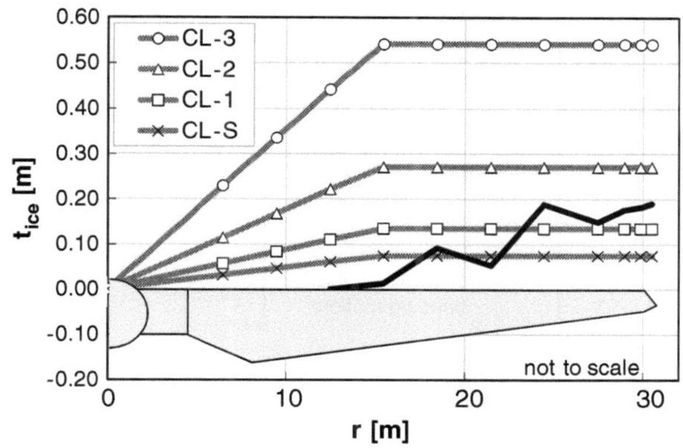

Fig. 3.50 Maximum ice thickness distributions

The first row contains the values of m_E for the trapezoidal ice mass distributions, the second row contains the moment of inertia of the blade respect to the root, the third row contains the value of $K_{bending}$ needed in FLEX5©, the fourth and the fifth row contains the maximum value of ice thickness in metres and normalised respect to the chord size. The columns contain the five contamination levels. CL-O is obtained with the ice mass distribution Observed in the numeric simulations, CL-S is a Special contamination level developed from the GL-standards that has the same blade root bending moment of CL-O CL-1 to 3 having progressively more severe contamination, have been developed in order to check the influence of the effects of the ice accretion.

In Fig. 3.50 for each blade station, the resulting ice thickness distributions for the five contamination levels are plotted.

It is worth mentioning two more effects that are not considered in the present model. The first has to do with the use of thermal ice prevention systems, in which case the blade material works at a higher temperature, with an effect on its mechanical properties (elastic modulus and stiffness). Secondly, ice contamination on sensors can reduce the control system's ability to prevent dangerous wind turbine operating conditions. Thus clean sensors have been adopted for this analysis.

The numerical simulation aimed to identify changes in the fatigue loads in icing conditions of the Tjæreborg turbine during operation. The extreme load assessment of IEC standard was not performed for the present study because the effect of operation during icing events is believed to mainly affect the cyclic loads. The load time series were post-processed with a rain flow count algorithm to achieve the equivalent load ranges for a reference frequency of 1 Hz. The linear damage theory was applied, assuming a given exponent for the Wöhler coefficient.

Figure 3.51 depicts a diagram of the wind turbine motions for the main monitored moments.

Two types of analyses were conducted: a sensitivity analysis, where the equivalent load ranges were compared directly, and a lifetime assessment, where the equivalent load ranges were calculated from a set of load cases, taking the probability of each load case into account.

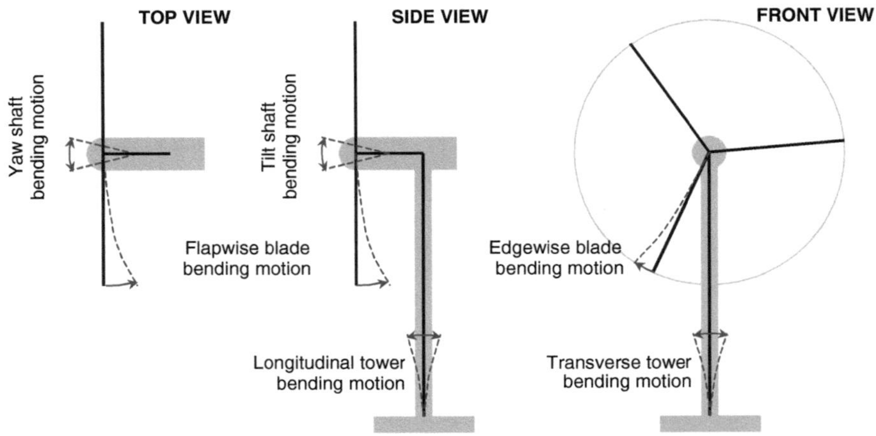

Fig. 3.51 Wind turbine motions for the main monitored moments

3.9.3 Sensitivity Analysis on the Physical Model

A first campaign of simulations was carried out to identify the effect of options chosen in the physical model for the wind turbine's aeroelastic behaviour. A single load case was simulated with different physical models of the iced rotor, i.e.

- changes in the airfoil aerodynamic performance data alone;
- changes in the mass distribution alone;
- simultaneous mass distribution and airfoil aerodynamic performance data changes.

The chosen load case corresponds to a power production characterised by a constant wind speed at hub height of 15 m/s and a wind gradient exponent of 0.14. Further details on the conditions adopted for the sensitivity analysis are in [65]. The sensitivity analysis led to the following main results:

1. the effect of icing on airfoil aerodynamics mainly influences the power output and has a lesser effect on the loads;
2. the main increase in the loads is determined on the tower root bending moments; some increase was observed in the load of the blade root bending moments and in the shaft bending moments. Only these sensors were therefore monitored in the subsequent analysis.

3.9.4 20-Year Fatigue Lifetime Assessment

Table 3.10 shows the set-up for the numerical simulations performed to assess the 20-year fatigue lifetime. The operating range of the Tjæreborg turbine ($V_{\text{cut,in}} = 5$ m/s, $V_{\text{cut,off}} = 25$ m/s; see Table 3.6) was divided into ten bins of 2 m/s of width.

Table 3.10 Set-up for calculating the 20-year lifetime

V range (m/s)	I (−) (%)	N_{Tot} (−)	N_{Clean} (−)	N_{CL-1} (−)	$N_{Deicing}$ (−)	N_{CL-2} (−)	$N_{Deicing}$ (−)	N_{CL-3} (−)	$N_{Deicing}$ (−)
5–7	18.0	35,697	28,558	5,711		1,142	286		
7–9	16.0	30,486	24,389	4,878		975	244		
9–11	14.5	22,089	17,671	3,534		707	177		
11–13	13.0	13,906	11,125	2,225		445	111		
13–15	13.0	7,708	6,166	1,234	6.2	246	1.2	62	0.3
15–17	13.0	3,793	3,034	607		122	30		
17–19	13.0	1,667	1,334						
19–21	13.0	657	526						
21–23	13.0	233	186						
23–25	13.0	74	59						

The iced rotor was simulated with 3 and with 2 iced blades for each contamination level and for each wind speed bin. Sudden de-icing was also simulated.

Operation during icing events was only simulated for the more populated wind speed bins from V = 5 m/s to 17 m/s. No power was generated during the tests in the wind speed bins from V = 5 m/s to 7 m/s with the iced rotor. These time series were not taken into account for the overall lifetime assessment because the power is very low during operation with contaminated airfoils.

In all, 10 simulations with a clean rotor and 11 for each contamination level were processed for the overall lifetime assessment.

All simulations except for those concerning sudden de-icing have a constant mean wind speed and a wind gradient exponent of 0.14. Further, a turbulent wind field is constructed on the basis of I which is given in the second column of Table 3.10. The light configuration of the Tjæreborg Turbine was selected for all these simulations. A Weibull distribution with a size parameter of 8 m/s and a shape parameter of 1.9 was chosen to calculate the load case probabilities. The number of hours of operation for each wind speed and contamination level is given in Table 3.10. The number of hours of operation during icing events amounts to 20 % of the total hour for a given wind speed bin N_{Tot}; this number of hours was subtracted from N_{Tot}, giving the number of hours of operation with a clean rotor N_{Clean}. The number of hours of operation in icing conditions for each wind speed bin was further divided into the three contamination levels. The percentage of hours of operation at the different contamination levels is a site-dependent parameter that can be estimated only by meteorological measurements. The chosen values are believed to be representative for a typical case.

Sudden de-icing caused by an extreme coherent gust has been simulated using a special approach.

A first simulation was run with three blades iced until the gust top. The values of the variables at the gust top were recorded and used as initialisation conditions for a second simulation with only one blade iced. The two time series were combined and the result was post-processed for the fatigue lifetime assessment.

Sudden de-icing was assumed only in the bin between V = 13 m/s and 15 m/s when the aerodynamic loads are highest. No turbulence was assumed for these simulations.

The load case probability was calculated assuming that 10 % of the 10-min events taking place between V = 13 m/s and 15 m/s determine a de-icing event.

Four types of load history were considered:

1. Clean: no icing occurs at any time, and hours of operation at each wind speed are given in the third column of Table 3.6. This load history was used as a reference standard case.
2. Balanced Icing: three iced blades are considered; the hours of operation with a clean rotor at each wind speed are given in the fourth column of Table 3.10, while the hours of operation during icing at each wind speed and contamination level are given in columns 5, 7 and 9 of Table 3.10.

3. Unbalanced Icing: this involves three-blade and two-blade icing, so the load history is the same as in the previous case, but two blades are contaminated during half the operating time with the iced rotor, (50 % 3-blade icing, 50 % 2-blade icing).
4. Unbalanced Icing with De-icing: three-blade icing, two-blade icing and sudden de-icing events take place. The sudden de-icing events (in terms of hours) listed in columns 6, 8 and 10 of Table 3.10 were added to the load history of the previous case.

The 20-year equivalent load ranges obtained for the blade root bending moments, shaft bending moments and tower root bending moments for the four above-described load histories are given in Table 3.11.

The equivalent load ranges in the cases involving operation during icing events are expressed in terms of the increase with respect to the standard case. Balanced icing increases the edgewise blade root bending moments by about 1 % (0.8 %) and the transverse tower root bending moment by about 21 % (21.4 %) (see for reference Fig. 3.51). Unbalanced icing increases the edgewise blade root bending moment by about 1 %, (0.8 %), the shaft bending moment by about 3 % and the transverse tower root bending moment by more than 400 %. For the load case probability considered in this study, sudden de-icing does not affect the 20-year fatigue lifetime.

The concept of temporal quantity event - Event Frequency Level (EFL)—was introduced in Chap. 2 to take into account the persistence of ice on the blades after the generic ice event. The temporal quantities can be, for example, the number of hours of operation with ice on one, two and three blades and the number of sudden shedding events from one or two blades.

The following procedure to integrate the aeroelastic analysis into the design path is proposed (this procedure is part of the more general system design path described in Chap. 5):

1. a set of arbitrary contamination levels (CL) and event frequency level (EFL) has to be defined;
2. the CL-EFL couples determine a matrix of arbitrarily load history cases for which the 20-lifetime assessment can be carried out;
3. a Damage level (DL) can be defined for every CL-EFL couple as the increase in the 20-year equivalent load range respect to those of the standard case;
4. the damage level matrix (DLM) can be built up for a set of chosen components of the WECS. The damage level matrix is turbine-dependent and it represents a "danger fingerprint" of a given machine that can be realised by the manufacturer or by the operator before knowing any information on the site;
5. any given site can then be fitted into one of the CL-EFL case based on its meteorological characteristics. In case the damage level exceeded a critical value $D_{critical}$, the data from the CL-EFL couple are fed as input to the design of the control system and of the anti-icing/de-icing system.

The previous procedure relays on the basic hypothesis that there is sufficient information about temporal and physical characteristics of icing of a given site

Table 3.11 20-year fatigue equivalent load ranges (reference frequency 1 Hz)

(kNm)	Clean M_{Clean}	Balanced icing $\Delta M/M_{Clean}$ (%)	Unbalanced icing $\Delta M/M_{Clean}$ (%)	Unbalanced icing + de-icing $\Delta M/M_{Clean}$ (%)
Flapwise blade root bending moment blade 1	727	−1.5	−1.4	−1.4
Edgewise blade root bending moment blade 1	1,209	0.8	0.6	0.6
Flapwise blade root bending moment blade 3	739	−1.5	−1.5	−1.5
Edgewise blade root bending moment blade 3	1,202	0.9	1.2	1.2
Yaw shaft bending moment	449	−2.5	3.1	3.1
Tilt shaft bending moment	453	−2.7	2.6	2.6
Longitudinal tower root bending moment	2,094	−2.7	−1.4	−1.2
Transverse tower root bending moment	685	21.4	482.4	482.6

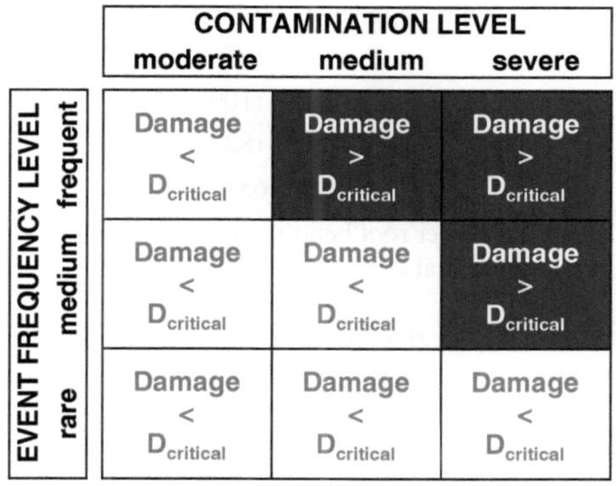

Fig. 3.52 Damage Level Matrix (DLM) representation

macro/micro area. They can be originated from direct measurements or approximate methods as the ones outlined in Chap. 2, and they can also be updated during the operational experience. In Fig. 3.52 an exemplification scheme for the Damage Level Matrix is represented.

The proposed procedure has been used for the Tjæreborg turbine processing the time series described in the previous sections. Contamination levels CL-1, CL-2 and CL-3 of Table 3.9 have been adopted. Three EFL with different percentages of operation during icing have been developed in a similar fashion as explained previously. The percentages of operation during icing for every event frequency level were: EFL-1 = 3, EFL-2 = 9 and EFL-3 = 27 %. 50 % 3-blade icing and 50 % 2-blade icing have been considered for every wind speed bin.

The damage level matrix for the edgewise blade root bending moment, yaw shaft bending moment and transverse tower root bending moment are reported in Tables 3.12, 3.13 and 3.14. The standard loads used to normalise the equivalent load ranges are shown in the caption of every table in the form of a matrix. The damage for the transverse tower root bending moment have also been normalised with respect to

Table 3.12 Damage level matrixes for the Tjæreborg turbine

	CL-1 (%)	CL-2 (%)	CL-3 (%)
EFL-3	0.9	2.8	9.4
EFL-2	0.3	0.9	3.9
EFL-1	0.1	0.3	1.4

Edgewise blade root bending moment, blade 3 (standard load $M_0 = 1{,}202\,\text{kNm}$)

Table 3.13 Damage level matrixes for the Tjæreborg turbine

	CL-1 (%)	CL-2 (%)	CL-3 (%)
EFL-3	5.8	1.4	2.6
EFL-2	1.9	0.5	0.8
EFL-1	0.6	0.1	0.3

Yaw shaft bending moment (standard load $M_0 = 449\,\text{kNm}$)

Table 3.14 Damage level matrixes for the Tjæreborg turbine

	CL-1 (%)	CL-2 (%)	CL-3 (%)
EFL-3	233 (109)	556 (214)	1213 (429)
EFL-2	151 (82)	391 (161)	883 (322)
EFL-1	92 (63)	272 (122)	644 (243)

Transverse tower root bending moment (standard transverse load $M_0 = 685$ kNm, in bracket standard longitudinal load $M_0 = 2{,}094$ kNm)

Table 3.15 20-year fatigue equivalent load ranges (reference frequency 1 Hz) for CL-O and CL-S

(kNm)	Clean M_{Clean}	CL-0 $\Delta M/M_{Clean}$ (%)	CL_0 $\Delta M/M_{Clean}$ (%)
Flapwise blade root bending moment blade 1	727	0.109	0.106
Edgewise blade root bending moment blade 1	1,209	0.076	0.091
Flapwise blade root bending moment blade 3	739	0.069	0.068
Edgewise blade root bending moment blade 3	1,202	0–078	0.096
Yaw shaft bending moment	449	2.185	2.463
Tilt shaft bending moment	453	1.787	2.047
Longitudinal tower root bending moment	2,094	−0.007	−0.001
Transverse tower root bending moment	685	50–196	50.870

the longitudinal tower root bending moment (values between brackets in Table 3.14) because the wind comes from different directions during the year.

The damage levels show a nonlinear trend with respect to the contamination level. The damage level for the edgewise blade root bending moment ranges from 0.1 to 9.4 %. The damage level for the yaw shaft bending moment ranges from ≤ 0.1 to 5.8 %. The damage level for the tower root bending moment ranges from 92 % to more than 1,200 % (from 63 to 429 %, by normalising respect to the tower root bending moment in the y-direction).

The proposed integrated procedure makes it possible to identify the cases where a redesign phase of some components is necessary for sites with a given icing risk.

3.9.4.1 Assessment of the Quality of the Input

Contamination levels CL-O and CL-S have been conceived to check the influence of the model used for the prediction of the amount of ice on the lifetime assessment. Both cases have the same mass magnification factor $K_{bending}$, but CL-O has a different ice thickness for every spanwise station, whereas CL-S has a trapezoidal ice thickness distribution (see Table 3.9 and Fig. 3.50).

The different distribution of $t_{ice}(r)$ determines two different sets of airfoil performances database: CL-S has three stations that follow the trapezoidal distribution while CL-O has five stations that follow the observed $t_{ice}(r)$ distribution.

The 20-year fatigue lifetime assessment was performed and compared for CL-S and CL-O; eight sensors were virtually monitored. An EFL with a percentage of operation during icing of 9 % has been developed in a similar fashion with that explained. The main results are reported in Table 3.15.

The increases in the equivalent load ranges with respect to the Clean case are small proving that CL-S and CL-O represent a moderate icing case. The change in the increases of the equivalent load ranges between CL-S and CL-O (third and fourth column of Table 3.15) and are extremely small suggesting that the trapezoidal ice mass distribution gives a consistent model of the dynamic behaviour of the iced rotor.

These results show relative insensitivity to the quality of the input conditions. This might be due to the limit of the aeroelastic code that accepts only a mass magnification factor. More difference is expected by assigning the whole mass distribution. The first three system exciting frequencies, estimated considering an intermediate rotational speed of 22.15 rpm, were: 1P = 0.37 Hz, 2P = 0.74 Hz, 3P = 1.11 Hz. Since the steel tower has a first eigenfrequency of about 0.60 Hz, the considerable increase in the equivalent load range of the transverse tower root bending moment could be due to resonance interaction.

The simple method outlined is able to capture major problems, whereas small increases of the equivalent load ranges fade into the background noise. The major problem identified with this methodology in the presented case concerned the dramatic increase in the transverse tower root bending moment due to operation with the unbalanced rotor. All the other components suffered minor changes in the equivalent load ranges. Light towers with a low first eigenfrequency value may have a resonance interaction with the exciting frequencies in the case of operation with an unbalanced rotor. The analysis showed the importance of simultaneously modelling the aerodynamics and mass distribution of the iced rotor. Since machine damage is limited to some components and conditions, the power generation during icing events is possible to some extent. The quantitative accuracy of the results depends also on the quality of the meteorological characteristics of the site. The most practical case of application for the presented method is the setting/tuning of the safety level for the control system (setting of the inertial sensors that decide when to possibly stop the power production). In prospective, this analysis can be integrated with the design of the forthcoming anti-icing/de-icing systems.

3.10 Simplified Analysis of Icing Rotor Unbalance

To complete the analysis of the previous paragraph on the mechanical effects of ice growth, a simplified method to determine the unbalance of iced rotors is described with the aim to discuss the possible effect of failure of the anti icing system on a single blade. This situation causes the rotor turning with only one blade iced and two clean. In this condition the iced blade experiences both an increase of the gravitational mass due to the accreted ice and a penalty in aerodynamic performance. The Tjæreborg turbine already used as benchmark in the preceding paragraph has been subjected to three contamination levels: the GL contamination level (CL-S) of Fig. 3.50, the CIRA computation outcome of Table 3.8, and the data from Fig. 3.24

Table 3.16 Mass of accreted ice and related degree of irregularity of the various contamination levels

Type of contamination	Total ice mass (kg)	DoI (−)
GL CL-0	83.09	0.0140
CIRA	118.95	0.0120
$k/c = 0.0059$	9.20	0.0012

(case 290, k/c = 0.0059). For each case the aerodynamic drop in performance has been adopted accordingly to the results shown previously. The inputs are given in Table 3.16 (columns 1 and 2). The inflow conditions are constant over the disc and the aerodynamic torque oscillation due to turbulence has been neglected.

The following relationships give the contributions to the shaft torque, the aerodynamic and mass contribution in the edgewise moments, for a single contaminated blade rotor.

$$M_{aero} = \frac{C_{L,actual}}{C_{L,clean}} C_{L,clean}\, c\, \frac{\rho\, V\, \Omega}{9} \left(R^3 - R_0^3\right)$$

$$M_{g,ice}(\theta) = m_{ice}\, R_G\, g\, \sin\theta$$

R_0 is the starting radius of the profiled blade planform and R_G is the ice mass barycentre radius.

The degree of irregularity (DoI) has been defined as:

$$DoI = \frac{\Omega_{max} - \Omega_{min}}{\Omega_{ave}} \tag{3.3}$$

These resulting moments have been processed by an ODE (Ordinary Differential Equations) solver in order to compute the fluctuation of the rotor angular speed. To make the computation more actual, the original inertia of the rotor of the Tjæreborg turbine has been reduced to take into account the light design of modern turbines. Thus 2,653 kg per blade have been used for the simulation, instead of the original 7963 listed in Table 3.7. The electrical generator model obeys to a classic $K \cdot \Omega^2$ law, where K has been chosen in order to get the rotational speed of 22.5 rpm at the nominal wind speed.

Results are given in Table 3.16 (columns 3 and 4), and in Figs. 3.53, 3.54, where the moments are normalised to the full clean blade ones as function of the azimuthal position.

The drop in aerodynamic performances is linked to the depletion of the aerodynamic efficiency. Penalty of about 73 % is observed in the CL-S contamination case together with large oscillations over the revolution. With regard to gravitational unbalance, oscillations as large as about 4 % occur for CIRA and CL-S contamination levels, while light surface roughness is causing only moderate effects on both aerodynamic and gravitational torques.

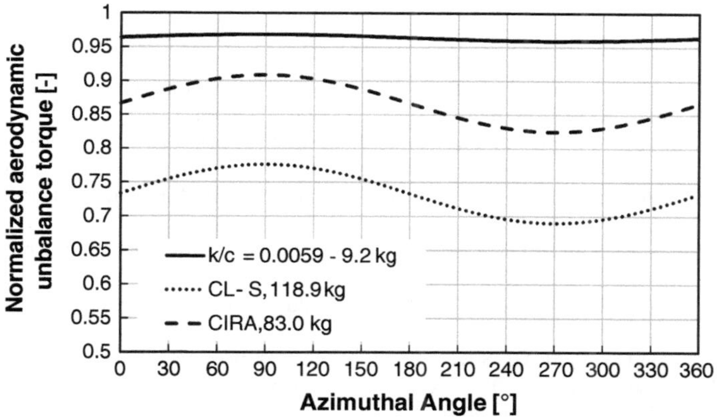

Fig. 3.53 Shaft torque normalized to the clean case as function of azimuthal position for different blade contamination

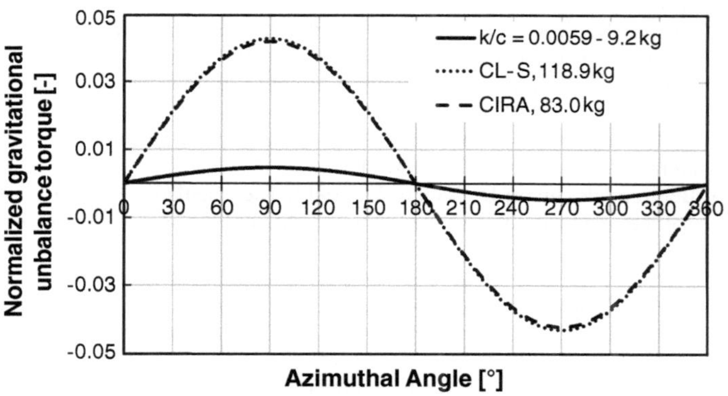

Fig. 3.54 Gravitational unbalance normalized to the clean case as function of azimuthal position for different blade contamination

The simplified model shows that also relevant levels of contamination determine negligible oscillation levels of the rotational speed, also when the lift coefficient is markedly reduced (−80 %) compared to clean one, and is probably comparable with that produced by sampled turbulence. This is nevertheless accompanied by a relevant drop in performances. This proves that the sole measurement of rotational irregularity is not a parameter sufficient to assess incipient icing of the rotor, but needs to be linked to some other performance indexes (i.e. power curve level, ambient temperature, etc.), as will be discussed in Chap. 5.

References

1. Miley SJ (1982) A catalog of low Reynolds number airfoil data for wind turbine applications. Department of Aerospace Engineering, Texas, College Station
2. Battisti L (2012) Gli impianti motori eolici. In: Battisti L (ed), ISBN 978-88-907585-0-8
3. Critzos CC, Heyson HH, Boswinkle RW Jr (1995) Aerodynamic characteristics of NACA 0012 airfoil section at angles of attack from 0 to 180 degrees. National Advisory Committee on Aeronautics. NACA-TN-3361
4. Timmer WA, van Rooij RPJOM (2004) Summary of the Delft University wind turbine dedicated airfoils. J Sol Energy Eng 125(4):488–496
5. Timmer WA (2008) Two-dimensional low-Reynolds number wind tunnel results for airfoil NACA0018. Wind Eng 32(6):525–537
6. Abbott IH, von Doenhoff AE, Stivers LS (1945) Summary of airfoil data. NACA Report 824
7. Lynch FT, Khodadoust A (2001) Effects of ice contamination on aircraft aerodynamics. Prog Aerosp Sci 37:669–767 (Pergamon)
8. Brumby RE (1988) The effect of wing ice contamination on essential flight characteristics. SAE aircraft ground de-icing conference, Denver, September 1988
9. Brumby RE (1991) The effect of wing ice contamination on essential flight characteristics. AGARD CP-496, paper 2
10. Timmer WA, van Rooij R (2003) Summary of the delft university wind turbine dedicated airfoils. Am Inst Aeronaut Astronaut (AIAA) J, AIAA-2003-0352
11. Thomas SK, Cassoni RP, MacArthur CD (1996) Aircraft anti-icing and deicing techniques and modeling. J Aircr 33(859):841–853
12. Al-Khalil KM, Keith TG, De Witt KJ (1993) New concept in runback water modeling for anti-iced aircraft surfaces. J Aircr 30(1):41–49
13. Neel CB, Bergrun NR, Jukoff D, Schlaff BA (1947) The calculation of the heat required for wing thermal ice prevention in specified icing conditions. NASA Langley Research Center, NACA TN 147
14. Gelder FT, Lewis JP (1951) Comparison of heat transfer from airfoil in natural and icing conditions. NASA Langley Research Center, NACA TN 2480
15. Makkonen L, Laakso T, Marjaniemi M, Finstad KJ (2001) Modeling and prevention of ice accretion on wind turbines. Wind Eng 25(1):3–21
16. Makkonen L, Autti M (1991) The effects of icing on wind turbines. Wind energy: technology and implementation (EWEC), pp 575–580
17. Makkonen L, Laakso T, Marjaniemi M, Wright J (2001) Results of Pori wind farm measurements. VTT, Energy Reports 42/2001
18. Bose N, Rong JQ (1990) Power reduction from ice accretion on a horizontal axis wind turbine. In: 12th wind energy conference Norwich. British Wind Energy Association, London
19. Seifert H, Richert F (1998) A recipe to estimate aerodynamics and loads on an iced rotor blades. In: Proceedings of the IV BOREAS conference, Enontekio, Hetta, Finland
20. Seifert H, Richert F (1997) Aerodynamics of iced airfoils and their influence on loads and power production. In: Proceedings of the European wind energy conference, Dublin, Ireland, pp 458–463
21. Wright WB (1995) Users Manual for the Improved NASA Lewis Ice Accretion Code LEWICE 1.6. NASA Langley Research Center, NASA CR 198355
22. Wright WB (2002) User Manual for the NASA Gleen Ice Accretion Code LEWICE-Version 2.2.2. NASA Langley Research Center, NASA/CR-2002-211793
23. Morency F, Tezok F, Paraschivoiu I (1999) Anti-icing system simulation using CANICE. J Aircr 36(6):999–1006
24. Morency F, Tezok F, Paraschivoiu I (2000) Heat and mass transfer in the case of anti-icing system simulation. J Aircr 37(2):245–252
25. Tran P, Brahimi MT, Pueyo A, Tezok F, Paraschivoiu I (1996) Ice accretion on aircraft wings with thermodynamic effects. J Aircr 32(2):444–446

26. Guffond D, Brunet L (1985) Validation du programme bidimensional de captation, ONERA, RT no. 20/5146 SY
27. Mingione G, Brandi V (1998) Ice accretion prediction on multielement airfoils. J Aircr 35(2):240–246
28. Dillingh JE, Hoeijmakers HWM (2003) Simulation of ice accretion on airfoils during flight. University of Twente, Faculty of Mechanical Engineering, Section Engineering Fluid Dynamics
29. Beaugendre H, Morency F, Habashi WG (2003) FENSAP-ICE's three-dimensional in-flight ice accretion module. J Aircr 40(2):239–247
30. Virk MS, Homola MC, Nicklasson PJ (2010) Effect of rime ice accretion on aerodynamic characteristics of wind turbine blade profiles. Wind Eng, 34(2):207–218. ISSN 0309-524X
31. Virk MS, Homola MC, Nicklasson PJ (2010) Relation between angle of attack and atmospheric ice accretion on large wind turbine s blade. Wind Eng 34(6):607–614
32. Virk MS, Homola MC, Nicklasson PJ (2012) Atmospheric icing on large wind turbine blades. Int J Energy Environ 3(1):18
33. Villalpando F, Reggio M, Ilinca A (2012) Numerical study of flow around iced wind turbine airfoil. Eng Appl Comput Fluid Mech 6(1):39–45
34. Sagol E, Reggio M, Ilinca A (2013) Issues concerning roughness on wind turbine blades. Renew Sustain Energy Rev 23:514–525. ISSN 13640321
35. Reid T, Baruzzi G, Ozcer I (2013) FENSAP—ICE simulation of icing on wind turbine blades, part 1: performance degradation. In: 51st AIAA aerospace sciences meeting including the new horizons forum and aerospace exposition, Grapevine, Texas, 7-10 January 2013
36. Mortensen K (2008) CFD simulation of an airfoil with leading edge ice accretion. Master thesis, Department of Mechanical Engineering, Technical University of Denmark
37. Fortin G, Perron J (2009) Spinning rotor blade tests in icing wind tunnel. In: AIAA 2009-4260, 1st AIAA atmospheric and space environments conference, San Antonio, TX, June, p 116
38. Hochart C, Fortin G, Perron J (2008) Wind turbine performance under icing conditions. Wind Energy 11:319–333
39. Jasinski WJ, Noe SC, Selig MS, Bragg MB (1997) Wind turbine performance under icing conditions. Trans ASME J Sol Energy Eng 120(1):60–65
40. Papadakis M, Yeong HW, Wei H, Wong SC, Vargas M, Potapczuk M (2005) Experimental investigation of ice accretion effects on a swept wing. U.S. Department of Transportation—Federal Aviation Administration, Final Report, DOT/FAA/AR-05/39, August 2005
41. Broeren AP, Bragg MB, Addy HE (2004) Effect of intercycle ice accretions on arifoil performance. J Aircr 41(1):165–174
42. Busch GT, Broeren AP, Bragg MB (2008) Aerodynamic simulations of a horn ice accretion on a subscale model. J Aircr 45(2):604–613. doi:10.2514/1.32338
43. Ashenden R et al (1996) Airfoil performance degradation by supercooled cloud, drizzle, and rain drop icing. Am Inst Aeronaut Astronaut J (AIAA) 33(6):1040–1046
44. Bragg MB, Broeren AP, Blumenthal LA (2003) Iced-airfoil and wing aerodynamics. SAE international paper 2003-01-2098
45. Bragg MB, Broeren AP, Blumenthal LA (2005) Iced-airfoil aerodynamics. Prog Aerosp Sci 41(5):323–418
46. Broeren AP, Bragg MB (2010) Effect of high-fidelity ice-accretion simulations on full-scale airfoil performance. J Aircr 47(1):240–254
47. Shin J (1994) Characteristics of surface roughness associated with leading edge ice accretion. AIAA paper, 94
48. Smith AM, Kaups K (1968) Aerodynamic surface roughness and imperfections. Society of Automotive Engineers, New York
49. Brumby RE (1991) Technical evaluation report on the fluid dynamics panel specialists meeting on effects of adverse weather on aerodynamic. AGARD advisory. Report 306
50. Lee S, Bragg MB (2003) Investigation of factors affecting iced airfoil aerodynamics. J Aircr 40(3):499–508. doi:10.2514/2.3123
51. Gurbacki HM (2003) Ice-induced unsteady flowfield effects on airfoil performance. Department of Aeronautical and Astronautical Engineering. University of Illinois, Urbana

52. Kim HS, Bragg MB (1999) Effects of leading-edge ice accretion geometry on airfoil aerodynamics. AIAA paper, 3150
53. Johnson SP (1936) Ice. Aviation 35:15–19
54. Lederer J (1939) Safety in the operation of air transportation. Norwich University, Norwich
55. Johnson CL (1940) Wing loading, icing and associated aspects of modern transport design. J Aerosol Sci 8(2):43–55
56. Burton T, Jenkins N, Sharpe N, Bossanyi E (2011) Wind energy handbook. Wiley, New York
57. Platt A (2012) NWTC design codes—WT perf, national wind technology center—NREL, Last modified 26-November-2012, http://wind.nrel.gov/designcodes/simulators/wtperf/
58. Seifert H, Scholz C (1990) Additional loads caused by ice on rotor blades during operation. In: Proceedings of the international conference, European community of wind energy, Madrid, 10-14 September 1990, pp 203–207
59. Jasinski WJ, Noe SC, Selig MS, Bragg MB (1997) Wind turbine performances under icing conditions. In: 35th aerospace science meeting and exhibit—AIAA, Reno
60. Volund P, Antikainen P (1997) Ice induced loads on wind turbines. In: European wind energy conference, Dublin, pp 664–667
61. Ganander H, Ronsten G (2003) Design load aspects due to ice loading on wind turbine blades. In: Proceedings of the VI BOREAS conference, Pyhatunturi, Finland
62. Øye S (1996) FLEX4 simulation of wind turbine dynamics. In: Proceedings of the international energy agency, annex XI, 28th meeting of experts, 11–12 April 1996, pp 71–77
63. Battisti L, Soraperra G (2003) Sistema antighiaccio per pale di turbine eoliche parte 2: sistemi a circolazione di aria. 58° Congresso ATI, Padova, September 2003, pp 8–12
64. Germanischer lloyd industrial services gmbh, business segment wind energy (2010), Guideline for the certification of wind turbines
65. Durstewitz M (2003) Windenergie in kalten klimaregionen. Erneuerbare Energ 12:3–34
66. Wright WB, Potapczuk MG (1998) Comparison of LEWICE 1.6 and LEWICE/NS with IRT experimental data from modern airfoil tests. NASA Langley Research Center, NTL Digital Repository, ID 1086
67. Broeren AP, Addy HE, Bragg MB (2004) Flowfield measurements about an airfoil with leading-edge ice shapes. AIAA paper, 59
68. Addy HE, Chung JJ (2000) A wind tunnel study of icing effects on a natural laminar flow airfoil. AIAA paper, 95

Chapter 4
Icing Process

Abstract This chapter discusses the models of the physics of the water impingement and ice formation mechanisms. Body discretization, external flow and temperature field, and body wetness have been analysed. The contents give the fundamentals of the design of anti-icing or de-icing systems. The icing process is described from the thermo-fluid-dynamic point of view. The aim is not to detail the ice growing process, but to give methods to determine the water mass flow captured by the aerodynamic profile, the impingement limits and the heat flows involved in the process on the surface, as ice prevention systems are designed to keep the surface reasonably clean of ice. To this aim the general theory for droplet trajectory includes the fixed cylinder case, the collision efficiency calculation for profiles at zero and other than zero AoAs. Calculation of the difference between translating and rotating blade on impinging water is presented and discussed. A numerical example for the profile of the Tjærborg wind turbine rotor is given. Finally, some relevant conclusions applied to wind turbines are drawn. The chapter analyses the water mass balance at the surface, and the thermo-fluid-dynamic processes at the iced surface by the concept of freezing fraction. Thus with the help of energy and mass conservation equations the problem of ice accretion and anti-ice design is presented and solved.

4.1 The Physics of the Ice Formation Mechanism

The freezing process is a complex phenomenon involving a large number of parameters. Only a few can be dealt with in controlled experiments. Air turbulence and climatic and meteorologic environment drifts introduce further variables in in-field testing.

Nevertheless, an understanding of the basis of icing mechanism is essential for estimating the intensity of the icing process on the wind turbine rotor and for developing adequate means for reducing the deriving hazards and inconveniences.

During the ice accretion process the following macro-phases are observed:

- The airborne super-cooled water droplets follow trajectories that will cause them partly to impinge upon a body and partly to deviate from it;
- The droplets impacting with an uncontaminated surface tend to merge into larger surface drops under the effects of surface tension and flow forces along the surface as imposed by the flow conditions evolving on the surface of the body;

L. Battisti, *Wind Turbines in Cold Climates*, Green Energy and Technology,
DOI 10.1007/978-3-319-05191-8_4

- A fraction of the colliding drops will either freeze on the surface or will run on the surface or will be shed from the surface because of the aerodynamic forces acting on them;
- The ice accretions formed by this initial freezing form a rougher body surface which enhances the convective heat transfer. This cools down the surface further, boosting the ice accretion process. As new rough shapes grow over the surface the problem gains in complexity due to the formation of complex boundary layer patterns.

The physical mechanisms associated with these phases can be described with the help of models describing:

1. Meteorological processes, which act in generation of liquid water within the air flow around the object (as rain and fog)—not treated here;
2. Mechanical processes associated with the motion of super-cooled water droplets that strike and accumulate water on the surface; these processes can be described with equations from fluid and particle mechanics which determine the trajectory pattern of the droplets within the air flow;
3. Thermodynamic mechanisms described by energy conservation principle (such as latent heat release, forced convection, evaporating heat transfer, radiative heat transfer, etc.) and mass conservation principles;
4. Mechanical balances of mass and surface forces, which tend to retain or shed part of the accreted ice and which play a role in defining the final ice shape.

On the whole incoming water that strikes the body, only a part will rush aft, since some will freeze, evaporate, shed and stand according to local thermal and fluid-dynamic conditions, see Fig. 4.1. The comprehension of the ice growth mechanism is therefore the basis for the selection and design of any ice prevention system.

At the surface, generally three thermodynamic phases are present, air and water vapours, liquid-water and ice. If air, vapour, water and ice over the surface are simultaneously present during ice accretion processes of an unheated body, the use of triple

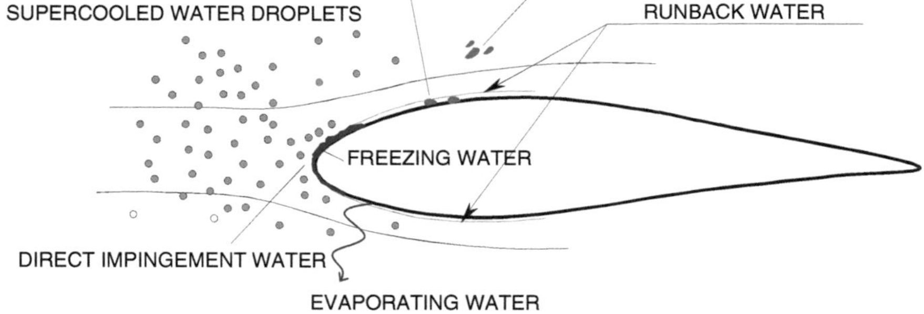

Fig. 4.1 Basic processes involving water in icing process

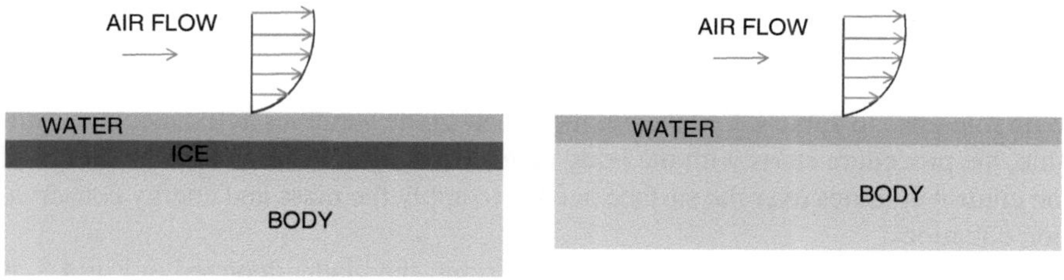

Fig. 4.2 Triple deck model (unheated body, *left*) and double deck model (heated body, *right*)

deck theory is necessary, while for un-iced heated bodies only double deck models (air, vapour and water) are of concern. In Fig. 4.2 triple deck and double deck models are depicted.

4.2 Ice Accretion/Ice-Free Conditions Simulations

The physical models and the relative computational procedure to determine the heat fluxes at the surface of the body and the ice growth, involves basically the steps indicated in Table 4.1. Ice growth and anti-ice calculation procedures adopt similar steps, although the finalisation of the models implies differences in the calculation path, which will be discussed in the following paragraphs.

Table 4.1 Computational steps

Step	Ice growth calculation	Anti-ice calculation
1	Site variables assessment	Site variables assessment
2	Body external geometry discretization	Body external geometry discretization
3	Not required	Body wall and internal geometry discretization
4	External flow field calculation	External flow field calculation
5	Not required	Internal flow field calculation
	Body wetness determination	Body wetness determination
6	a. Particles trajectory calculation b. Particles impingement calculation	a. Particles trajectory calculation b. Particles impingement calculation
	Mass and energy conservation	Mass and energy conservation
7	Ice growth calculation	Mass and energy flows calculation
		Conjugate heat transfer
8	Geometry modification by adding the ice growth	Not required
9	Not required	Required local heat flows for anti-icing design strategy
10	Not required	External/internal heat supply and power assessment
11	Iteration from steps 5–9 until icing duration	Not required

4.2.1 Body Discretisation–Geometry Domain

With reference to Table 4.1, as the output of step 1 (site variables assessment) is available, the procedure starts with the body geometrical discretisation and the design of the control volumes over the surface needed to apply the mass and energy conservation equations.

With reference to a generic wind turbine rotor, the blade, depicted in Fig. 4.3, is built up by a series of spanwise adjacent blade elements (or stations) identified by the index i variable from 1 to N. Each station is thus comprised between the radii r^i and r^{i+1}.

Each blade element is then divided on a series of chordwise adjacent panels identified by the index j, variable from 1 to J. The width of each j-panel is given by the length or line segments, which are variable along the airfoil surface, so that shorter segments can be concentrated in regions where the radius of curvature is the smallest. The physical properties such as temperature, pressure, thermal conductibility, etc., are lumped in correspondence to the centroid of each panel.

Straight-line segments constituting the lower surface of the control volumes replace the body surface. The control volumes are located on the surface of the body and protrude outside of it as shown in Fig. 4.4.

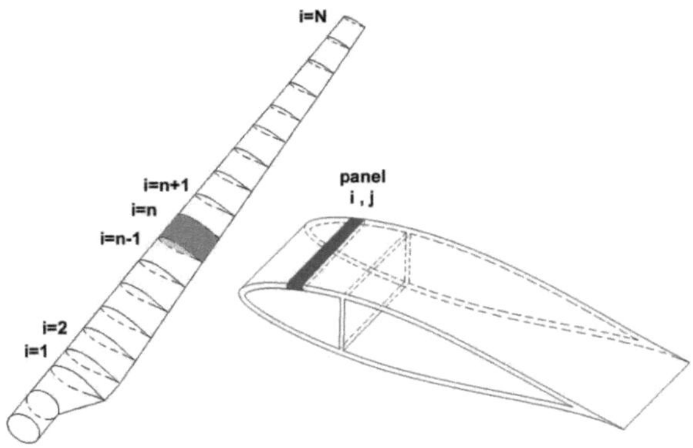

Fig. 4.3 Blade discretization scheme

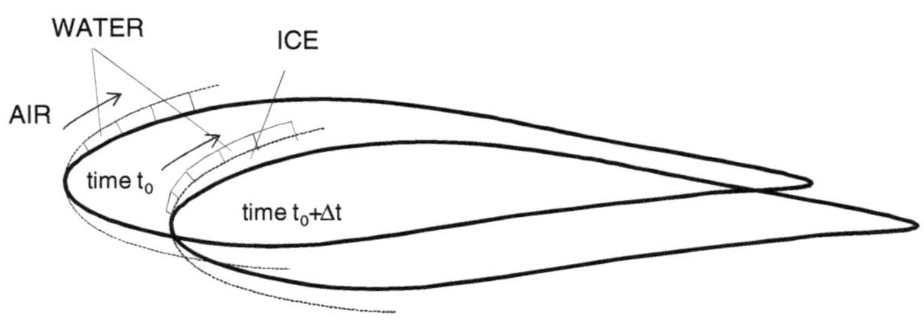

Fig. 4.4 Control volumes on the surface

The lower boundary of the control volume is initially on the surface of the not contaminated geometry and is displaced outward as the ice accretes. Successively, the control volume is always located on either the clean or iced surface. The procedure is therefore marching in time and theoretically uses a fresh, updated geometry to account for the ice layer growth time to time.

A time-stepping procedure is applied to simulate ice accretions. Initially, the flow field and droplet impingement characteristics are determined for the clean geometry. The ice growth rate on each control volume is then determined by a suitable model (described later). When a time increment is specified, this growth rate is converted into an ice thickness and the body coordinates are adjusted to account for the accreted ice. Although the most desirable option is to repeat the entire procedure, calculating each time the new flow field about the ice to update the water local collection efficiency and thermodynamic data, this option leads to an increase in the computational time to accrete a layer of ice. Most of the computational time is actually spent in calculating the flow field.

4.3 External Flow and Temperature Field

From late 1940, the aircraft research community has been developing analytical methods to calculate the heat flows at the surface [1]. The introduction of computer aided computation and the decrease in computer calculation cost have led to the rapid development of ice accretion simulation tools in the early 1980s [2]. Usually, such methods are based on either 2D or quasi-3D inviscid flow codes (panel method) to compute the airflow solution. A potential flow code is desirable due to speed of execution. These codes produce an approximate flow solution by using a distribution of sources, sinks and/or vortices along the body geometry that simulates the flow outside of the viscous boundary layer. This is a reasonable assumption for high-Reynolds number flows where the boundary layer is sufficiently thin. In calculating the flow field, contributions from all the sources, sinks and/or vortices are summed. One limitation to this approach is that the panel flow codes are generally limited to cases where the angles of attack (AoA) are low (separation is not handled) and this hypothesis is acceptable for only limited operational ranges of wind turbines.

The codes are coupled usually with a Lagrangian particle tracking techniques for droplet impingement calculations, and on a 1-D mass and heat transfer balance at the surface to predict ice shapes. The best-known codes using this structure are NASA's LEWICE [3, 4], ONERA [5] and Bombardier Aerospace's CANICE [6]. Another important output from the flow solver is the location of the stagnation point. This is the point at which the local tangential velocity is zero, or most nearly so. For a clean aerofoil, selection of the stagnation point is a trivial procedure, as the surface is smooth, but it exhibits severe problems in the evaluation of velocities near the surface at panel edges. Especially due to the irregular shapes associated with ice formation, multiple stagnation points can develop on the iced surface. This can lead to difficulty in selecting the point at which to start the boundary layer calculations for the upper and lower surfaces. This is a common problem occurring for instance in ice "horns" formations. For near-freezing temperatures, much of the incoming water does not freeze on impact, and prediction of surface water flow is critical. Viscous/non-viscous coupling methods are difficult to apply to separated flow, being usually the norm in the glaze ice accretion cases.

Some authors use, for this reason, Navier-Stokes over a Euler code instead of a potential flow code. The baseline module of LEWICE is for instance [3, 4], an ice accretion prediction code that applies a time-stepping procedure to calculate the shape of an ice accretion. The potential flow field can be calculated in LEWICE using the Douglas Hess-Smith 2-D panel code. From 2.2 version the potential flow module can be bypassed by setting a flag in the user input file. In this mode, the user has the option to call a grid generator and grid-based flow solver (Euler or Navier-Stokes) or to read in the solution file from this flow solver. Current CFD technologies definitely overcome some of the limitations of the cited methods, and open new possible uses in a fully 3D approach coupling aerodynamics, particle motion and thermal processes. FENSAP for instance, can work in viscous (Navier-Stokes) mode to handle clean and degraded flows via 3D compressible turbulent Navier-Stokes equations [7]. Water capture at the profile is made by a 3D Eulerian method [8] instead of the Lagrangian one. 3D mass balance and heat transfer at surface are solved using partial differential equations for prediction of 3D ice accretion shapes [9]. Additionally, the anti-ice heat fluxes across the body wall are determined via a conjugate heat transfer problem [10].

If the amount of ice accreted during the time step is small, it is possible to use the same flow field and thermodynamic data from the previous time step. It is worth to mention that this option does not produce results that are as accurate as those in the first option, especially for glaze ice accretions. But the advantage is that the computational time required is significantly reduced.

One of the advantages of wind turbine aerodynamics is the use of incompressible solution of the Navier-Stokes solvers. This simplification reduces the computational time to calculate the flow field and enables more frequent calculations of the flow field about the iced geometries in a given time. Thus the accuracy of the ice accretion prediction for low Mach number flows will be enhanced by the use of the incompressible Navier-Stokes solver.

A step back in the complexity of the flow field simulation is operating in the application of anti-icing systems design. Here since a ice-not-tolerant strategy is of concern, the surface would remain reasonably clean of ice, allowing in most cases (i.e. wind turbine production) the use of panel solvers. The de-icing strategy could instead be more carefully approached, since small ice growth is admitted, asking possibly for more realistic flow solvers.

Despite the method used for solving the flow field, for each volume, some basic variables have to be computed to characterise the flow field and temperature for the successive heat transfer computations.

As the pressure at the edge of the boundary layer is known, the flow velocity v_e is evaluated in correspondence to the edge of the boundary layer from:

$$p_e - p_\infty + \frac{1}{2}\rho_\infty W^2\left[1 - \left(\frac{v_e}{W}\right)^2\right] = 0 \tag{4.1}$$

The ambient total pressure p^0_∞ is defined as

$$p^0_\infty = p_\infty + \frac{1}{2}\rho_\infty W^2 \tag{4.2}$$

The temperature at the edge of the boundary layer is thus evaluated through the isentropic process relationship:

$$T_e = T^0_\infty \left(\frac{p_e}{p^0_\infty}\right)^{\left(\frac{k-1}{k}\right)} = T^0_\infty \left(\frac{p_e}{p^0_\infty}\right)^{\varepsilon} \tag{4.3}$$

Being the ambient total temperature T^0_∞ is defined as:

$$T^0_\infty = T_\infty \left(\frac{p^0_\infty}{p_\infty}\right)^{\left(\frac{k-1}{k}\right)} = T_\infty \left(\frac{p^0_\infty}{p_\infty}\right)^{\varepsilon} \tag{4.4}$$

The recovery temperature, T_{rec} at any point on the blade surface is the surface temperature that would be obtained if the blade were an adiabatic body. This temperature is given from:

$$T_{rec} = T_e + r\frac{v_e^2}{2\,c_p} \tag{4.5}$$

where r is the recovering factor defined as (incompressible flow):

$$r = 1 - \left(\frac{v_e}{W}\right)^2 \left[1 - Pr^z\right]$$

At the edge of the boundary layer, $r = Pr^z$ with $z = 1/3$ in the turbulent flow field and $z = 1/2$ in the laminar one. The specific heat at constant pressure c_p appearing in Eq. (4.5) and the isoentropic coefficient ε of Eqs. (4.3) and (4.4) are considered constant and equal to that of the dry air.[1]

[1] A rigorous analysis reveals that the isoentropic coefficient ε within the boundary layer is an equivalent coefficient, taking into account that the viscous layer is formed on dry air, vapour and water. This means that it should be computed as weighted average of the mixture components in the volume according to the following:

$$c_{p,eq} = \frac{\sum m_i c_i}{\sum m_i} = \frac{c_{p,air} + \dfrac{m_{vap}}{m_{air}} c_{p,vap} + \dfrac{m_w}{m_{air}} c_w}{1 + \dfrac{m_{vap}}{m_{air}} + \dfrac{m_w}{m_{air}}} \tag{4.6}$$

After substituting the numerical values, it results that $c_{p,eq} \sim c_{p,air}$, so the specific heat of air at constant pressure and hence ε will be used for the following analysis within the boundary layer.

4.4 Modelling the Body Wetness

The quantity of water present in clouds, which could possibly wet the profile, is evaluated through:

- The liquid water content (LWC): is a measure of the content of the liquid water in a given amount of air (g/m^3);
- The size of droplets: d (μm) or the distribution of droplets diameter (MVD).

In in-cloud icing, the profile wetting is a crucial factor to evaluate the rate of ice accretion. Cloud droplets are very small, typically 5–400 μm in diameter, and have correspondingly small terminal velocities (0.007 m/s for 15 μm diameter droplets). Although in windless conditions there are essentially no droplet fluxes, the motion of the wind turbine rotor causes a relevant relative velocity at which the droplet strikes on the surface (at tip about 70 m/s). Not only the relative velocity of the body but also the motion, i.e. the rotation, plays a role in the droplets trajectory interaction.

Modelling the body wetness allows determining:

- The mass rate of water impingement on the airfoil (water intercepted);
- The area of impingement;
- The distribution of the water over this area.

The mass rate of impinging water gives the indication of the quantity of the water which must be maintained at a liquid state until it either evaporate or runs off the trailing edge when the formation of ice aft the area of impingement, normally termed *runback*, has to be avoided.

The area of impingement influences the extent of heated region to be provided at the leading edge.

The local distribution of water impingement is required in the calculation of the heating requirement in areas where water is striking.

The mass of water impinging on the surface is a complex function dependent on different parameters. When there is wind, the wind drag on the droplets carries them along and the droplets follow the wind streamlines around any obstacle in their path (Fig. 4.5). Only the inertia of the droplets, which tends to keep them moving in a straight-line path, makes them diverge from the wind streamlines.

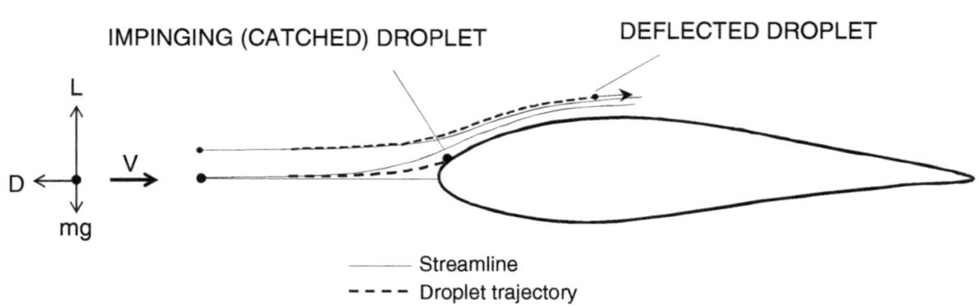

Fig. 4.5 Representation of the droplets motion past a body

The amount of water droplets in the air collected from a structure results from a balance between their drag, the tendency to follow the streamlines and their inertia, i.e. the tendency to continue in a straight-line path, see Fig. 4.5.

A trajectory calculation is thus needed to determine the amount of water impinging on the surface and the extent of the surface directly wetted.

The *collision efficiency* is defined as the fraction of water in a mass of air travelling within two adjacent streamlines that is captured by the airfoil. Its local value is obtained by analysing the water particles trajectories past a body.

If a streamtube enclosed by two trajectories (2D steady flow) and two terminal sections is identified, one far upstream (0) and the other onto the surface location of interest (p) as illustrated in Fig. 4.6, the assumptions are that:

1. The droplets do not affect the velocity field;
2. Droplets are spheres and the droplet diameter distribution is described by the Median Volume Diameter (MVD);
3. The initial droplet velocity equals the free stream velocity.

The mass conservation equation for the streamtube becomes:

$$LWC_0\ V_0\ A_0 = LWC_p\ V_p\ A_p \tag{4.7}$$

In this frame, the local collection efficiency, β_{2D}, is defined as:

$$\beta_{2D} = \frac{LWC_p\ V_p}{LWC_0\ V_0} = \frac{A_0}{A_p} \tag{4.8}$$

and represents the fraction of the liquid water content captured by that location on the aerofoil.

With reference to the 2D scheme of Fig. 4.6, by taking a unitary length of the spanwise element, this leads to the differential form as:

$$\beta_{2D} = \frac{\Delta y_0}{\Delta s} = \frac{dy_0}{ds} \tag{4.9}$$

The *stagnation line collision efficiency*, β_0, is defined as the value registered between two streamlines enclosing the stagnation line.

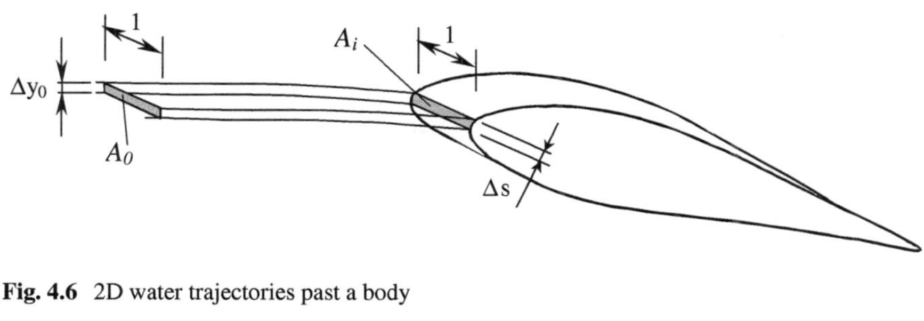

Fig. 4.6 2D water trajectories past a body

The *total collection efficiency* represents the total fraction captured by the airfoil trajectories:

$$E = \frac{1}{y_{max}} \int_{S_u}^{S_l} \beta_{2D} \, ds \tag{4.10}$$

If the droplets followed straight-line trajectories, it could be written as the ratio of the actual mass impinging the water to the maximum transverse size of the body:

$$E = \frac{y_0}{y_{max}} \tag{4.11}$$

where y_0 is the vertical distance between the droplet release points of the upper and lower surface tangent trajectories (called upper and lower impingement limits) and y_{max} is the maximum transverse dimension of the body.

The following important definitions can be given with reference to Fig. 4.7:

- The upper and lower impingement limits (s_u, s_l) are the last points aft of the stagnation point, which are hit by a water droplet.
- The local collection efficiency (also called catch efficiency) is a measure of the particle's deflection from the free stream and represents the fraction of the liquid water content captured by that location on the airfoil (useful to determine the accretion shape for ice).
- The total collection efficiency (overall collection efficiency) represents the total fraction captured by the airfoil (useful for determining overall ice accretion rates) [11].

The local mass rate of water impingement for unit surface area in mass per second, square metre, for a fixed body is:

$$\dot{m}_{w,imp} = LWC \cdot V \cdot \beta \tag{4.12}$$

In order to establish the possibility of runback forming aft of the heated area of the body, it is necessary to know the total quantity of water droplet intercepted per unit of body span, computed as

$$\dot{m}_{w,imp,\,t} = \int_{s_l}^{s_u} \dot{m}_{w,imp} \, ds \tag{4.13}$$

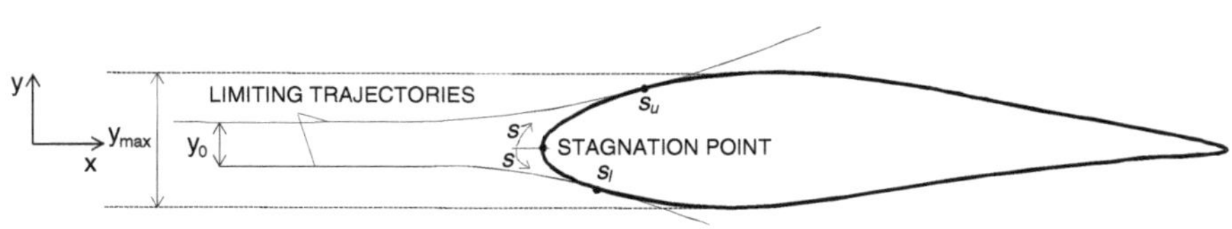

Fig. 4.7 Profile section and impingement quantities

thus

$$\dot{m}_{w,imp,\,t} = LWC \cdot V \cdot E \cdot y_{max} \tag{4.14}$$

In the case of a cloud, where the water drops are not of uniform size, but have pattern of size distributions, the rate of impingement can be computed if the distributions are known or assumed. The rate of water impingement at any point is the sum of all rates of impingement of the volume of water contained in each drop size.

$$\dot{m}_{w,imp,\,t} = \sum_{i=1}^{n} n_i \beta_i \cdot LWC \cdot V \tag{4.15}$$

where n_i is the mass fraction of the liquid water contained in drops of a particular size and β_i is the collision efficiency for the corresponding to that drop size. The MVD (Median Volume Diameter) of the droplets can be used to model a droplet diameter distribution (Finstad et al. [12]).

4.4.1 Droplets Impinging Upon the Fixed Cylinder

The study of the water particle trajectory and the impact on the cylinder is essential to understand the mechanics of motion and the interaction of the water drops with the bodies. Therefore, before passing to the analysis of droplet interaction with airfoils, it is convenient to work with cylinders in cross flow, to derive and discuss quantities as local collection efficiency, stagnation collection efficiency, overall stagnation efficiency, and the resulting quantity of water captured on the body. This case is relevant because the cylinder resembles the leading edge area of some profiles (LE inscribed cylinder).

Basic studies for the design of equipment aimed at the protection of aircraft components against ice formation were undertaken in the early 1930s, developing theory on droplet trajectories for right circular cylinders [13], for bodies of revolution [14], for elbows [15] and for airfoils (NACA and NASA). Recently, the calculations of water droplet impingement on cylinders are most useful in connection with flight instruments used in the study of LWC, droplet size and their distributions in icing clouds (rotating cylinders), icing accretions rate measurements devices, and bodies as suspended electrical cables.

The hypothesis is that at a large distance ahead of the body (free stream conditions) the droplets move with the same velocity as the air, and the droplets always remain spherical and rigid (this approximation is valid for drop radii less than about 500 μm). The motion of the particles results from the action of the inertia, the frictional force and the gravitational forces (see Fig. 4.5). Application of the momentum conservation to the particle gives:

$$m_d \frac{d\bar{V}}{dt} = -D + (\rho_w - \rho_{air})\bar{V} g \tag{4.16}$$

Drag force is given by:

$$D = \frac{1}{2}\rho_{air} C_D \pi \frac{d^2}{4}\left(\frac{d\bar{s}_d}{dt}\right)^2 \tag{4.17}$$

$d\bar{s}_d/dt$ represents the velocity of the water particle tangent to the trajectory line s. The related Reynolds number is:

$$Re_d = \frac{\rho_{air}\frac{d\bar{s}_d}{dt}d}{\mu_{air}} \tag{4.18}$$

Therefore Eq. (4.17) becomes

$$D = \frac{Re_d}{8} C_D \pi d \mu_{air} \frac{d\bar{s}_d}{dt} \tag{4.19}$$

If the gravitational forces can be neglected, Eq. (4.16) can be written as:

$$\frac{4}{3}\pi \frac{d^3}{8}\rho_w \left(\frac{d\bar{s}_d}{dt}\right)^2 = \frac{Re_d}{8} C_D \pi d \mu_{air}\left(\frac{d\bar{s}_d}{dt} - \bar{V}\right) \tag{4.20}$$

and by non-dimensionalizing (by dimension L, and velocity V_∞),

$$\frac{1}{18}d^2 \frac{\rho_w}{\mu_{air}}\frac{V_\infty}{L}\left(\frac{d\bar{s}_d}{dt}\right)^2 \frac{L}{V_\infty^2} = C_D \frac{Re_d}{24}\frac{1}{V_\infty}\left(\frac{d\bar{s}_d}{dt} - \bar{V}\right) \tag{4.21}$$

The Stokes number (inertia parameter) is:

$$K_{st} = \frac{\rho_w d^2 V}{9L\,\mu_{air}} \tag{4.22}$$

The dimensionless form of the motion in x direction component becomes:

$$\frac{dx^2}{dt^2} = C_D \frac{Re_d}{24}\frac{1}{K_{st}}\left(\frac{V_x}{V_\infty} - \frac{dx}{dt}\right) \tag{4.23}$$

where $L_{ref} = D$, $V_{ref} = 2V_\infty$ $t_{ref} = D/V_\infty$, $x = \bar{s}/L_{ref}$, $\tau = t/L_{ref}$ and in the y component

$$\frac{dy^2}{dt^2} = C_D \frac{R_d}{24}\frac{1}{K_{st}}\left(\frac{V_y}{V_\infty} - \frac{dy}{dt}\right) \tag{4.24}$$

The Reynolds number can be conveniently expressed in terms of the free stream Reynolds number as:

$$Re_\infty = \frac{\rho_{air} V_\infty d}{\mu_{air}} \tag{4.25}$$

so that:

$$\left(\frac{Re_d}{Re_\infty}\right)^2 = \left(\frac{dx}{dt} - \frac{V_x}{V_\infty}\right)^2 + \left(\frac{dy}{dt} - \frac{V_y}{V_\infty}\right)^2 \tag{4.26}$$

4.4.1.1 The Drag of Spheres

In very viscous flows, i.e. $Re_d < 2$ (Stokes flows) the drag force exerted on a sphere is:

$$D = 6\pi \mu d V/2$$

In these conditions:

$$C_D = \frac{D}{\frac{1}{2}\rho v^2 \frac{d^2}{2}} = \frac{24}{Re_d}$$

In the flows relevant to ice accretion the Reynolds numbers based on particle diameter, see Eq. (4.25) are actually not very low, (although generally not exceeding 200) and the Stokes formulation is no longer accurate.

At the relative velocity of the airfoils (and of the wind turbine blades), therefore, one has to rely on other formulations (generally semiempirical). One of the most famous, being also one of the first made available, is the Langmuir and Blodgett one (hereafter called LB) [13, 16]. They computed a series of water drop trajectories about cylinders, spheres and ribbons, taking into consideration deviations from Stokes law. The formulation proposed for Reynold number below 1,000 is:

$$C_D \frac{Re_d}{24} = 1.0 + 0.197 Re_d^{0.63} + 2.6 \cdot 10^{-4} Re_d^{1.38} \tag{4.27}$$

A new integration procedure proposed by Finstad et al. [20] increased the accuracy of original LB by a higher order integration scheme, revealing differences of up to 10 % on same ranges on stagnation collision efficiency. The drag formulation proposed by Wright for LEWICE is based on the following relationships [3]:

$$\begin{aligned} C_D &= \frac{24}{Re_d} + 0.4 + \frac{6}{1 + Re_d^{0,5}} \\ C_D &= \frac{24}{Re_d} + 0.3 + \frac{6}{1 + Re_d^{0,5}} \end{aligned} \tag{4.28}$$

for $Re_d < 100$ and $Re_d \geq 100$ respectively.

And the one from Beart and Pruppacher [17], for Reynolds numbers below 200:

$$\frac{D}{D_s} = 1 + 0.102 Re_d^{0.995}$$

$$\frac{D}{D_s} = 1 + 0.115 Re_d^{0.802} \tag{4.29}$$
$$\frac{D}{D_s} = 1 + 0.189 Re_d^{0.632}$$

valid for $0.2 < Re_d \geq 2$, $2 < Re_d < 21$ and $21 < Re_d \geq 200$ respectively. These relationships are input in the following expression to get the drag force:

$$C_D = \frac{D}{D_s}\frac{24}{Re_d} \tag{4.30}$$

The outputs of these models are plotted in Fig. 4.8. It is worth to recall that the LB results, since the experimental conditions are used to obtain the correlation, cannot be in principle applied to precipitation droplets. The terminal velocity of drizzle and rain drops, typically from 200 μm to 1 or 2 mm in diameter, is between about 2 and 7 m/s [18].

Thus, even in no wind, there is a downward flux of water that is a balance between gravity and the drag of the still air on the drops. When there is wind, the drops gain horizontal velocity in addition to their vertical velocity.

4.4.1.2 The Modified Langmuir Parameter

If the Reynolds number is based on the body dimension D:

$$Re_\infty = \frac{\rho_{air} V_\infty D}{\mu_{air}} \tag{4.31}$$

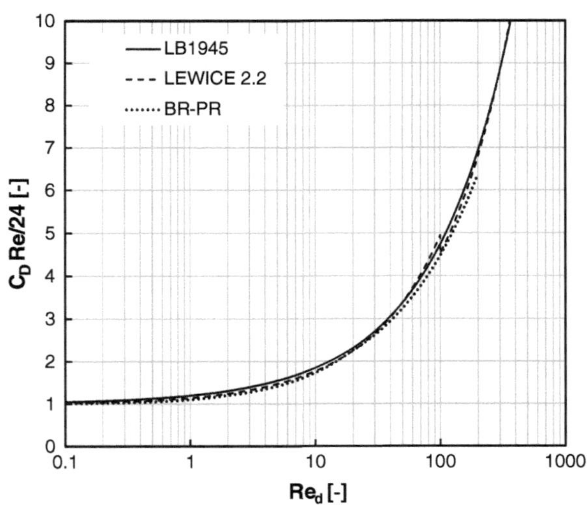

Fig. 4.8 LB model for droplet drag [13] compared to NASA LEWICE drag [3] and Beart and Pruppacher relationships [17]

by inputting the Stokes number Eq. (4.22) into Eq. (4.23) one gets:

$$\frac{dx^2}{dt^2} = \frac{\rho_{air}}{\rho_w} C_D \frac{Re_d}{24} \frac{1}{Re_D} 9 \frac{D^2}{d^2} \left(\frac{V_x}{V_\infty} - \frac{dx}{dt} \right) \tag{4.32}$$

or:

$$\frac{dx^2}{dt^2} = K_L \left(\frac{V_x}{V_\infty} - \frac{dx}{dt} \right) \tag{4.33}$$

where K_L is called the *modified Langmuir parameter*:

$$K_L = \frac{\rho_w}{\rho_{air}} \frac{1}{C_D} \frac{24}{Re_d} \frac{1}{9} \left(\frac{d}{D} \right)^2 Re_D \tag{4.34}$$

The modified Langmuir parameter K_L allows to synthetically describe the inertia effects of the water drops with regard to the body they are interacting on: for small values of K_L, the droplet velocity will tend to the air velocity V and will follow the streamlines. This implies that relatively small droplets $(d/D)^2 << 1$, i.e. for large bodies, the icing will not be an important issue, while for small bodies or for smaller parts of larger bodies, icing will be more and more important as the body size decreases.

4.4.1.3 Some Relevant Conclusions Applied to Wind Turbines

When we refer to wind turbines, Eq. (4.34) states that for large wind turbine blades the modified Langmuir parameter K_L, assuming the droplet diameter is spanwise constant along the blade, will be proportional to:

$$K_L \propto \frac{W}{c} \approx \frac{W}{t} \tag{4.35}$$

Since for large wind turbines the ratio W/t can vary as much as 20–30 times along the blade span, Eq. (4.34) reveals that the outer blade sections will be more wetted than the inner ones.

For small wind turbine blades K_L can be written as:

$$K_L \propto \frac{d}{c} \approx \frac{d}{t} = \text{cost.} \tag{4.36}$$

Since the chord (and the corresponding thickness of the profile is small), the ratio $(d/t)^2$ for small wind turbines is relatively high, and this implies that the blade of a small turbine will be evenly wetted on the whole span and icing will be in general important for this class of machines. The picture in Fig. 4.9 is an example of full freezing of a small turbine at Summit Station (Greenland, 3,200 m.a.s.l.), which results to be completely blocked by ice.

Fig. 4.9 Example of freezing on small size turbine. presented at HS seminar (Swedish Wind Energy Association), Stockholm, March 26, 2012, reproduced in the Presentation of Gran Ronsten (WindREN) with courtesy of Polar Field Services

4.4.2 The Determination of the Stagnation Collision Efficiency

LB also give tabulated values of the total collision efficiency E (essential for determining overall ice accretion rates) and β_0 (used to determine the accretion shape for rime ice) as function of two non-dimensional parameters, the Stokes number K_{st} and the Langmuir parameter Φ.

$$K_{st} = \frac{\rho_w d^2 V}{9 D \mu_{air}} \tag{4.37}$$

$$\Phi = \frac{Re_d^2}{K_{st}} \tag{4.38}$$

The main advantage of using Φ is that it is not a function of the particle diameter. The case $\Phi = 0$ represents the Stokes flow.

Matching the Reynolds number and the inertia parameter for the scaled and reference case will satisfy also drop trajectory similarity. Unfortunately, in practical situations it is not always possible to satisfy both parameters simultaneously. The modified inertia parameter K_0, introduced by Langmuir and Blodgett in 1946 [16], overcomes this problem. It combines the effects of the inertia parameter and the Reynolds number and thereby provides a single similarity parameter to satisfy.

One starts defining the drag range λ_D as the average drag ratio over the Reynolds numbers from zero to Re_d that the particle is subjected to during its trajectory:

$$\frac{\lambda_d}{\lambda_{d,ST}} = \frac{1}{Re_d} \int_0^{Re_d} \frac{dRe}{C_D \frac{Re}{24}} \tag{4.39}$$

where λ_d is the drop range in absence of gravity, and $\lambda_{d,ST}$ is the drop range in absence of gravity within Stokes law validity. Since LB noted that for K = 1/8 drop

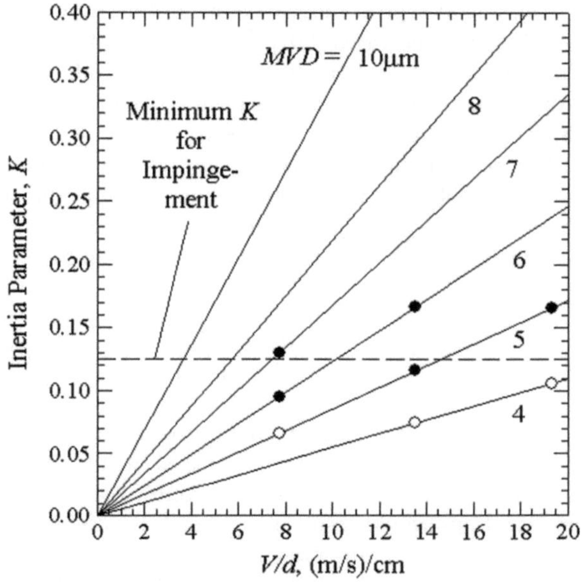

Fig. 4.10 Comparison of simplified impingement theory of LB and computation from LEWICE 2.0 code [19]

impingement would not occur, the following experimentally deduced relationship holds:

$$K_0 = \frac{1}{8} + \frac{\lambda_d}{\lambda_{d,ST}}\left(K - \frac{1}{8}\right) \tag{4.40}$$

for

$$K_0 > \frac{1}{8}$$

Figure 4.10 shows some results that compare this theory with the computation of LEWICE code. Drop impingement for small values of inertia parameter are shown. Solid symbols are the result of LEWICE 2.0 calculations showing some impingement points. Open symbols are the result of LEWICE 2.0 calculations for which no impingement occurred. All LEWICE calculations were for a 1.83 m chord NACA 0012 airfoil (d = 5.78 cm) at zero AoA.

The range parameter was tabulated by LB as a function of the drop Reynolds number, Re_d. A fit to the LB tabulation can be found in Anderson [19].

$$\frac{\lambda_d}{\lambda_{d,ST}} = \frac{1}{0.8388 + 0.0014385 \cdot Re_d + 0.1847 \cdot Re^{0.5}} \tag{4.41}$$

or the original formulation of Langmuir and Blodgett giving directly $K = 0$:

$$K_0 = \frac{1}{8} + \frac{K - 1/8}{1 + 0.0967 Re_d^{0.6367}} \tag{4.42}$$

This equation gives range parameters within ±0.2 % of LB values for $14 = Re_d = 600$. The error increases to 0.4 % for a Re_d of 800. For a static temperature T of 0 °C, Re_d for icing encounters range from 18 to 760, for airspeeds from 45 to 180 m/s, drop median-volume diameters from 10 to 50 μm and air pressure from 101,325 to 48,671 Pa (sea level to 6,000 m a.s.l.)

It is possible now to derive an expression for the local collision efficiency. LB published tables of the stagnation-point collision efficiency as a function of the inertia parameters, K_{st} and K_0 for cylinders. They showed that fairly accurate values could be calculated from the following equation valid for $K_{st} = 7.5$.

$$\beta_0 = \frac{1.40\left(K_0 - \frac{1}{8}\right)^{0.84}}{1 + 1.40\left(K_0 - \frac{1}{8}\right)^{0.84}} \tag{4.43}$$

A sample set of conditions which would produce a value of $K_{st} = 7.5$ at T = 0 °C, is D = 2.5 cm, V = 147.5 m/s and d = 20 μm. For a NACA 0012 airfoil, it corresponds to a chord of 31.6 in (80.3 cm). Larger model sizes, lower velocities or smaller drops would result in smaller values of K_{st}. However, this equation gives values of β_0 that are consistent with LEWICE calculations over a wide range of conditions. Figure 4.11 gives the LE collection efficiencies that closely match those found by the more detailed procedure of the LEWICE code. The equation value for β_0 ranged from 4.7 % lower than the LEWICE value at $K_0 = 0.4$–0.23 % higher at $K_0 = 34$. The conditions considered in Fig. 4.11 provide a range of inertia parameter, K_{st}, of 0.72 to 202, suggesting that LB's upper limit of $K_{st} = 7.5$ is very conservative.

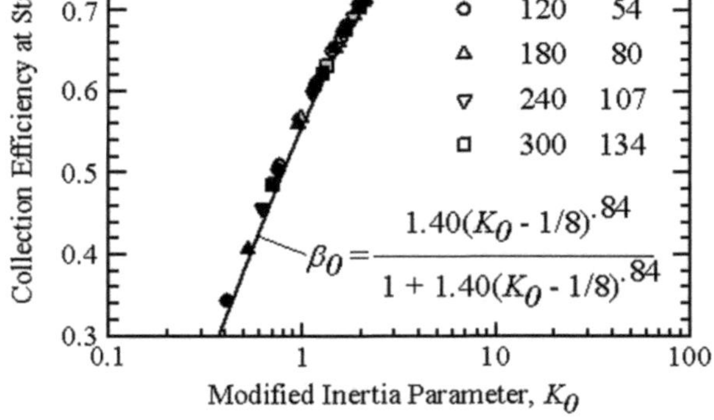

Fig. 4.11 Stagnation collection efficiency for NACA 0012 airfoils at 0 °C AoA. Static temperature, –12.2 °C. Static pressure, 100 kPa; airspeed, 54–134 m/s, MVD, 10–50 μm; LWC, 1 g/m^3. Open symbols, 7-in chord; shaded symbols, 21-in chord; solid symbols, 31.5 in chord. Data represented by symbols are from LEWICE predictions [19]

Finstad [20] deduced the expression for the local collision efficiency at the cylinder stagnation point β_0, and the overall collision efficiency E:

$$\beta_0 = [1.218K_{st}^{-0.0067}e^{(-0.551K_{st}^{-0.643})} - 0.17] - [0.00305(\phi - 100)^{0.43}] \cdot [2,220K_{st}^{-0.45}e^{(-0.767K_{st}^{-0.806})} - 0.068] \quad (4.44)$$

and

$$E = [1.030K_{st}^{-}0.00168e^{(-0.796K_{st}^{-}0.780)} - 0.040] - [0.00944(\phi - 100)^{0.344}] \cdot [2.657K_{st}^{-}0.519e^{(-1.06K_{st}^{-}0.842)} - 0.029] \quad (4.45)$$

with the following conditions:

$$0.17 < K_{st} < 1000$$

$$\phi = \frac{Re_D^2}{K_{st}}$$

$$100 < \phi < 10000$$

$$Re = \frac{\rho_{air} w_\infty d}{\mu}$$

Comparison of the methods is given in Fig. 4.12 for $\Phi = 10^3$.

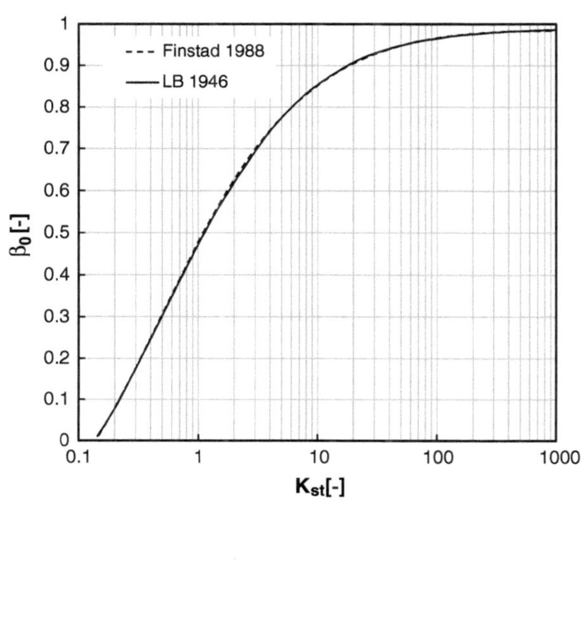

Fig. 4.12 Stagnation collection efficiency β_0 as function of Stokes number according to LB and Finstad approximations for $\Phi = 10^3$

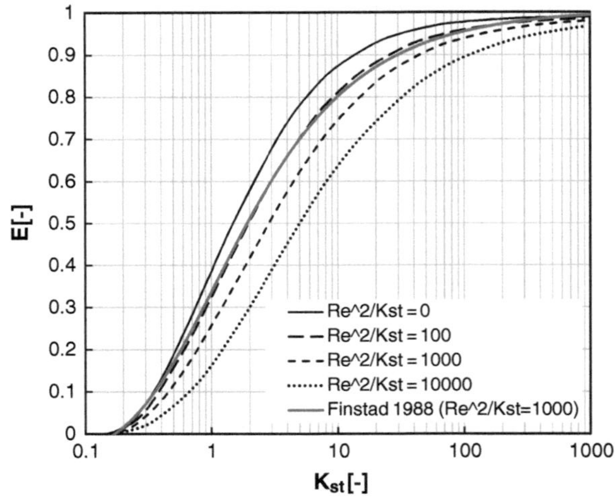

Fig. 4.13 Overall collision efficiency E as function of Stokes number according to LB for Φ ranging from 0 to 10^4, and Finstad

The overall collision efficiency is given in Fig. 4.13 for Φ ranging from 0 to 10^4. It is worth to recall that this range would apply to most physical applications, encompassing the following parameter ranges:

- $10 \leq \text{MVD} \leq 4000\,\mu\text{m}$
- $0.01 \leq \text{D} \leq 1.0\,\text{m}$
- $5 \leq \text{W} \leq 200\,\text{m/s}$
- $-20\,°\text{C} \leq \text{T} \leq 0\,°\text{C}$

The works of LB [16], and Finstad [20] were carried out under the hypothesis of potential flow field about an infinitely long, circular cylinder. The applicability of this assumption on real cases depends on the cylinder Reynolds number and the resulting boundary layer and wake. At high Reynolds number (cylinder diameter based), the cylinder boundary layer tends to be thin and, therefore, it would expected to have a marginal effect on the droplet trajectories. At high K_{st}, that is, at high velocities, the trajectories tend to become ballistic, assuming straight-line paths. The wake at these high numbers will resemble a kind of 'solid afterbody' to the cylinder, having a small effect on the collision efficiency, as experimentally proved by Finstad. At low Reynolds numbers, the cylinder boundary layer is thicker and eventually the potential flow approximation is no longer valid.

4.4.3 2D Scheme for the Particles Trajectory Calculation

The following 2D model allows determining, by a time Marching integration, the trajectories of water droplets past a body of arbitrary shape in two-dimensional flow, provided the streamline velocity components are known or can be computed. The method considers the deviation of the water droplets' behaviour from Stokes law because of velocity and drop size range. In general, water drops are not sufficiently small to obey Stokes' law of resistance as previously shown and the force acting on the drop can be determined only by modelling the drag coefficient for spheres.

The procedure starts from the cited Langmuir Blodgett equations (4.23), (4.24) and (4.26) valid for the stationary cylinder:

$$\frac{dx_d}{dt}\frac{d}{dx}\left(\frac{dx_d}{dt}\right) = C_D\frac{Re_d}{24}\frac{1}{K_{st}}\left(V_x - \frac{dx_d}{dt}\right) \tag{4.46}$$

and in the y component

$$\frac{dy_d}{dt}\frac{d}{dy}\left(\frac{dy_d}{dt}\right) = C_D\frac{Re_d}{24}\frac{1}{K_{st}}\left(V_y - \frac{dy_d}{dt}\right) \tag{4.47}$$

where

$$K_{st} = \frac{2}{9}\frac{\rho_w}{\rho_{air}}\left(\frac{r}{c}\right)^2\frac{\rho_{air}\,V_\infty\,c}{\mu_{air}} \tag{4.48}$$

Equations (4.46) and (4.47) can be rewritten as:

$$\frac{dx_d^2}{dt^2} = -C_D\frac{Re_d}{24}\frac{1}{K_{st}}\left(\frac{dx_d}{dt} - V_x\right) \tag{4.49}$$

$$\frac{dy_d^2}{dt^2} = -C_D\frac{Re_d}{24}\frac{1}{K_{st}}\left(\frac{dy_d}{dt} - V_y\right) \tag{4.50}$$

The average velocity of a droplet over a small time interval is:

$$\bar{v}_d = \frac{v_{n+1} + v_n}{2} \tag{4.51}$$

and the average acceleration:

$$\bar{a}_d = \frac{a_{n+1} + a_n}{2} \tag{4.52}$$

The velocity components of a droplet after a time of travel Δt while it covers the distance from n to $n+1$ are given by:

$$\left(\frac{dx_d}{dt}\right)_{n+1} = \left(\frac{dx_d}{dt}\right)_n - \left\{C_D\frac{Re_d}{24}\frac{1}{K_{st}}\left[\left(\frac{dx_d}{dt}\right)_n - (V_x)_n\right]\right\}\Delta t \tag{4.53}$$

$$\left(\frac{dy_d}{dt}\right)_{n+1} = \left(\frac{dy_d}{dt}\right)_n - \left\{C_D\frac{Re_d}{24}\frac{1}{K_{st}}\left[\left(\frac{dy_d}{dt}\right)_n - (V_y)_n\right]\right\}\Delta t \tag{4.54}$$

and the position is:

$$(x_d)_{n+1} = (x_d)_n\left(\frac{dx_d}{dt}\right)_n\Delta t - \frac{1}{2}\left\{C_D\frac{Re_d}{24}\frac{1}{K_{st}}\left[\left(\frac{dx_d}{dt}\right)_n - (V_x)_n\right]\right\}\Delta t^2 \tag{4.55}$$

$$(y_d)_{n+1} = (y_d)_n\left(\frac{dy_d}{dt}\right)_n\Delta t - \frac{1}{2}\left\{C_D\frac{Re_d}{24}\frac{1}{K_{st}}\left[\left(\frac{dy_d}{dt}\right)_n - (V_y)_n\right]\right\}\Delta t^2 \tag{4.56}$$

It is assumed the acceleration of the drops is constant over the interval Δt. The set of equations presented can be used to compute the trajectory of a particle by Marching in time. As initial boundary condition far upstream one can state that the velocity of the drop is that of the air $V_\infty = V_0$. After a Δt increment one observes a finite difference $(dx_d/dt - V_x)$ and $(dy_d/dt - V_y)$. Therefore, far upstream at position 0:

$$\frac{dx_{d,0}}{dt} = V_{x,0} \tag{4.57}$$

$$\frac{dy_{d,0}}{dt} = V_{y,0} \tag{4.58}$$

the new position of the particle is:

$$x_1 = x_0 + \Delta x \tag{4.59}$$

$$y_1 = y_0 + \Delta y \tag{4.60}$$

being:

$$\Delta y = \frac{V_{y,0}}{V_{x,0}} \Delta x$$

The droplet velocity is thus:

$$v_{x,1} = \frac{dx_{d,1}}{dt} = \frac{dx_{d,0}}{dt} + \Delta v_x \tag{4.61}$$

$$v_{y,1} = \frac{dy_{d,1}}{dt} = \frac{dy_{d,0}}{dt} + \Delta v_y \tag{4.62}$$

Equation (4.49) becomes:

$$v_x dv_x = C_D \frac{Re_d}{24} \frac{1}{K_{st}} (v_x - V_x) dx \tag{4.63}$$

and substituting Eqs. (4.61) and (4.57) into Eq. (4.63) one gets:

$$v_{x,1} = (v_{x,1} - V_{x,0}) = C_D \frac{Re_d}{24} \frac{1}{K_{st}} (v_{x,1} - V_{x,1}) \Delta x \tag{4.64}$$

and solving for $v_{x,1}$:

$$v_{x,1} = \frac{V_{x,0} + C_D \frac{Re_d}{24} \frac{1}{K_{st}} \Delta x - \left[\left(V_{x,0} + \frac{Re_d}{24} \frac{1}{K_{st}} V_{x,1} \Delta x \right)^2 - 4 \frac{Re_d}{24} \frac{1}{K_{st}} V_{x,1} \Delta x \right]^{\frac{1}{2}}}{2} \tag{4.65}$$

Analogously, one obtains for the vertical velocity component:

$$v_{y,1} = \frac{V_{y,0} + C_D \frac{Re_d}{24} \frac{1}{K_{st}} \Delta y - \left[\left(V_{y,0} + \frac{Re_d}{24} \frac{1}{K_{st}} V_{y,1} \Delta y \right)^2 - 4 \frac{Re_d}{24} \frac{1}{K_{st}} V_{y,1} \Delta x \right]^{\frac{1}{2}}}{2} \tag{4.66}$$

Assuming that y_0 is the initial ordinate of the droplet far upstream, the mass flow of water carried is:

$$V \cdot LWC \cdot L \cdot y_0$$

being L the width of the streamtube. This mass flow impinges over the leading edge area, thus, since:

$$V \cdot LWC \cdot L \cdot y_0 = \dot{m}_{w,imp} \cdot L \cdot s$$

for an infinitesimal span length over the body, the impingement mass $\dot{m}_{w,imp}$ results as:

$$\dot{m}_{w,imp} = V \cdot LWC \cdot \frac{dy_0}{ds} \tag{4.67}$$

The total mass of impinging water over the body is:

$$\dot{m}_{w,imp,\,t} = \int_{s_l}^{s_u} \dot{m}_{w,imp}\, ds = V \cdot LWC \cdot y_{max} E \tag{4.68}$$

where:

$$E = \frac{y_{0,limit}}{y_{max}}$$

$y_{0,limit}$ is the ordinate of the upstream starting point of the trajectory which impinges the body and y_{max} is the maximum ordinate of the body.

At this end the velocity of the air past the body is used to find the intersection between particle trajectories and body streamlines.

4.4.3.1 Practical Implementation of the Calculation

The MVD approximation is used as per the droplet diameter. Note that if a droplet distribution that varies from d_{min} to d_{max} is used, the impingement limits have to be established for each droplet size. The maximum impingement limits are defined by the impingement limits of the largest droplet diameter in the distribution. To initiate the computation, the number of particles to be released (i.e. 400), and the minimum temporal step given by the following relationship have to be specified:

$$\Delta t_0 = \frac{c}{V_0\ N_{\Delta t_c}}$$

where c is the chord length, V_0 is the far upstream velocity and $N_{\Delta t_c}$ is the number of steps to cover the chord distance.

Furthermore, a constant k_t to define the variation in the temporal step as departing from the profile LE, has to be set (i.e. 10)

$$t = \Delta t_0 \left(1 + k_t \frac{\bar{x}_p}{c}\right)$$

The N particles of water released into the flow are placed upstream at the edge of the x coordinate of the grid. The minimum and maximum y coordinates of the particles at the initial time are evaluated by calculating three values of the ordinate along the border of minimum x:

- From the line streamline passing by the leading edge (0, 0);
- The straight line passing through the point (0, 0) and with a slope equal to the tangent of the angle of attack;
- The straight line passing through the point (c, 0) and with a slope equal to the tangent of the angle of attack.

At the minimum value among these ordinates the maximum distance of the lower part of the profile from the x axis is subtracted. This gives the minimum ordinate of the particles. Similarly, at the maximum value of these the maximum distance of the upper part of the profile from the x axis is added. This provides the maximum ordinate of the particles.

The angle γ of the relative velocity with respect to the abscissa is:

$$\gamma = \arctan\left(\frac{\frac{dy_d}{dt} - V_y}{\frac{dx_d}{dt} - V_x}\right)$$

The flow velocity $(V_x,\ V_y)$ is calculated by interpolation on the finer grid provided for the point in question, for instance, by using the method Piecewise Cubic Hermite Interpolating Polynomial (PCHIP) of MATLAB®.

The particle velocity is given by:

$$V = \left[\left(\frac{dx_d}{dt} - V_x\right)^2 + \left(\frac{dy_d}{dt} - V_y\right)^2\right]^{0.5}$$

The differential equations of the first order can be solved by an explicit Runge-Kutta 4th order method, by considering however the drag force to the initial point of each time step for simplicity.

The collision is detected if the particle penetrates the perimeter of the profile. In that case the collision point closest to the contour is found. The collision point is approximated by the intersection of the straight line passing through the point ith and the ith –1 of the particle trajectory and the straight line tangent to the profile at the closest point. The simulation is terminated when a particle passes over the collision zone on the profile.

Once the points of collision are known, the curvilinear abscissa along the profile is calculated. The collection efficiency is approximated with the ratio between the spacing of the particles and the spacing between two collision points along the curvilinear abscissa, therefore:

$$\beta \approx \frac{dy\,\cos\alpha}{ds}$$

4.4.4 The Solution for the Fixed Cylinder

The above model has been implemented to discuss the effect of particle size, wind speed and cylinder dimension on droplet trajectories, collision efficiency, impingement limits and stagnation collision efficiency. To accomplish this purpose, the potential flow past a cylinder was used to create the flow field. This can obtained by superimposing a sink and a source dipole at uniform velocity W, thus getting the following stream function:

$$\psi\left(x,\ y\right) = y \cdot W \left[1 - \frac{\left(\frac{D}{2}\right)^2}{x^2 + y^2}\right] \tag{4.69}$$

The size and thermo-fluid-dynamic boundary conditions were selected by replicating the diameter of the inscribed cylinder in four radial sections of the Tjæreborg wind turbine blade, as shown in Table 4.2. Ambient input data for the simulations are given in Table 4.3.

Figures 4.14, 4.15, 4.16 and 4.17 show the trajectories of the droplet and the impingement points for the four cylinders used in the simulations.

The points have been then superimposed in Figs. 4.18 and 4.19 to the Finstad correlation [20], previously shown. The correspondence is quite good, and the value of Section 0 does not appear, since the K_0 parameter is less than the critical value of 1/8.

As expected, by moving from root Section 0 to tip Section 13, the limits of impingement increase, due to the conspiring effect of the smaller diameter and higher wind velocity, and more surface will be wetted. As one can further note, the 0 Section tends to remain completely dry, since the large body is in transverse dimension.

Table 4.2 General blade data of Tjæreborg wind turbine based cylinder analysis

Section	r/R (–)	Radial station (m)	Profile type	c (m)	Equivalent LE cylinder D (m)	W (m/s)
0	0.09	2.75	Circular	1.8	1.8	15.19
8	0.61	18.46	NACA 4416	2.1	0.105	44.8
10	0.80	24.46	NACA 4414	1.5	0.060	58.25
13	0.98	29.86	NACA 4412	0.96	0.032	70.45

Table 4.3 Input data for the simulations

Parameter	Value
p_∞ (Pa)	101 325
T_∞ (°C)	–2.0
LWC (g/m^3)	0.2
MVD (μm)	20.0
Humidity (–)	0.99

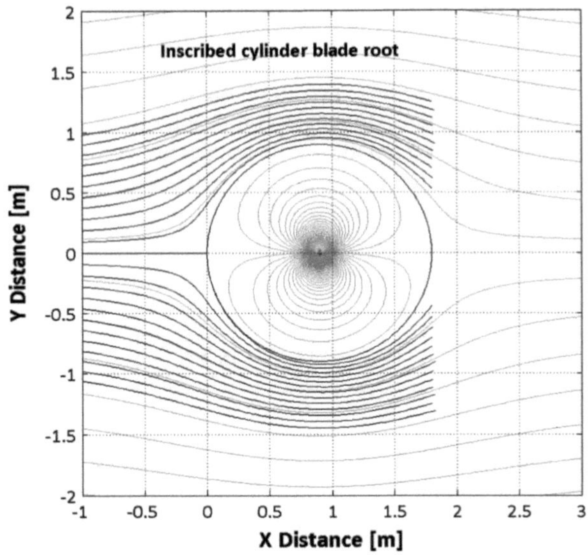

Fig. 4.14 Droplet trajectories and impingement points of the inscribed cylinder of Section 0 of Table 4.2

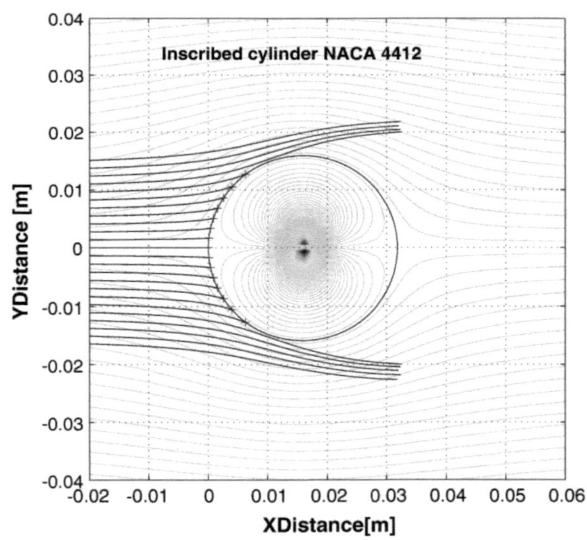

Fig. 4.15 Droplet trajectories and impingement points of the inscribed cylinder of Section 8 of Table 4.2

The stagnation point collision efficiency and the total collision efficiency for the four cases are presented in Table 4.4. The last column of Table 4.4 indicates the minimum droplet diameter causing a value of the stagnation collision efficiency bigger than zero, i.e. the threshold ambient condition for the surface, to become wetted by the droplet impingement.

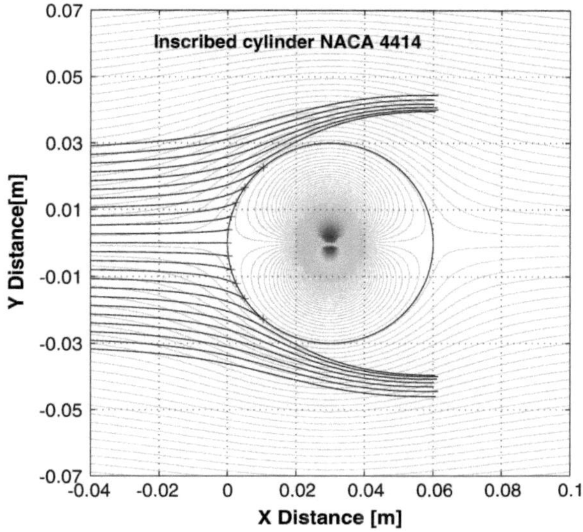

Fig. 4.16 Droplet trajectories and impingement points of the inscribed cylinder of Section 10 of Table 4.2

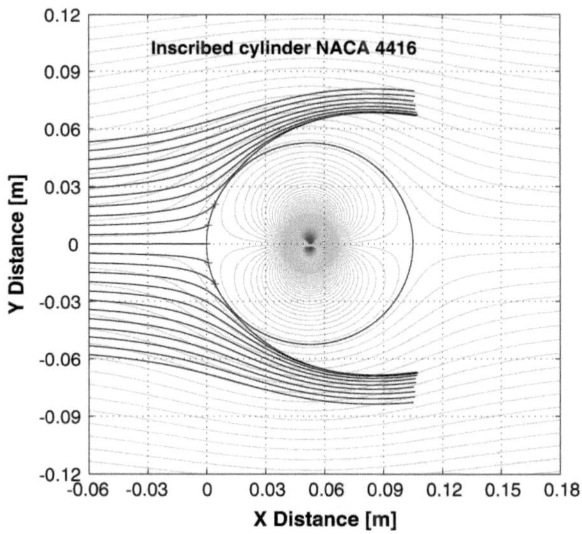

Fig. 4.17 Droplet trajectories and impingement points of the inscribed cylinder of Section 13 of Table 4.2

Lastly, as criterion to set up the number of particles released in the space field, the accuracy in predicting the collision efficiency has been computed by varying the number of particles released. Figure 4.20 shows the convergence of the stagnation and the total collision efficiency as the number of droplets is increased, and the parameter N/D for the case of the cylinder inscribed in the NACA 4412 profile (the other cases replicate these results). While for the stagnation collection efficiency β_0 a good

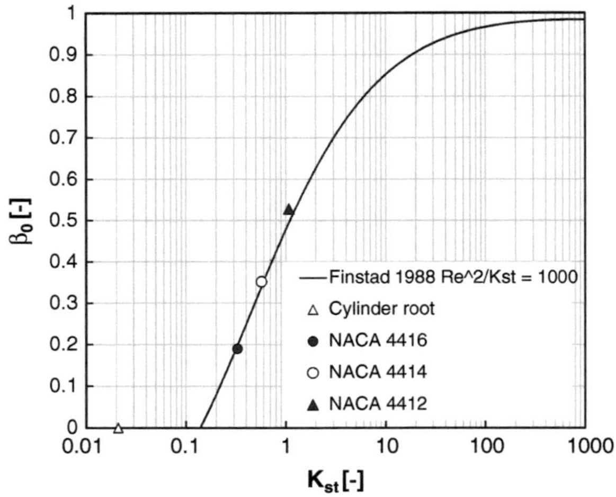

Fig. 4.18 Stagnation points collision efficiency and Finstad correlation

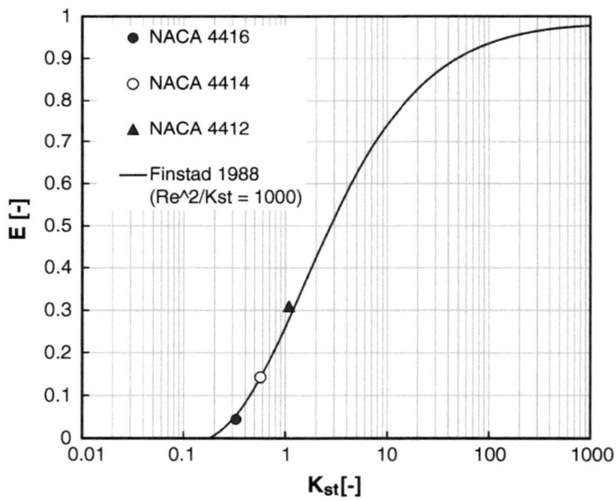

Fig. 4.19 Total collision efficiency E and Finstad correlation

Table 4.4 Re_d, K_{st}, ϕ, β_0, E, MVD_{min} for the simulation involving data of Tables 4.2 and 4.3

Section	Re_d	K_{st}	ϕ	β_0	E	MVD_{min} (μm)
0	15.41	0.021	13,717	0.001	0.000	68
8	15.41	0.326	728.3	0.190	0.044	9
10	15.41	0.571	416.2	0.351	0.142	7
13	15.41	1.080	220.0	0.528	0.310	4

numerical convergence is achieved already for N/D of about 1,570 (≈50 particles released), the convergence on E is achieved for N/D of about 7,300 (≈235 particles). This last value has been proved to be consistent for a large range of parameters (MVD, LWD, D), and had been set as control parameter for the simulations.

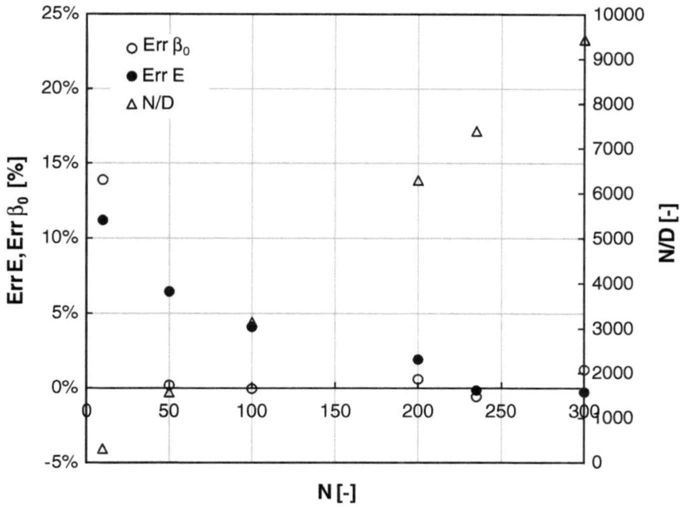

Fig. 4.20 Convergence and number of particles released

4.4.5 Collision Efficiency at Zero AoA Airfoil LE

During the Second World War and early after that period, calculations performed on NACA airfoils indicated that, for large values of drop size and wind speed:

- The assumption of the equivalent cylinder would not hold for blade airfoils;
- There is an additional variable which is the angle of attack.

Since experimental tests are difficult and costly, extensive calculations were undertaken in the past to determine the drop trajectories for most aerodynamic profiles. A very comprehensive study on a 6-foot NACA 23012 airfoil at 0°AoA (LWC = 0.82 g/m^3, T = 20°F, V = 195 mph) is reported here and briefly commented [21] upon because of its exemplar picture of the physics of the problem

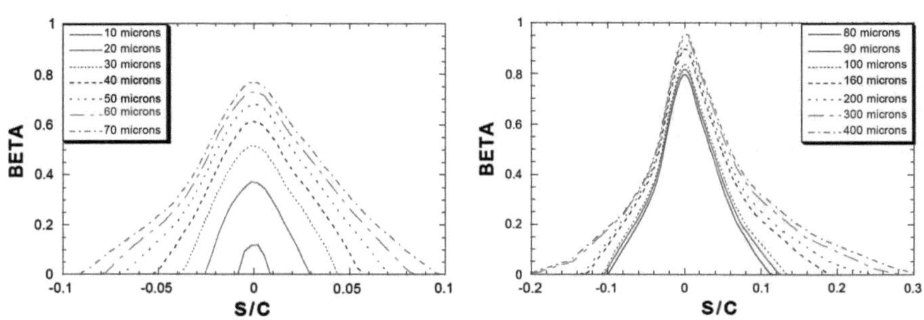

Fig. 4.21 Local collection efficiency, β, for drop size from 10 to 400 μm for NACA 23012 airfoil [21]

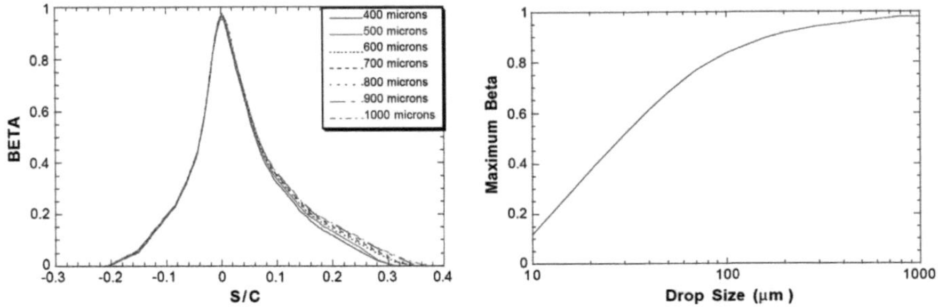

Fig. 4.22 Local collection efficiency, β, for drop size from 400 to 1,000 μm for NACA 23012 airfoil [21]

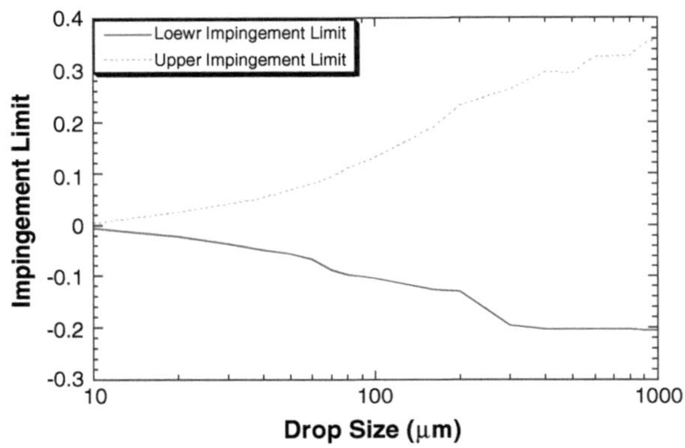

Fig. 4.23 Maximum collection efficiency as function of the drop size [21]

treated. Figures 4.21 and 4.22 show the local collection efficiency, β, for drop size from 10 to 1000 μm. These results are rearranged in Figs. 4.23 and 4.24, where the local collection efficiency, the impingement limits and the total collection efficiency are shown as functions of the droplet size.

The maximum local collection efficiency increases with drop size, as the upper and lower limits, which move accordingly downstream. Smaller droplets, which have less inertia, have lower collision efficiencies. The total collection efficiency increases with drop size. There is a large variation over the drop size range. The graph of Fig. 4.21 shows a rapid increase in total collection efficiency up to 100–200 μm range, which gradually approaches the value of 1 at 1,000 μm (see Fig. 4.22).

The larger a drop gets, the more ballistic its trajectory will be and the local collection efficiency becomes simply a reflection of the curvature of the airfoil geometry. The degree to which the drops are deflected by the airfoil is shown in Fig. 4.23.

The upper and lower impingement limits have as theoretical limits the maximum and minimum thicknesses of the airfoil. If the body/droplet interaction causes no deflection, the total collection efficiency is one (Fig. 4.24). The 1,000 μm drop size impingement limit is very close to this value.

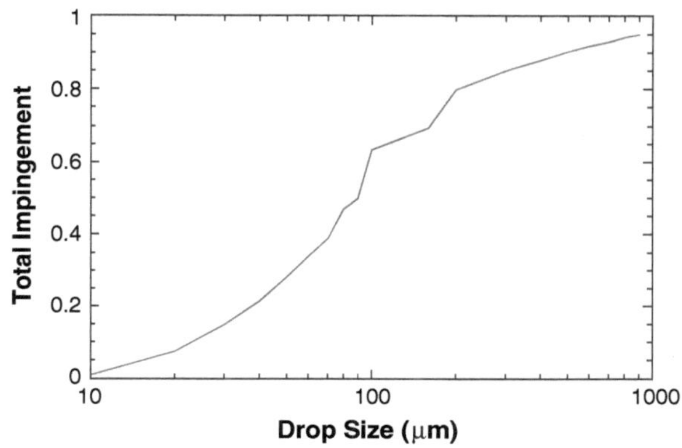

Fig. 4.24 Impingement limits and total impingement as function of the drop size [21]

The results of the investigation are that:

(a) Smaller droplets, which have less inertia, have lower collision efficiencies;
(b) The collision efficiency of droplets is smaller for large obstacles than for small ones, because streamlines diverge relatively farther in front of large obstacles than small ones, and;
(c) In high winds, droplets have more momentum so their collision efficiency is higher.

4.4.6 Collision Efficiency at AoA Other than Zero and Scaling Effects

A closer look at the water trajectories when the AoA is different from zero, is illustrated in Fig. 4.25, where the asymmetric droplet release with respect to the leading edge is clearly shown.

A comparison of LEWICE output of LE collection efficiency β_0 at angles of attack of 0° and 10° with those from Eq. (4.43) is given in Fig. 4.26. For the NACA

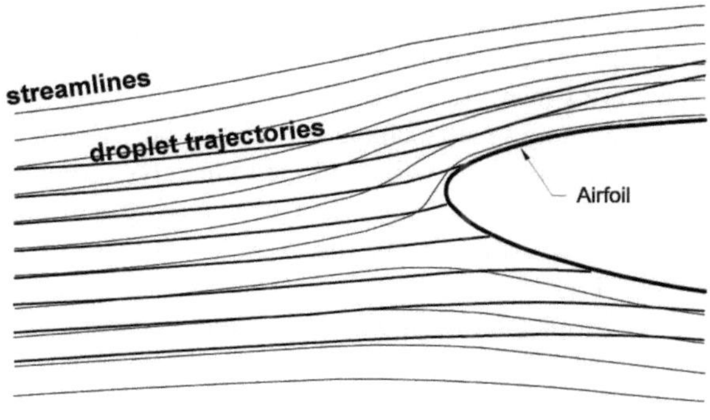

Fig. 4.25 Qualitative effect of AoA other than zero of droplet trajectories

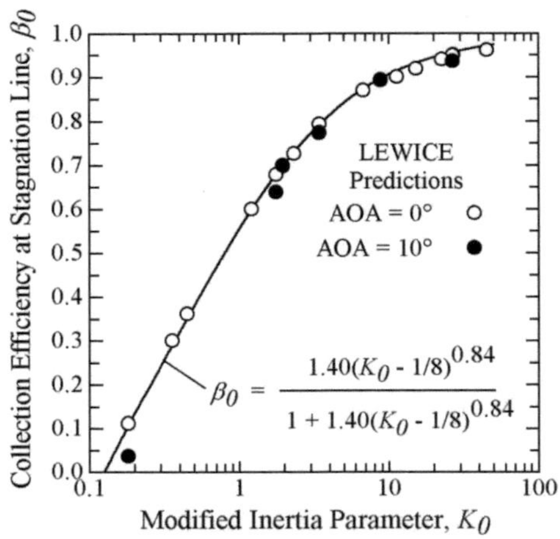

Fig. 4.26 Comparison of LEWICE determinations of leading-edge collection efficiency at angles of attack of 0 and 10° with those from Eq. (4.43) [19]

0012 airfoil used for these computations β_0 varies little from 0° to 10° AoA. If the scale model is set at the identical AoA as the reference, matching of K_0 will still produce the correct drop trajectories for the scale test although there may be a small error in β_0 if Eq. (4.43) used. As a consequence, in order to match β_0 of scale and reference values, it is only necessary to match K_0. Being K_0 quite easy to calculate, this parameter has been universally adopted in scaling methods similarity parameter to ensure similarity of drop trajectories.

This consideration leads to a very important conclusion on the total amount of water collected. The above considerations have stated that at the stagnation line of a clean surface the collision efficiency is the same for scaled model and reference profile. Therefore, if proper scaling rules are adopted (scaled and reference models are geometrically similar) aft of the stagnation line the collection efficiencies must vary in the same way for both models. This agreement of scaled and reference collection efficiencies is shown in Fig. 4.27, which gives LEWICE predictions for two NACA 0012 airfoils of different sizes.

The conditions for the smaller model were scaled from those of the larger in such a way that the stagnation (K_0 and therefore β_0) for the two matched. The two curves are indistinguishable over the range for which accretion occurs. Therefore, to help identify the curve for the reference case the area under that plot has been shaded. Consequently, it is only necessary to match the stagnation K_0 to properly scale drop trajectories over the entire clean airfoil. It is therefore assumed that as ice accretes, because the geometry changes in the same way for both models, the time-varying collection efficiency will continue to match everywhere.

Figure 4.28 shows the effect of changing AoA from 0° to 4° on the collection efficiency distribution for the NACA 0012 airfoil. The results are obtained with LEWICE [22].

The droplet diameter was 20 μm and the air speed was 100 mph (44.7 m/s). While the β_0 varies not remarkably from 0° to 4° AoA, the extension of the wet limits widen as the AA increases, especially over the pressure side of the profile.

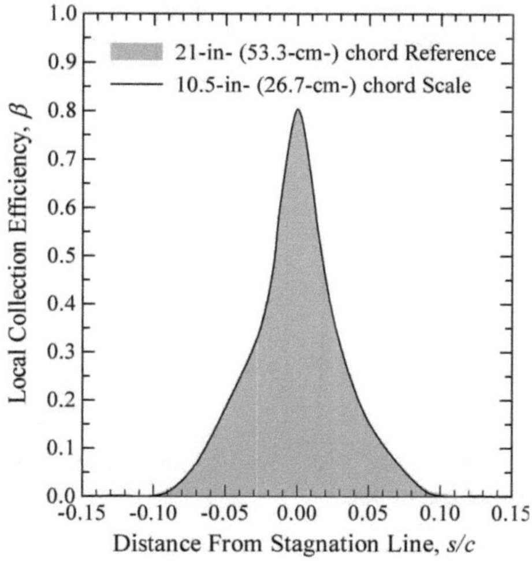

Fig. 4.27 Scaled and reference collection efficiencies: LEWICE predictions for two NACA 0012 airfoils of different sizes. NACA 0012 airfoils at 0° AoA. reference conditions: c_R, 21 in (53.3 cm); V_R, 67 m/s, d_R, 30.0 μm, scaled model: c_S, 10.5 in (26.7 cm); V_S, 117 m/s, d_S 15.6 μm [19]

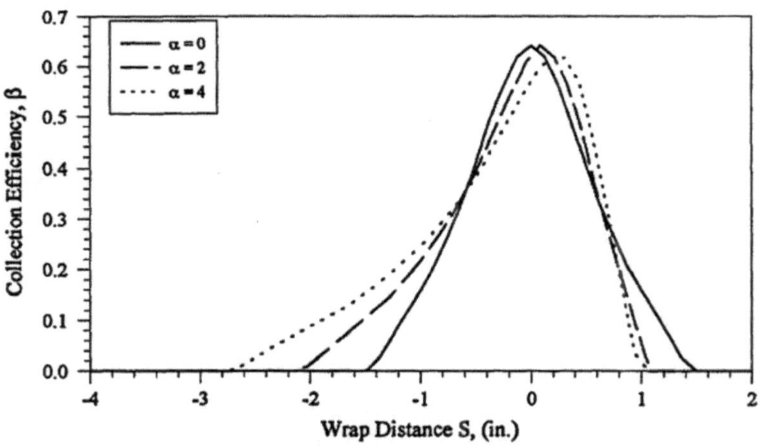

Fig. 4.28 Determinations of leading-edge collection efficiency and impingement limits at angles of attack of 0°, 2° and 4° for the NACA 0012 [22]

4.4.7 Example

On the basis of validation carried out in Sect. 4.4.4, the 2D local and total water impingement for the radial sections of the Tjæreborg wind turbine blade have been calculated. The conditions for the simulations are listed in Tables 4.3 and 4.5. These results will be used in the simulations of Chap. 5 for the anti-icing system. The

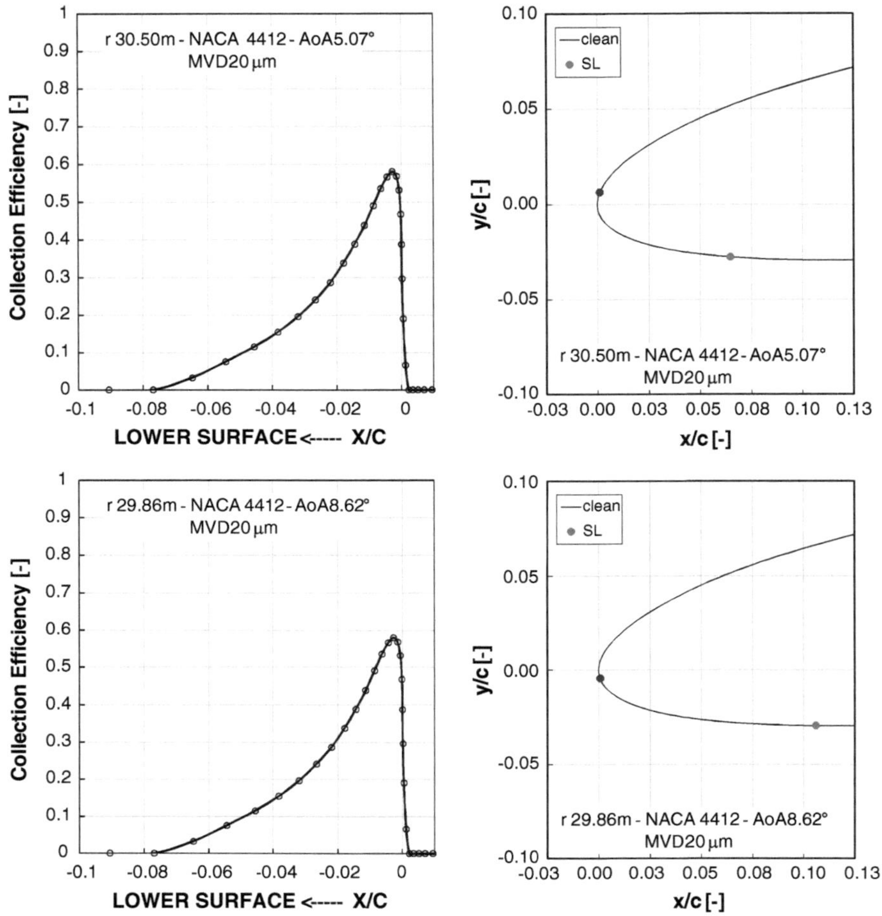

Fig. 4.29 r/R = 1 (*up*), 0.98 (*down*)

results of the calculation of the impingement are given in Figs. 4.29, 4.30, 4.31 and 4.32. Ambient static temperature was 270.15 K, pressure, 100 kPa, LWC = 0.1 g/m^3, MVD = 20 μm.

At the tip blade section, the maximum 2D collision efficiency β (see curve for r/R = 1 in left graph of Fig. 4.33) and the water mass flow collected (left graph of Fig. 4.34) are reached. No water is collected below r/R = 0.4.

The impingement limits change moving from tip to hub sections, as shown in the right graph of Fig. 4.33), achieving the maximum wet length between r/R = 0.95–0.98 (Section 12 and 13). As a consequence, a peak is observed at r/R = 0.95–0.98 in the right Fig. 4.34 where the water collected per metre of blade length is plotted. The very last tip section exhibits a smaller amount of water captured, due to the drop in the relative speed caused by the tip losses.

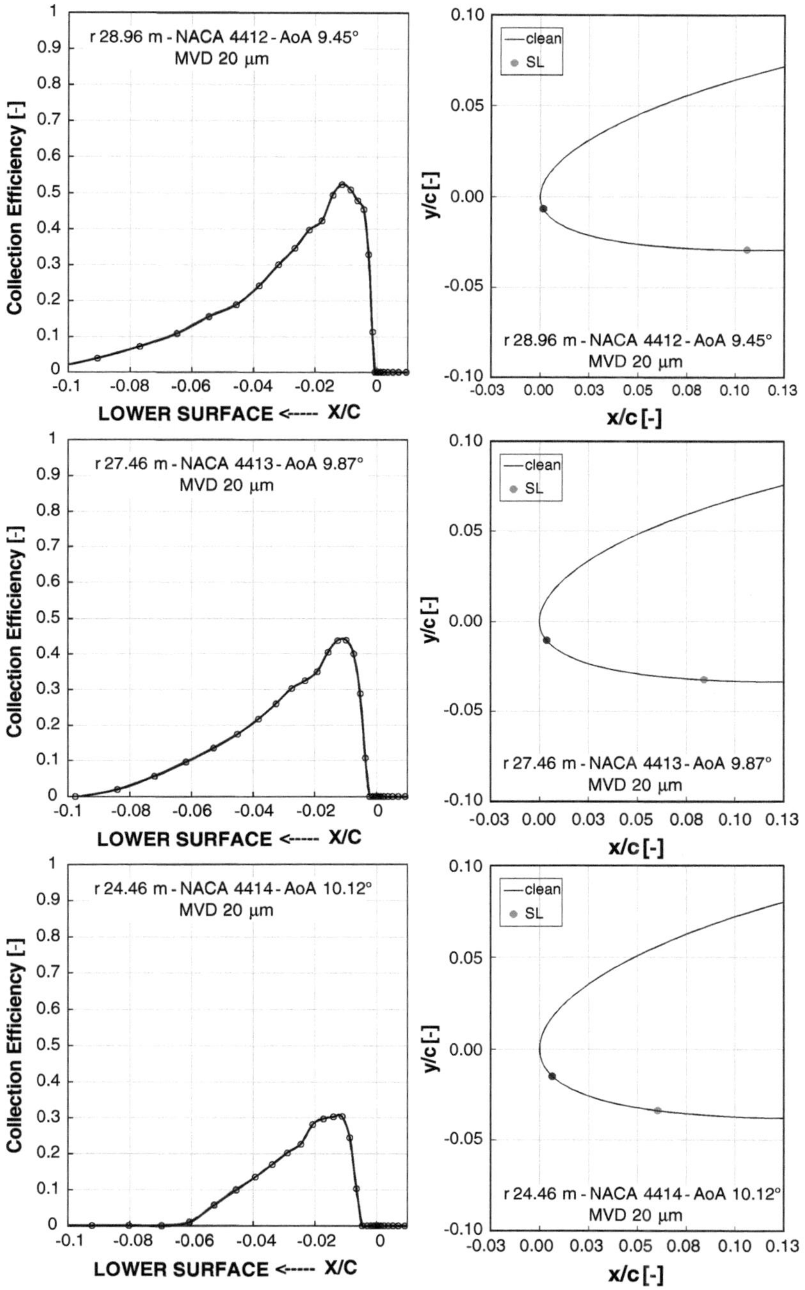

Fig. 4.30 r/R = 0.95 (*up*), 0.90 (*centre*) and 0.80 (*down*)

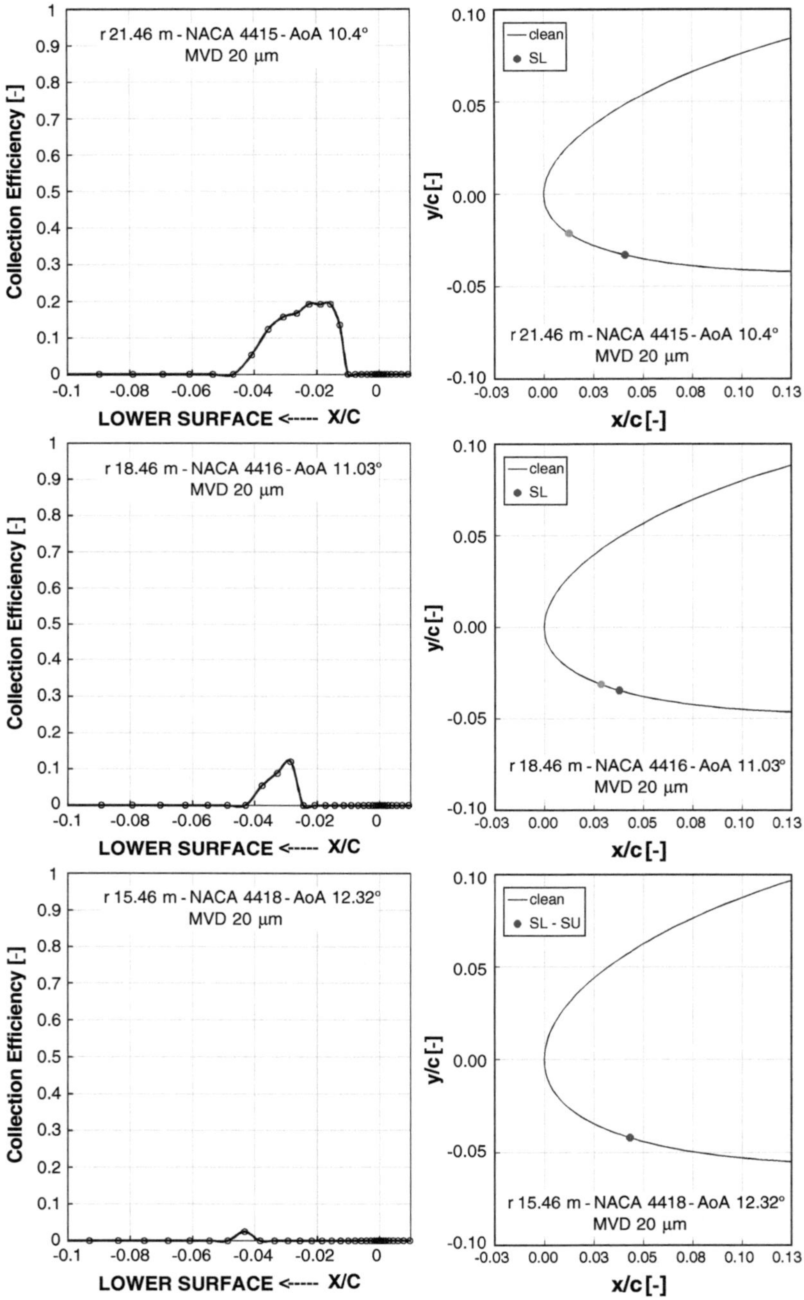

Fig. 4.31 r/R = 0.70 (*up*), 0.61 (*centre*) and 0.51 (*down*)

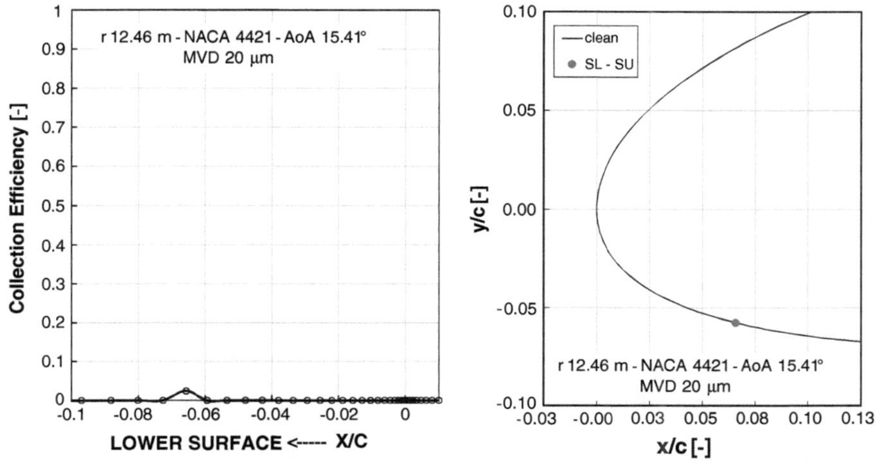

Fig. 4.32 r/R = 0.41

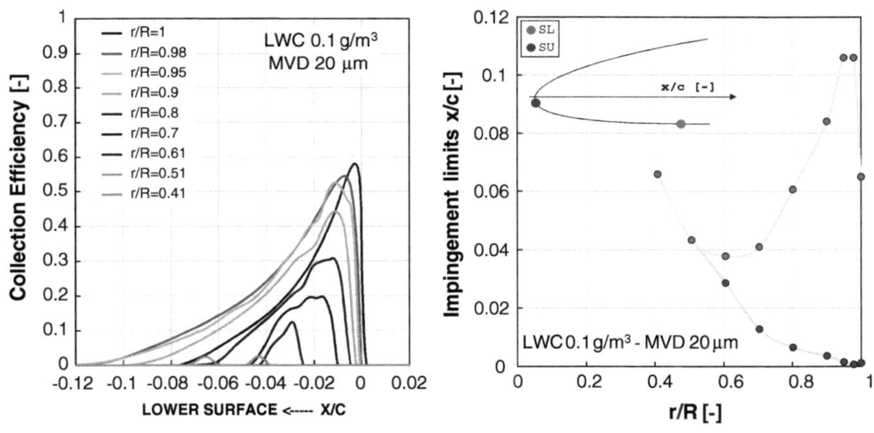

Fig. 4.33 Impinging mass distribution (*left*) and wet blade coordinate x/c (*right*) in each blade section

4.4.8 The Rotating Airfoil

If the body is rotating, the water impingement process can no longer be considered as two-dimensional. In fact, while if the two-dimensional flow conditions holds, two impingement limits, an upper and lower, are required (see Fig. 4.6); for three-dimensional flow, the limits of impingement may vary spanwise along the surface of the body (i.e. the blade). The problem requires the adoption of four trajectories and no longer two, as depicted in Fig. 4.35.

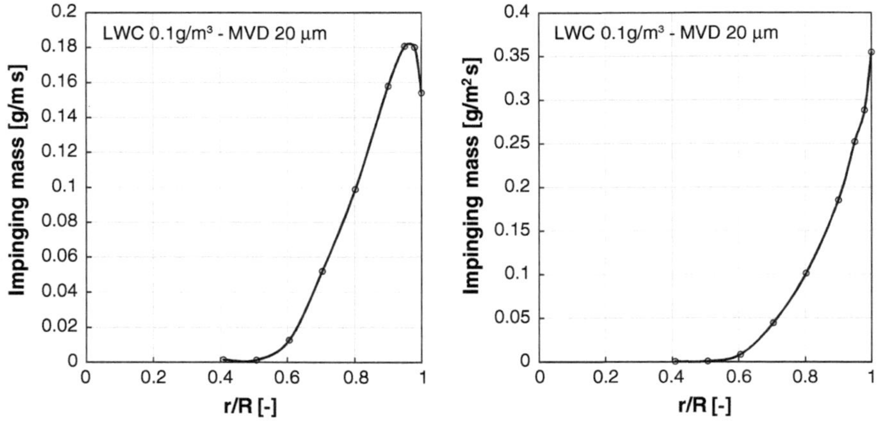

Fig. 4.34 Impinging mass distribution per m^2 of blade surface (*left*) and per m of blade length (*right*) versus r/R

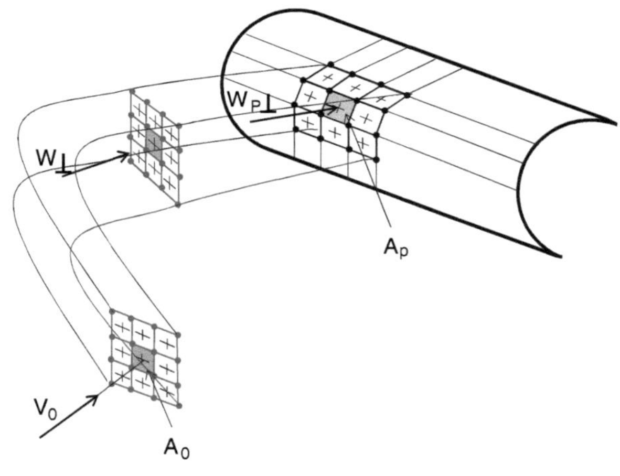

Fig. 4.35 Impingement limit of rotating body and water trajectories

4.4.8.1 Limits of Impingement

Figure 4.35 indicates that the identification of the impact area is not straightforward. If the limits of impingement fall inside a panel, these limits need an interpolation procedure to be determined. Some numerical codes take into account the 3-D approach (NASA, ONERA, DRA, FENSAP-ICE).

4.4.8.2 3D Impingement

The rotation determines a *rotating sampling* effect that increases markedly the amount of water captured by the body. As previously stated, a part of extreme events, as super large droplets rain, the effect of the forces due to gravity and interparticle interactions are negligible compared to those due to aerodynamic drag and centrifugal forces [23].

To analyse 3D swirling flows, the most used form of the equations of motion of particles immersed in a fluid are written in a rotating frame of reference using cylindrical polar coordinates as follows:

$$\frac{d^2 r_p}{dt^2} = F_r + r_p \left(\frac{d\theta_p}{dt} + \Omega \right)^2 \tag{4.70}$$

$$r_p \cdot \frac{d^2 \theta_p}{dt^2} = F_\theta - 2 \cdot \frac{dr_p}{dt} \left(\frac{d\theta_p}{dt} + \Omega \right)^2 \tag{4.71}$$

$$\frac{d^2 z_p}{dt^2} = F_z \tag{4.72}$$

where r_p, θ_p, and z_p define the particle location in cylindrical polar coordinates, and Ω is the rotor blade angular velocity. The centrifugal force and Coriolis acceleration are represented by the last term on the right-hand side of Eqs. (4.70) and (4.71). The forces per unit mass of particles of interaction between the two phases are given by the first term on the right-hand side of Eqs. (4.70)–(4.72). The force of interaction between the two phases is dominated by the drag due to the difference in velocity between the solid particles and the flow and is given by:

$$\bar{F} = \frac{3}{4} \frac{C_D}{d} \left[\left(W_r - \frac{d\, r_p}{dt} \right)^2 + \left(W_\theta - \frac{d\, r_p \theta_p}{dt} \right)^2 + \left(W_z - \frac{d\, z_p}{dt} \right)^2 \right] \cdot (\bar{W} - \bar{W}_p) \tag{4.73}$$

where W_r, W_θ , and W_z are the relative wind velocities in the radial, circumferential, and axial directions, respectively, d is the drop diameter, and C_D the drop drag coefficient. The numerical integration of Eqs. (4.70)–(4.72) are carried out in the three-dimensional flow field starting with the particle initial position and velocity.

The only experimental experiences have been made on propellers. They have shown that the rotation induces important variations in local and total efficiency evaluation, namely:

- For constant rotational speed, β increases as the radial location moves outboard;
- When the rotational speed increases, β increases;
- There is the effect of inflow (by varying thrust). For what concerns WTs the streamlines diverge as they approach the rotor (deformation of the stream tube).

The local efficiency of the surface is defined in three dimensions as the ratio of the cross-sectional area of the mass tube far upstream of the body to the surface area of the body bounded to the impinging stream tube, as shown in Fig. 4.35. Assuming that the droplets have negligible velocity with respect to freestream, the four trajectories are released at the corner of the area A_0 upstream and impinge on the corner points of a surface panel covering the area A_p.

The problem of determining the 3D collision efficiency can be solved only numerically. However, a pseudo 3D collection efficiency can be deduced by using some simplifying assumptions.

The 2D collision efficiency is given by Eq. (4.74):

$$\beta_{2D} = \frac{LWC_p\, V_p}{LWC_0\, V_0} = \frac{A_0}{A_p} \tag{4.74}$$

As the blade moves with the relative velocity W in its moving frame, the mass conservation principle written along the streamtube depicted in Fig. 4.29, yields:

$$\beta_{3D} = \frac{(LWC)_p\, W_p}{(LWC)_0\, V_0} \tag{4.75}$$

By combining the two equations we get:

$$\beta_{3D} = \beta_{2D} \frac{W_p}{V_p} \tag{4.76}$$

On a wind turbine blade, the relative velocity changes either due to radial variation along the blade or due to varying operating conditions of the rotor (variation of rotational speed). The local speed ratio x is defined as:

$$x = \frac{\Omega\, r}{V_0} \tag{4.77}$$

The streamlines diversion (expansion) can be taken into consideration by an induction factor a, therefore:

$$V_p = V_0\,(1 - a)$$

Since

$$W^2 = [V_p^2 + (\Omega r)^2]^{0.5}$$

or

$$\frac{W}{V_p} = \left[1 + \left(\frac{x}{1-a}\right)^2\right]^{\frac{1}{2}}$$

In terms of tip speed ratio the preceding equation becomes:

$$\frac{W}{V_p} = \left[1 + \left(\frac{TSR}{1-a}\right)^2 \left(\frac{r}{R}\right)^2\right]^{\frac{1}{2}}$$

the 3D collection efficiency becomes:

$$\beta_{3D} = \beta_{2D}\left[1 + \left(\frac{TSR}{1-a}\right)^2 \left(\frac{r}{R}\right)^2\right]^{\frac{1}{2}} \tag{4.78}$$

4.4.8.3 3D Collision Efficiency Calculation

The data of Table 4.5 are assumed to perform calculation of the collision efficiency at the leading edge. The inscribed cylinder in the leading edge blade D(r) varies according to the blade profile and thickness from root to tip with the obtained relationships, interpolating the data of Table 4.2.

Figure 4.36 shows the 2D collision efficiency as function of the Stokes number and the radius position. The 2D collision efficiency increases towards the blade tip due to the thinner profile (smaller inscribed diameter D). The water mass impinging on the leading edge is shown in Fig. 4.37.

The 3D collision efficiency at leading edge point is now computed by the aid of Eq. (4.78) and in Fig. 4.38 it is compared to the 2D value found in Fig. 4.36.

It is evident from this simple model that the rotating system is leading the system to a paramount accumulation of water mass flow towards the tip of the blade compared to the equivalent pure translating 2D boundary condition. The collision efficiency excesses 4 times the 2D case and the corresponding water mass scales correspondingly.

Now it is interesting to analyse what happens when only the wind speed changes, while environment conditions are invariant. This calculation generates for each blade radius a different situation, showing that for 3D condition each blade station exhibits a maximum in collision efficiency at a given tip speed ratio, which is not necessarily the same at any radius.

Figure 4.39 depicts the situation for mid span section obtained for MVD = 20 μm, V_0 = 3–12 m/s. The maximum collision efficiency is achieved at about $TSR = 6$, and while for 2D model β_0 drops as TSR grows, the 3D model shows instead a more complex feature. This example shows how, from the design point of view,

Table 4.5 Input parameters

Parameter	Value
ρ_{air} (kg/m^3)	1.20
ρ_d (kg/m^3)	1 000
μ_{air} (Pa s)	0.0000018
d (μm)	20
LWC (g/m^3)	0.2
V_0 (m/s)	12
TSR (–)	6
a (–)	0

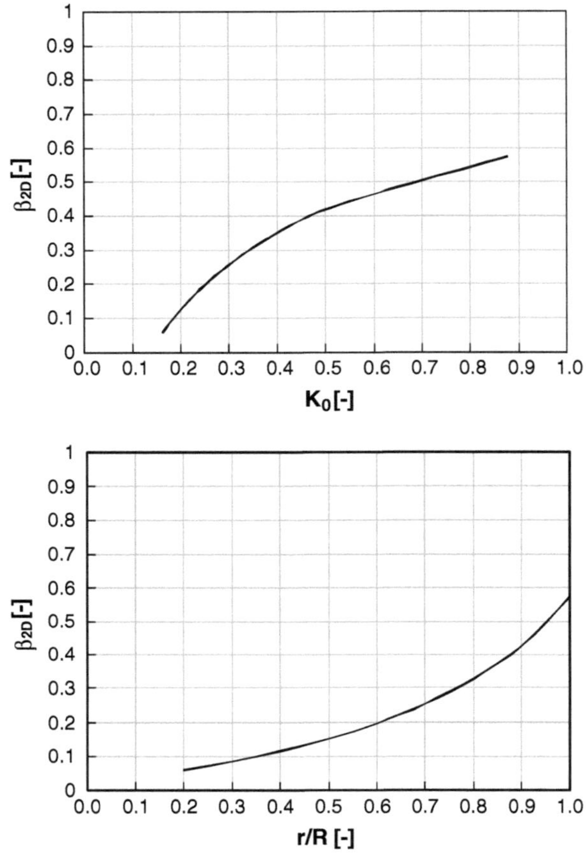

Fig. 4.36 Collision efficiency as function of the Stokes number (*top*) and the blade radius (*bottom*)

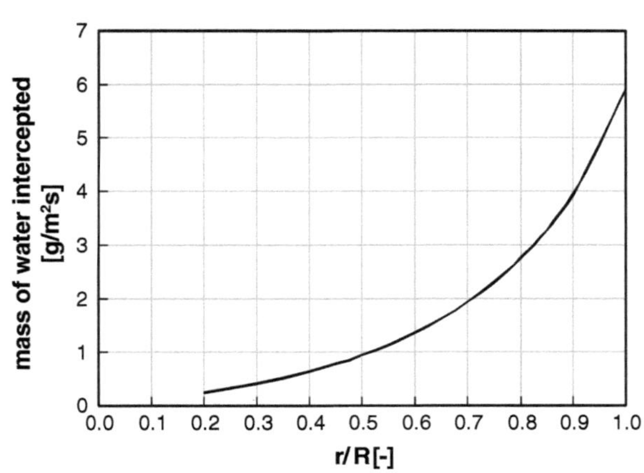

Fig. 4.37 Water mass impinging on the leading edge, LWC = $0.2\,\mathrm{g/m^3}$

Fig. 4.38 3D collision efficiency as function of the blade radius compared to 2D collision efficiency (*top*) and ratio β_{3D}/β_{2D} (*bottom*)

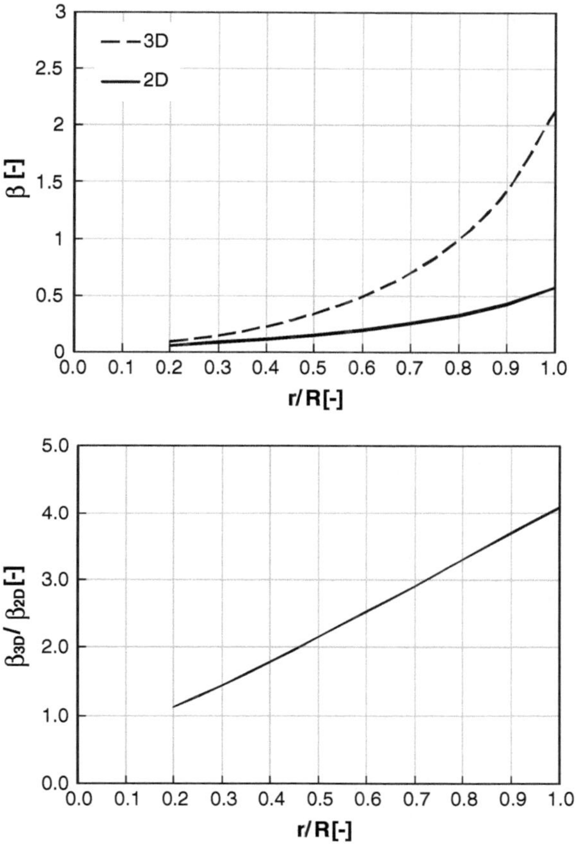

Fig. 4.39 3D and 2D stagnation collision efficiency as function of the tip speed ratio for blade mid span section

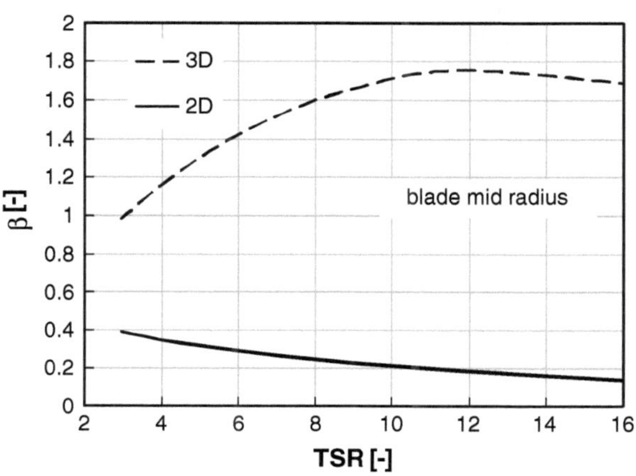

the more risky situation for the condition leading to potential icing over the whole operating conditions needs to be carefully investigated. This is essential to set the proper boundary condition for the anti-icing system design.

4.5 Mass Conservation Equation

In Fig. 4.40 the mass fluxes crossing a given control volume at the surface are depicted. The mass conservation equation is written as:

$$\sum_{i}^{n} \dot{m}_i - \sum_{out}^{m} \dot{m}_{out} = \frac{m}{dt} \qquad (4.79)$$

m/dt is the accumulation term.

If a permanent flow is assumed (i.e. $m/dt = 0$) Eq. (4.79) gives:

$$\dot{m}_{w,imp} + \dot{m}_{w,in} - \dot{m}_{w,out} - \dot{m}_{w,ev} - \dot{m}_{ice} - \dot{m}_{w,sh} - \dot{m}_{w,st} = 0 \qquad (4.80)$$

where

$\dot{m}_{w,imp}$ = mass flux of impinging water
$\dot{m}_{w,in}$ = mass flux of runback water entering a given section
$\dot{m}_{w,out}$ = mass flux of runback water leaving a given section
$\dot{m}_{w,ev}$ = mass flux of evaporating/sublimating water
$\dot{m}_{ice}$ = mass flux of water freezing in a given section
$\dot{m}_{w,sh}$ = mass flux of water shedding
$\dot{m}_{w,st}$ = mass flux of standing water due to surface tension.

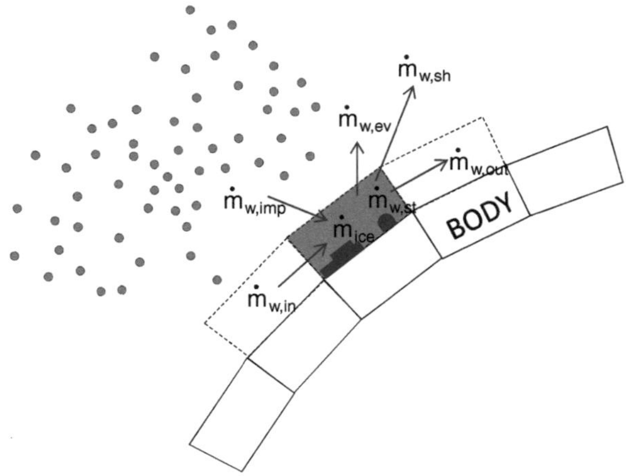

Fig. 4.40 Mass balance at the control volumes

4.5.1 Analysis of the Elementary Mass Fluxes

4.5.1.1 Impinging Water Mass Flow

The quantity of water flow per unit area that impinges on each panel is:

$$\dot{m}_{w,imp} = LWC\,\beta_{3D}\,V \qquad \text{for } \beta > 0 \tag{4.81}$$

While, of course $\dot{m}_{w,imp} = 0$ for $\beta = 0$.

The collection efficiency relative to each panel β_{2D} can be determined according to the simplified approach presented in [24] or the one described in the preceding paragraphs.

The actual collection efficiency for rotating airfoils, can be obtained by correcting the two-dimensional collection efficiency as follows:

$$\beta_{3D} = \beta_{2D}\,F\left(TSR, \frac{r}{R}\right) \tag{4.82}$$

where F is a function defined as:

$$F\left(TSR, \frac{r}{R}\right) = \left[1 + \frac{TSR^2}{1-a}\left(\frac{r}{R}\right)^2\right]^{1/2} \tag{4.83}$$

Equation (4.82) implies that $\beta_{3D} \geq \beta_{2D}$ and the factor F defined by Eq. (4.83), takes into account the effect of rotation.

4.5.1.2 Evaporating Water Mass Flow

The water can either evaporate or sublimate, the latter condition arising when the water film no longer covers the iced surface. Despite the condition involved, the potential water mass flow leaving the control volume via evaporation or sublimation is given by:

$$\dot{m}_{ev}^{pot} = h_m \Delta\rho_v$$

where h_m is the vapour mass transfer coefficient. The latter can be expressed as:

$$h_m = \frac{h_c}{\rho_{air} c_{p,air} L^{2/3}}$$

$\Delta\rho_v$ is the water vapour density difference across the boundary layer

$$\Delta\rho_v = \rho_{v,s} - \rho_{v,e}$$

being $\rho_{v,s}$ is the water vapour density at the surface and $\rho_{v,e}$ is water vapour density at the boundary layer edge.

L is the Lewis number given by:

$$L = \frac{k_{air}}{\rho_{air} c_{p,air} D_{ev}}$$

D_{ev} is the diffusivity of water vapour in air.

The vapour density of the air is given by the state equation of ideal gases is given by:

$$\rho_v(T) = \frac{e_v(T)}{\frac{R}{mm_w} T}$$

therefore, the potential evaporating water mass flow becomes:

$$\dot{m}_{ev}^{pot} = \frac{h_c}{\rho_{air} \, c_{p,air} \, L^{2/3}} \frac{mm_w}{R} \left(\frac{e_{v,s}}{T_s} - \frac{e_{v,e}}{T_e} \right) \tag{4.84}$$

The vapour pressure at the surface is the saturated vapour pressure

$$e_{v,s} = e_{v,s}^{sat}$$

The potential evaporating water mass flow thus becomes:

$$\dot{m}_{ev}^{pot} = \frac{h_c}{\rho_{air} \, c_{p,air} \, L^{2/3}} \frac{mm_w}{R} \left(\frac{e_{v,s}^{sat}}{T_s} - \frac{e_{v,e}}{T_e} \right) \tag{4.85}$$

The vapour pressure at the edge of the boundary layer is related to the vapour pressure in the free stream via the Dalton law, and assuming no condensation or evaporation while the drop is travelling toward the airfoil, therefore:

$$\frac{e_{v,e}}{p_e} = \frac{e_{v,\infty}}{p_\infty} \tag{4.86}$$

$$e_{v,\infty} = Rh \; e_{v,\infty}^{sat}$$

Equation (4.85) becomes:

$$\dot{m}_{ev}^{pot} = \frac{h_c}{\rho_{air} \, c_{p,air} \, L^{2/3}} \frac{mm_w}{R} \left(\frac{e_{v,s}^{sat}}{T_s} - \frac{Rh \; e_{v,\infty}^{sat} \; p_e}{T_e \; p_\infty} \right) \tag{4.87}$$

The relationships for the saturation vapour pressure over water is commonly used both in the air and at the accretion surface. The following empirical equations are adopted:

for $T < T_0 = 273.15\,K$

$$e^{sat} = 6894.7 \exp\left\{ 20.15247167 - \frac{11097.16963}{1.8 \; T} \right\}$$

and for T > $T_0 = 273.15\,\text{K}$

$$e^{sat} = 6894.7 \exp\left\{14.56594634 - \frac{7129.219482}{1.8\,T - 72}\right\}$$

The density and temperature at the boundary layer edge can be replaced using the ideal gas and isoentropic relationships:

$$T_e = T_\infty \left(\frac{p_e}{p_\infty}\right)^{\frac{\gamma-1}{\gamma}}$$

$$\rho_e = \rho_\infty \left(\frac{p_e}{p_\infty}\right)^{\frac{1}{\gamma}}$$

Here since $c_{p,eq} \approx c_{p,air}$, the specific heat of air at constant pressure and hence γ will be used for the following analysis within the boundary layer.

If the model assumes the thickness of the water film being negligible, all control volumes take instantaneously the temperature of the underlying layer.

The mas flux of evaporating water finally (standing neglected) is given by the conditions:

$$\dot{m}_{w,ev} = \dot{m}_{w,ev}^{pot}$$

for

$$\dot{m}_{w,ev} < \dot{m}_{w,imp} + \dot{m}_{w,in}$$

and

$$\dot{m}_{w,ev} = \dot{m}_{w,imp} + \dot{m}_{w,in}$$

for

$$\dot{m}_{w,ev} > \dot{m}_{w,imp} + \dot{m}_{w,in}$$

4.5.1.3 Freezing Water Mass Flow

The mass flux of water turning to ice in the control volume can be determined by the following equation:

$$\dot{m}_{ice} = f\,\left(\dot{m}_{w,imp} + \dot{m}_{w,in}\right) \tag{4.88}$$

The quantity f appearing in Eq. (4.88) is called the *freezing fraction*, introduced by Messinger [1] and its significance and use will be discussed in the next paragraph.

4.5.1.4 Shed Water Mass Flow

The mass flux of shed water is evaluated on the basis of qualitative observations of icing physics. The shedding phenomena is dominated by the Weber number.

The *Weber number* is given by

$$We = \frac{\rho_e \, W_e^2 \, d_b}{\sigma}$$

where ρ_e is the air density (kg/m^3) at a given location at the edge of the boundary layer, W_e is the relative air velocity (m/s) at a given location at the edge of the boundary layer, σ is surface tension between water and air (kg/s^2) and d_b is the diameter of a bead of surface water (m).

If the Weber number is below a critical value (a value of 500 is used currently) no water is shed. Above this point, the per cent mass lost is equal to the per cent difference in Weber number. The critical Weber number (We_{cr}) calculation is usually determined by the empirical equation [22]:

$$\begin{cases} \dot{m}_{w,sh} = \dot{m}_{w,out} \left(1 - \dfrac{We_{cr}}{We}\right) & \text{for } We > We_{cr} \\ \dot{m}_{w,sh} = 0 & \text{for } We < We_{cr} \end{cases}$$

with $\text{We}_{cr} = 500$.

It should be emphasised that this relation is not based on any quantitative measurements of mass loss. The amount of mass lost using this criterion is low, which actually matches the qualitative experimental observations.

4.5.1.5 Standing Water Mass Flow

The mass flux of standing water is computed through the empirical equation [22]:

$$\dot{m}_{w,st} = \frac{h_b \, \rho_w}{\Delta t} \left\{\dot{m}_{w,st}\right\}_{t=t-\Delta t} \tag{4.89}$$

where h_b is the height of water bead 0.5 d_b.

This mass flow contribution takes into account the amount of unfrozen water which is not allowed to leave the control volume due to surface tension (*Weber number*) effects. This amount is determined independently, as long its value does not exceed the amount of unfrozen water available.

4.5.2 Water Film Continuity and Layer Break-Up

It is worth to mention that the film of water flowing on the surface is going to experience rupture somewhere along on the surface because of a complex interaction between surface tension and mass forces as exemplified in Fig. 4.41.

In this case, some portions of the surface are no longer wetted. This situation has paramount effects on the cooling process of the surface, as will be evident in the next paragraphs, as evaporation is one of the major contributions in the heat transfer.

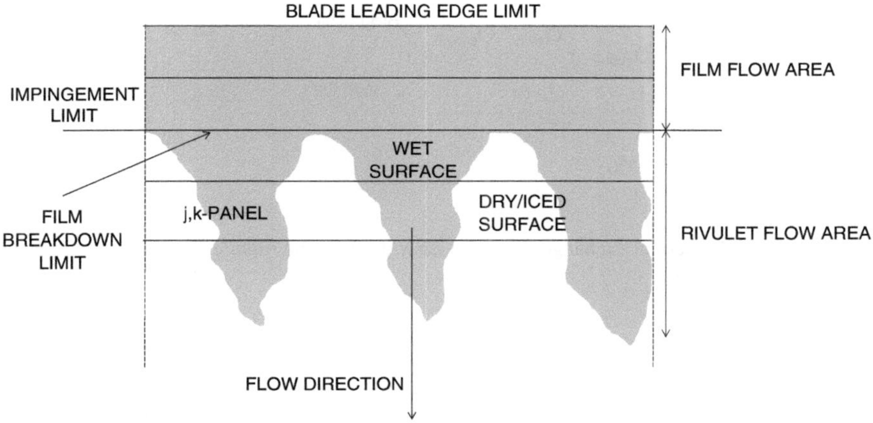

Fig. 4.41 Schematic of the film rupture on the surface

It is thus necessary to define the area ratio K_A which defines the fraction of the area covered by water (beads or rivulets) A_w to the area of the control volume (panel) A_p:

$$K_A = \frac{A_w}{A_p}$$

In area of impingement, measurements indicate a value for K_A of 1. The water film breakdown occurs only downstream of the impingement limits, where the water runs back in rivulets. Flight and tunnel data shows a rapid decrease from 1 to about 0.3, with a gradual decline in the value of K_A. Typical variations in wetness faction aft the area of impingement is shown in Fig. 4.42, taken from [25].

The film height varies with evaporation, temperature, pressure gradient and shear stress. The rivulets will form when the film reaches a critical height.

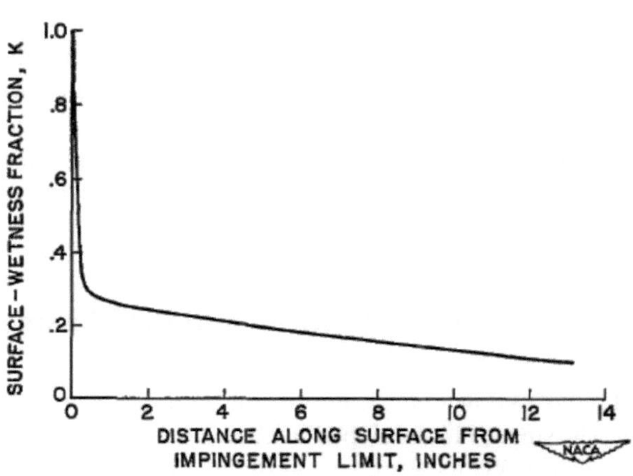

Fig. 4.42 Variation of wing-surface-wetness fraction aft of area of water impingement with distance from limit of impingement [25]

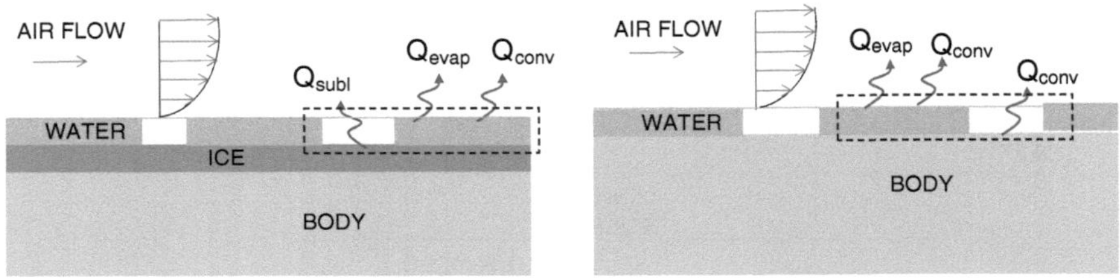

Fig. 4.43 Possible thermal exchange scenarios deriving from the rupture of water layer over the iced and not iced surface

From the heat transfer point of view, two situations can generate at the surface:

1. The existence of surface/ice/water/air interfaces for unheated surfaces;
   The fraction of iced surface that is no longer covered by the water undertakes a sublimation process. The control volume experiences both evaporation and sublimation at the air-water and air-ice interfaces respectively. The specific heat of sublimation, Δh_{subl} is equal to the sum of the specific heats of fusion Δh_f and vaporisation Δh_{ev}:
   $$\Delta h_{subl} = \Delta h_f + \Delta h_{ev}$$
2. The existence of surface/water/air interfaces for heated surfaces.
   The fraction of the surface that is not covered by the water undertakes a cooling process due to evaporation. The control volume experiences thus convection and evaporation at the surface-air and water-air interfaces respectively.

These situations are shown in Fig. 4.43.

In the most simple approach the film of water over the surface is supposed sufficiently thin to be contained into the boundary layer. This hypothesis is confirmed by either calculations and direct observations. The assumption that the water film assumes a constant temperature on the considered panel, equal to the blade external wall temperature, follows as a consequence and can be considered valid for the analysis of unheated surfaces. A more complete approach can be found at the end of the chapter.

Experimental tests on aircraft wings indicate that the water film moves with constant velocity along the wall driven from the air friction. It has been considered that some radial water migration is also present for some wind turbine blade sections due to centrifugal effects. This effect removes actually the no slip assumption at the air boundary layer interface. A model taking into account this condition is developed in [26].

4.6 The Freezing Fraction and the Messinger Model

The most known physical model used in ice accretion codes is based on the work of Messinger [1] developed to calculate the heating requirements for icing protection and LWC measurement systems. The model is based on the temperature of an

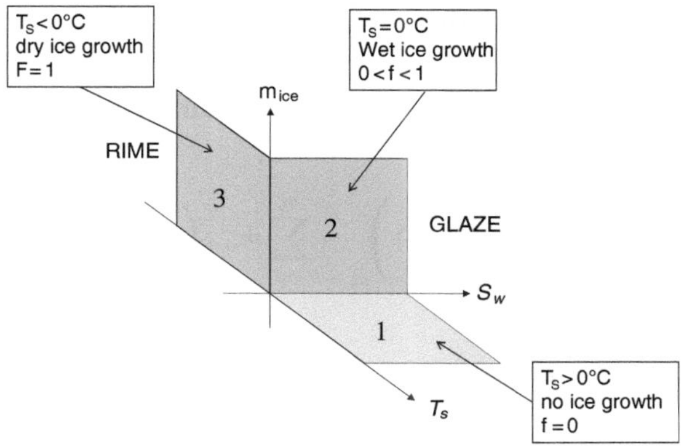

Fig. 4.44 Graphical representation of the freezing fraction

unheated surface in icing conditions for three surface temperature regimes, i.e. less than 273.15 K, equal to 273.15 K and above 273.15 K. In general, this type of model works reasonably well in dry or rime icing conditions (i.e. air temperature well below freezing and low LWC values. In glaze icing conditions (i.e. air temperature close to freezing and high LWC values) the results are much less satisfactory. The poor agreement derives from neglecting the surface water dynamics in the models, which lead to not accurate run-back flow phenomena, and heat conduction through the water-ice interface.

Numerically, the freezing fraction, according to the original definition of Messinger, is the fraction of impinging liquid that freezes within the region of impingement.[2] Although it can assume any values between $-\infty$ and $+\infty$, physically only the following range of figures are of interest:

- $f = 1$: occurrence of wet ice growth, all the water entering the control volume freezes within the control volume, leading to rime ice formation;
- $f = 0$: no ice formation;
- $0.3 \geq f \geq 0$: occurrence of wet ice growth, leading to glaze ice formation.

The situation is depicted in Fig. 4.44, taken from [27], where the three-axis representation allows qualitatively evaluating the effect on ice mass, water film thickness and surface temperature as function of the three freezing fraction ranges mentioned before.

Ice accretions are often composed by glaze, rime and intermediate regions. Local value of the freezing fraction therefore varies along the surface, and shall be calculated using the mass and energy balances for each control volume at the surface.

[2] Another definition (for numerical implementation purposes) considers the fraction of the total liquid entering the control volume that freezes within the control volume.

4.7 Energy Conservation Equation

The formulation of the energy conservation applied at the control volume enveloping the water layer gives:

$$\sum_{i}^{n} \dot{m}_{in}\left(h_{in}+\frac{W_{in}^{2}}{2}\right)_{in}-\sum_{out}^{m} \dot{m}_{out}\left(h_{out}+\frac{W_{out}^{2}}{2}\right)_{out}+\sum_{k}^{p} \dot{q}_{k}=\frac{d(m_{w}\cdot u)}{dt} \tag{4.90}$$

$d(m_w \cdot u)/dt$ is the accumulation term, and can be written as:

$$\frac{d(m_w\cdot u)}{dt}=\frac{dm_w}{dt}u+\frac{du}{dt}m_w$$

The first right-side term is null according to the permanent flow conditions stated by the mass conservation principle.

In order to couple the energy conservation with models developed under the assumptions of equilibrium temperature (the Messinger model) the variation in the internal energy of the water control volume is also null:

$$\frac{du}{dt}m_w=\frac{c_w\left(T_w-T_{ref}\right)}{dt}m_w$$

and therefore Eq. (4.90) is used in its steady state formulation:

$$\sum_{i}^{n} \dot{m}_{in}\left(h_{in}+\frac{W_{in}^{2}}{2}\right)_{in}-\sum_{out}^{m} \dot{m}_{out}\left(h_{out}+\frac{W_{out}^{2}}{2}\right)_{out}+\sum_{k}^{p} \dot{q}_{k}=0 \tag{4.91}$$

Equation (4.91) can now be expanded by the following contributes:

$$\begin{aligned}
&\dot{m}_{w,in}\left(h_{w,in}+\frac{W_{w,in}^{2}}{2}\right)-\dot{m}_{w,out}\left(h_{w,out}+\frac{W_{w,out}^{2}}{2}\right)\\
&+\dot{m}_{w,imp}\left(h_{w,imp}+\frac{W_{w,imp}^{2}}{2}\right)-\dot{m}_{w,sh}\left(h_{w,sh}+\frac{W_{w,sh}^{2}}{2}\right)\\
&+\dot{m}_{w,st}\,h_{w,st}-\dot{m}_{w,ev}\,\Delta h_{w,ev}-\dot{m}_{ice}\,\Delta h_{ice}+\sum_{k}^{p}\dot{q}_{k}=0
\end{aligned} \tag{4.92}$$

The heat fluxes as shown in Fig. 4.45 are:

$$\sum_{k}^{p}\dot{q}_{k}=-\dot{q}_{conv}+\dot{q}_{cond}-\dot{q}_{lr}+\dot{q}_{sr} \tag{4.93}$$

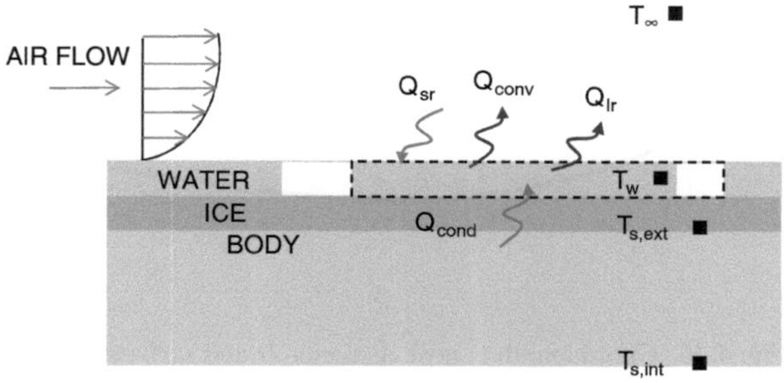

Fig. 4.45 Schematic of the heat fluxes involved in energy conservation

being:

$$\dot{q}_{conv} = h_c\,(T_s - T_{rec})$$
$$\dot{q}_{cond} = \frac{k_{ice}}{t_{ice}}\left(T_{ice} - T_{s,ext}\right) = \frac{k_b}{t_b}\left(T_{s,ext} - T_{s,int}\right)$$
$$\dot{q}_{lr} = \varepsilon\,\sigma\left(T^4_{s,ext} - T^4_{\infty}\right)$$
$$\dot{q}_{sr} = S_r \tag{4.94}$$

k_{ice}/t_{ice} is the thermal resistance of the ice layer.

These heat fluxes are analysed in detail in the following section.

4.7.1 Analysis of the Heat Fluxes Contributes

4.7.1.1 Convective Heat Transfer Coefficient

The convective heat transfer coefficient can be evaluated using either empirical correlations of aeronautical origin, or by more complex approaches, appropriate for the airfoil geometry and the flow regime.

With reference to Fig. 4.46, the following correlation [28, 29] for the very leading edge can be used:

$$Nu_c = Re_c^{0.5}\,Pr^{0.4}\left[1.14 + \left(\frac{c}{D}\right)^{0.5} - 2.353072\left(\frac{c}{D}\right)^{3.5}\left(\frac{s}{c}\right)^{3}\right]$$

and for the after region of the airfoil it is:

$$\begin{cases} Nu_c = 0.286\,Re_c^{0.5}\left(\dfrac{u_e}{W}\right)^{0.5}\left(\dfrac{c}{s}\right)^{0.5} & \text{for laminar flow} \\ Nu_c = 0.296\,Pr^{1/3}\,Re_c^{0.8}\left(\dfrac{u_e}{W}\right)^{0.8}\left(\dfrac{c}{s}\right)^{0.2} & \text{for turbulent flow} \end{cases}$$

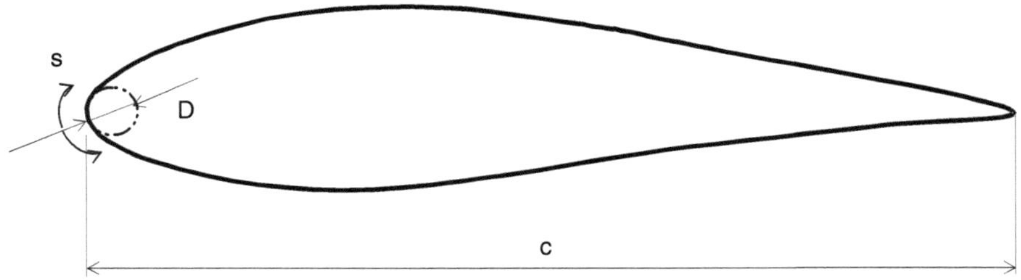

Fig. 4.46 Chord length c, nose diameter D and surface coordinate s of airfoil

The convective heat transfer coefficient is locally deduced from the definition of the *Nusselt number*:

$$Nu_c = \frac{h_c \cdot c}{k_a} \tag{4.95}$$

The heat transfer convection coefficient can be more accurately computed through integral boundary layer calculation methods, where the boundary layer is laminar and the convection coefficient is computed by means of the Smith and Spalding [30] relationship:

$$h_{c,e} = Pr \frac{c_1 \, K_t \, u_e^{1+c_2}}{\left(\upsilon \int_o^s u_e^{c_3} ds\right)^{0.5}} \tag{4.96}$$

where the c_1, c_2, e, c_3 coefficients are proportional to the Prandtl number.

The most uncertain factor influencing the convective heat transfer coefficient is the location of transition from laminar to turbulent flow in the boundary layer and the chordwise extent of the transition region.

Disturbances of the boundary layer by the presence of water on the blade surface cause premature transition. Experimental evidences from NASA indicate that the start of transition under such conditions occurs near the rear extremity of water drop impingement.

No certain criterion has been stated for the transition in such conditions, therefore its location is conventionally assumed just after the first panel at the stagnation point. For the turbulent boundary layer, the following relationship is employed:

$$h_{c,t} Pr \frac{\frac{c_f}{2} \rho \, u_e \, c_p}{Pr_t + 0.52 \sqrt{\frac{c_f}{2}} \left(u_e \frac{c_f}{2} \frac{K_s}{\upsilon}\right)^{0.45} Pr^{0.8}} \tag{4.97}$$

The skin friction coefficient is calculated as a function of the boundary layer momentum thickness and the surface sand grain roughness:

$$\frac{c_f}{2} = \left[\frac{0.41}{\ln\left(\frac{864\,\theta_t}{K_s} + 2.568\right)} \right]^2 \tag{4.98}$$

The boundary layer momentum thickness is evaluated extending the Thwaite's method to turbulent velocity profiles:

$$\theta_t = \frac{0.0263\,\upsilon^2}{u_e^{3.4}} \int_0^s u_e^4 ds + \theta_l \tag{4.99}$$

Finally, regarding the turbulent Prandtl number Pr_t, Wright suggests using a value of 0.9. However, Schilchting shows that this value is representative of the ratio between mass and energy diffusion only deep inside the turbulent boundary layer (i.e. $\delta_t < 0.2$), while at the borders of the boundary layer, the turbulent Prandtl number can decrease up to 0.5. For this reason it seems to be more correct to use an average Pr_t; as air is considered, this states that $Pr_t = Pr$.

Most advanced models use the rivulets and bead height to obtain a local surface equivalent roughness to be used for a critical Reynolds number in order to assess the transition.

4.7.1.2 Longwave Radiation

Longwave radiation is computed by assuming that an iced foil is radiating at 0 °C and the clouds and surrounding rain-filled air are radiating at the air temperature T. The emissivity is assumed to be 1 (emissivity of ice is 0.98 and of water is 0.96 [31]).

$$\dot{q}_{lr} = \varepsilon\,\sigma\left(T_s^4 - T_\infty^4\right)$$

4.7.1.3 Shortwave Radiation

$$\dot{q}_{sr} = S_r$$

The incoming global diffuse radiation flux S_r is either measured or modelled.

At some weather stations the global diffuse radiation flux is measured as an average value for each hour. The derivation of this value is based on the assumption that:

- The albedo of the iced profile is zero (the ice is clear);
- The incoming radiation to the accretion is absorbed in either the ice or the underlying body;
- There is no radiation reflected back from the ground to the body.

- The albedo of the ice is close to, but greater than zero, and depends on how much air and/or snow is incorporated in the accretion;
- The shorter-wavelength visible radiation may not be absorbed in the ice, but it is probably absorbed in the underlying body, which is assumed to be in thermal equilibrium with the ice accretion;
- The ground surface reflects some incoming radiation, depending on the ground cover and the amount and condition of snow and ice on the ground.

Although these assumptions are only approximately correct, they are somewhat offsetting in their simplicity.

The shortwave and longwave fluxes are opposite in sign and comparable in magnitude, but there is no shortwave flux at night, when the air temperature may be relatively cold and the longwave flux relatively large.

For anti-icing design general purpose, a part of specific systems designed to have a given surface emissivity (black painted coatings for blades) the radiative contributions can be discarded.

4.8 The Solution of the Problem

In order to simplify the problem, the following assumption can be made.

- The kinetic energy associated to the velocity of the water film over the surface is negligible, therefore:
$$\frac{W_{w,in}^2}{2} \approxeq 0, \qquad \frac{W_{w,out}^2}{2} \approxeq 0$$
- The water flow shed and the water flow standing are neglected:
$$\dot{m}_{w,sh} \approxeq 0, \qquad \dot{m}_{w,st} \approxeq 0$$
- The radiant heat flows are small compared to the others:
$$\dot{q}_{sr} \approxeq 0, \qquad \dot{q}_{lr} \approxeq 0$$

The physical properties of water and ice are invariant with temperature. Rime ice density is assumed, as the model cannot in principle describe glaze icing. The phase chance occurs only at $T = T_{ref} = 273.15\,\text{K}$. Impinging water, runback water and ice assume instantaneously for each panel the surface temperature, therefore $T_{w,in}^i = T_{w,out}^{i-1} = T_w = T_{ice} = T_{s,ext}$.

Thus the energy conservation Eq. (4.92) simplifies to:

$$\dot{m}_{w,in}\, c_w \left(T_{w,in} - T_{ref}\right) - \dot{m}_{w,out}\, c_w \left(T_{w,out} - T_{ref}\right) + \beta\, LWC\, W_{w,imp} \left[c_w \left(T_{w,imp} - T_{ref}\right) + \frac{W_{w,imp}^2}{2}\right]$$

$$
\begin{aligned}
& - K_A \left[\frac{h_c}{\rho_{air}\, c_{p,air}\, L^{2/3}} \frac{mm_w}{R} \left(\frac{e^{sat}_{v,s}}{T_s} - \frac{Rh\, e^{sat}_{v,\infty}\, p_e}{T_e\, p_\infty} \right) \right] \Delta h_{ev}^{T=T_{ref}} \\
& - \dot{m}_{ice} \left[c_{ice} \left(T_{ice} - T_{ref} \right) + \Delta h_f^{T=T_{ref}} \right] \\
& - h_c \left(T_w - T_\infty \right) - h_c \left(T_e + \frac{r\, W_e^2}{2\, c_{p,air}} - T_\infty \right) + \dot{q}_{cond} = 0
\end{aligned} \tag{4.100}
$$

The variable inventory gives for each control volume:

- Thermodynamical variables: T_{ref}, c_w, c_{ice}, $\Delta h_{ev}^{T=T_{ref}}$, ρ_{air}, $c_{p,air}$, L, mm_w, R, $\Delta h_f^{T=T_{ref}}$, $e^{sat}_{v,s}$;
- Climatic variables: T_∞, p_∞, LWC, Rh;
- Turbine and site (mixed) variables: $T_{w,in}$, $T_{w,out}$, $T_{w,imp}$, T_{ice}, T_s, K_A, β, W_e, p_e, T_e, h_c, $W_{w,imp}$, $\dot{m}_{w,in}$, $\dot{m}_{w,out}$, $\dot{m}_{w,imp}$, $\dot{m}_{ice}$, $\dot{q}_{ice}$.

The unknown variables are T_s, $\dot{m}_{w,out}$ and $\dot{m}_{cond}$.

Assumed variables are $T_{w,in}$, $T_{w,out}$, $T_{w,imp}$, T_{ice}, $w_{w,imp}$, $\dot{m}_{w,in}$, $\dot{m}_{w,imp}$, $\dot{q}_{cond}$, all others being known. The two auxiliary equations needed to solve the problem in closed form are given by mass conservation and the freezing fraction:

$$
\dot{m}_{w,imp} + \dot{m}_{w,in} - \dot{m}_{w,out} - \dot{m}_{w,ev} - \dot{m}_{ice} = 0 \tag{4.101}
$$

$$
\dot{m}_{ice} = f \cdot \left(\dot{m}_{w,imp} + \dot{m}_{w,in} \right) \tag{4.102}
$$

The assumptions are:

$$
\begin{aligned}
T^i_{w,in} &= T^{i-1}_{w,out} = T_{s,ext} \\
T_{w,imp} &= T_\infty \\
T_{ice} &= T_{s,ext} \\
W_{w,imp} &= W \\
\dot{m}^i_{w,in} &= \dot{m}^{i-1}_{w,out} \\
\dot{q}_{cond} &= 0
\end{aligned}
$$

$\dot{m}_{imp}$ is derived from particle trajectory analysis.

4.8.1 CASE A: The Ice Accretion Solution $T_w < T_s \leq 0\,°C$

A time-stepping procedure is applied to accreted ice. Initially, the flow field and droplet impingement characteristics are determined for the clean geometry.

It is first assumed that the equilibrium surface temperature, T_s, equals 273.15 K.

The terms of the energy equation are then evaluated at this temperature, and the expression is solved to determine the freezing fraction, f. For $0.0 < f < 1.0$, $T_s = 273.15$ K, and the initial assumption was correct. A value of $f < 0.0$ indicates that the surface temperature is greater than 273.15 K. Therefore, the solution is obtained by setting $f = 0.0$ and solving for T_s. Note that an iterative procedure is required since many of the terms are functions of T_s.

Similarly, $f > 1.0$ indicates that T_s is less than 273.15 K, and f must be set equal to 1.0.

Again, an iterative procedure has to be applied to determine the thermodynamic characteristics of the control volume. The mass balance is used to determine the mass flow rate of runback water out of the control volume. Any water flow exiting the control volume will flow away from the stagnation point and into the next control volume.

The procedure for the computational solution is illustrated in Fig. 4.47.

The scheme is repeated for the adjacent downstream control volume and continued along the upper and lower surfaces of the body.

When a time increment is specified, this growth rate is converted into an ice thickness and the body coordinates are adjusted to account for the accreted ice.

The ice thickness in the time step Δt is given by:

$$\Delta t_{ice} = \frac{\left(\dot{m}_{w,imp} + \dot{m}_{w,in}\right) f}{\rho_{ice}} \Delta t \tag{4.103}$$

The ice mass is deduced from Eq. (4.102).

After calculating this initial ice layer, two options are available. The most desirable option is to repeat the entire procedure, beginning with the calculation of the

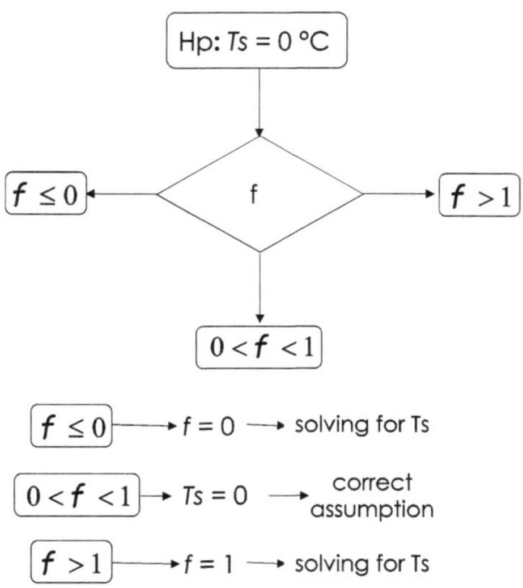

Fig. 4.47 Schematic procedure of computational solution

flow field about the iced geometry, to obtain revised local collection efficiency and thermodynamic data.

Unfortunately, since the majority of the computational time is spent in calculating the flow field, this option also increases the computational time to accrete a layer of ice. Ice section is added in the normal direction of the surface with the bisection line between two adjacent panels.

4.8.2 CASE B: The Ice-Free Surface Solution $T_s > 0\,^{\circ}C$

In this case the energy equation becomes:

$$\begin{aligned}
&\dot{m}_{w,in}\, c_w \left(T_{w,in} - T_{ref}\right) - \dot{m}_{w,out}\, c_w \left(T_{w,out} - T_{ref}\right) \\
&\quad + \beta\, LWC\, W_{w,imp} \left[c_w \left(T_{w,imp} - T_{ref}\right) + \frac{W_{w,imp}^2}{2} \right] \\
&\quad - K_A \left[\frac{h_c}{\rho_{air}\, c_{p,air}\, L^{2/3}} \frac{mm_w}{R} \left(\frac{e_{v,s}^{sat}}{T_{s,ext}} - \frac{Rh\, e_{v,\infty}^{sat}\, p_e}{T_e\, p_\infty} \right) \right] \Delta h_{ev}^{T=T_{ref}} \\
&\quad + \left(\dot{m}_{w,in} + \dot{m}_{w,imp}\right) c_w \left(T_{ref} - T_{s,ext}\right) \\
&\quad - h_c \left(T_w - T_\infty\right) - h_c \left(T_e + \frac{r\, W_e^2}{2\, c_{p,air}} - T_\infty \right) + \dot{q}_{gen} = 0
\end{aligned} \tag{4.104}$$

The variable inventory gives for each control volume:

- Thermodynamical variables: T_{ref}, c_w, c_{ice}, $\Delta h_{ev}^{T=T_{ref}}$, ρ_{air}, $c_{p,air}$, L, mm_w, R, $e_{v,s}^{sat}$;
- Climatic variables: T_∞, p_∞, LWC, Rh;
- Turbine and site (mixed) variables: $T_{w,in}$, $T_{w,out}$, $T_{w,imp}$, $T_{s,ext}$, K_A, β, R, W_e, p_e, T_e, h_c, $w_{w,imp}$, $\dot{m}_{w,in}$, $\dot{m}_{w,out}$, $\dot{m}_{w,imp}$, $\dot{q}_{gen}$.

The unknown variables are $\dot{q}_{gen}$, $\dot{m}_{w,out}$.

Figure 4.48 shows the main heat contribution involved in the process.

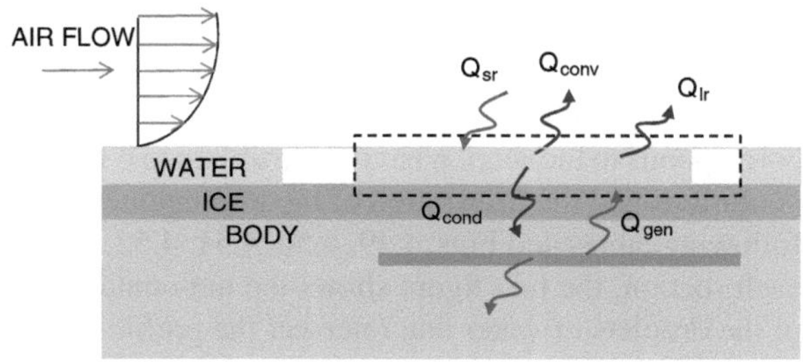

Fig. 4.48 Physical scheme of the ice free heated surface

In general for internal hot air circulation IPS systems, we can write:

$$\dot{q}_{gen} = \dot{q}_{IPS} = \frac{k_b}{t_b}\left(T_{s,int} - T_{s,ext}\right)$$

while for electric embedded electrical resistance

$$\dot{q}_{gen} = \dot{q}_{IPS} = \eta_{IPS}\, I^2\, R$$

being η_{IPS} the thermal efficiency of the system.

Assumed variables are $T_{w,in}$, $T_{w,out}$, $T_{w,imp}$, $T_{s,ext}$, $W_{w,imp}$, $\dot{m}_{w,in}$, $\dot{m}_{w,imp}$ all others being known.

One auxiliary equation is needed and is given by mass conservation:

$$\dot{m}_{w,imp} + \dot{m}_{w,in} - \dot{m}_{w,out} - \dot{m}_{w,ev} = 0 \tag{4.105}$$

The assumed variables are:

$$\begin{aligned} T^{i}_{w,in} &= T^{i-1}_{w,out} = T_{s,ext} \\ T_{w,imp} &= T_{\infty} \\ W_{w,imp} &= W \\ \dot{m}^{i}_{w,in} &= \dot{m}^{i-1}_{w,out} \end{aligned} \tag{4.106}$$

$T_{s,ext}$ depends on IPS strategy and technology.

4.8.3 Worked Example on Blade Icing

Typical results of the application of the icing model is given in Fig. 4.58, where some results obtained in collaboration with the University of Trento and CIRA (Italian Aerospace Research Centre) are shown. MULTI-ICE code was used.

The code MULTI-ICE calculates the flow field through a method in panels, and estimates the growth of ice through a predictor-corrector method. Table 4.6 reports the data of the Tjæreborg turbine blade used for the simulation: LWC = 0.8 g/m^3 and d = 20 μm have been set as meteorological parameters. The static temperature has been set at 270.15 K, the static pressure at 100 kPa and the ice growing time was 45 min.

For the plan shape of the blade of the wind turbine sections from 14 (which corresponds to the section next to the tip) up to 6 have been considered. The operative conditions and the illustration of the impingement and ice accretion are given in the following tables and Figs. 4.49, 4.50, 4.51, 4.52, 4.53, 4.54, 4.55, 4.56 and 4.57. For each section, the first figure shows the not contaminated profile and the trajectories of the droplets of water that intersect the profile calculated with respect to the clean profile, and the second the ice shape.

Table 4.6 Geometric and fluid-dynamic details of the blade sections considered for simulation

Blade geometry								
Section	R (m)	Profile	c(m)	t/c (%)	β_{twist} (°)	ϕ_{flow} (°)	AoA(°)	W(m/s)
0	0.00	Circular	1.80					
1	1.46	Circular	1.80					
2	2.75	Circular	1.80					
3	2.96	Circular	1.80					
4	6.46	NACA 4430	3.30	30.58	8.00	36.16	30.16	19.35
5	9.46	NACA 4424	3.00	24.10	7.00	26.23	21.23	25.32
6	12.46	NACA 4421	2.70	21.15	6.00	19.41	15.41	31.66
7	15.46	NACA 4418	2.40	18.71	5.00	15.32	12.32	38.19
8	18.46	NACA 4416	2.10	16.81	4.00	13.03	11.03	44.80
9	21.46	NACA 4415	1.80	15.44	3.00	11.40	10.40	51.50
10	24.46	NACA 4414	1.50	14.40	2.00	10.12	10.12	58.25
11	27.46	NACA 4413	1.20	13.33	1.00	8.87	9.87	65.04
12	28.96	NACA 4412	1.05	12.76	0.50	7.95	9.45	68.43
13	29.86	NACA 4412	0.96	12.74	0.20	6.82	8.62	70.45
14	30.50	NACA 4412	0.94	12.59	0.16	3.23	5.07	71.92

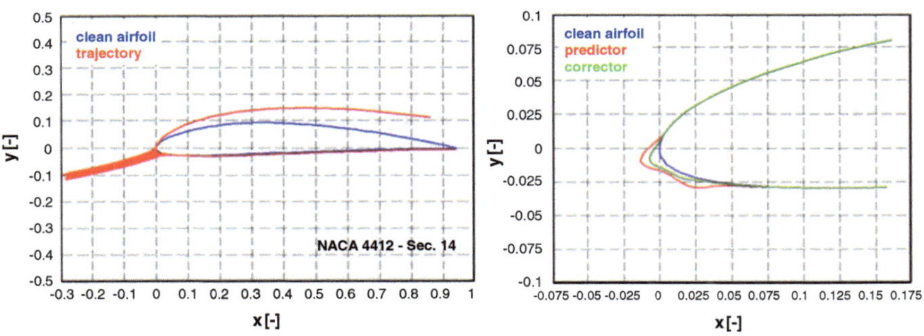

Fig. 4.49 Section 14 results (see Table 4.6): the trajectories of the water droplets on the clean profile (*left*) and the ice accretion shape (*right*)

From Section 9 to the Section 6, it can be seen that there is no substantial production of ice on the profiles. This is also verified by the coefficient of impingement, which is very limited in these sections.

On the experience gathered by these simulations, a second set of computations have been made extending the icing time to 180 min. Examples of ice shapes obtained for a very long time of ice exposition are shown for three significant sections: Sections 7, 9 and 14 (Fig. 4.58) are presented.

As the thickness of the blade decreases, moving from the root to the tip, the impingement limits increases and the accreted ice mass becomes important. Twenty percentage of the blade chord is obtained after 180 min on the tip section. The

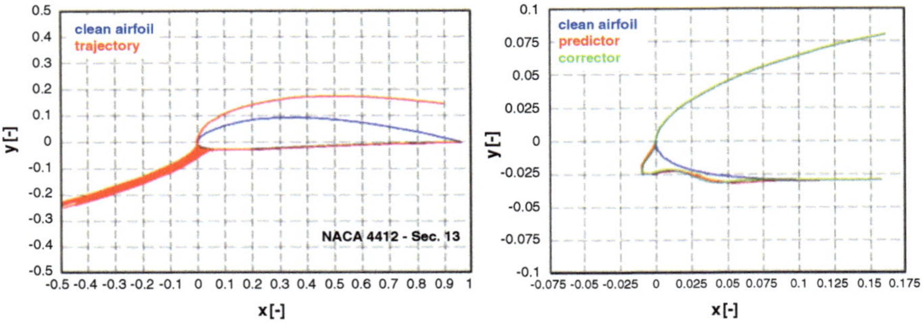

Fig. 4.50 Section 13 results (see Table 4.6): the trajectories of the water droplets on the clean profile (*left*) and the ice accretion shape (*right*)

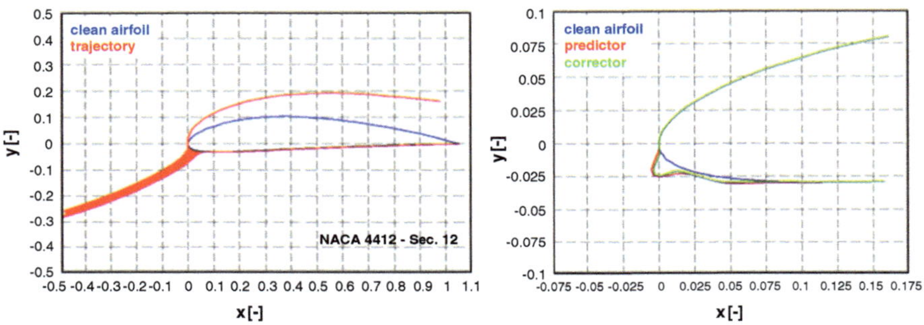

Fig. 4.51 Section 12 results (see Table 4.6): the trajectories of the water droplets on the clean profile (*left*) and the ice accretion shape (*right*)

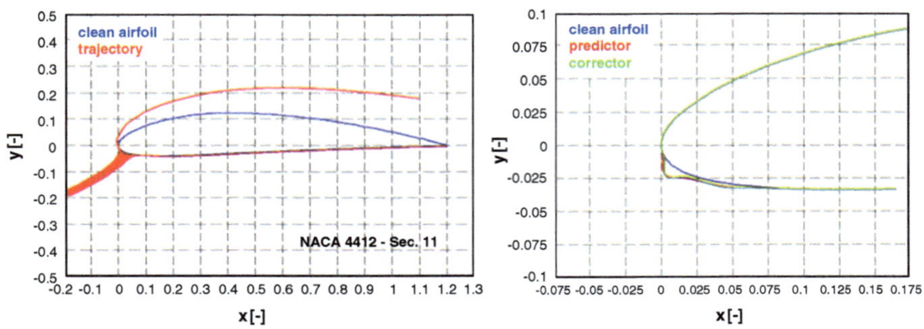

Fig. 4.52 Section 11 results (see Table 4.6): the trajectories of the water droplets on the clean profile (*left*) and the ice accretion shape (*right*)

maximum collection efficiency increases from root to tip. The position of the maximum collection efficiency moves to s/c = 0 position because of the progressive reduction of the flow angle. From Section 9 to Section 6, it can be seen, for fixed operating conditions, a substantial production of ice in the profiles of the wing is not

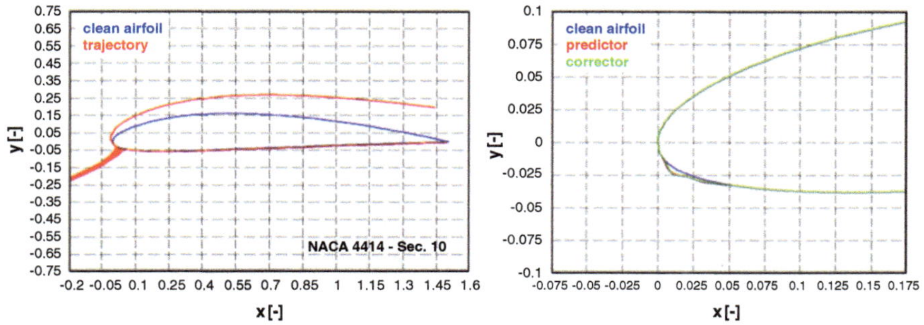

Fig. 4.53 Section 10 results (see Table 4.6): the trajectories of the water droplets on the clean profile (*left*) and the ice accretion shape (*right*)

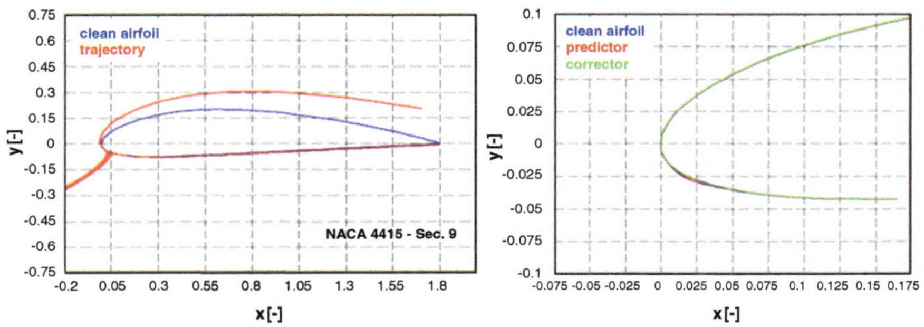

Fig. 4.54 Section 9 results (see Table 4.6): the trajectories of the water droplets on the clean profile (*left*) and the ice accretion shape (*right*)

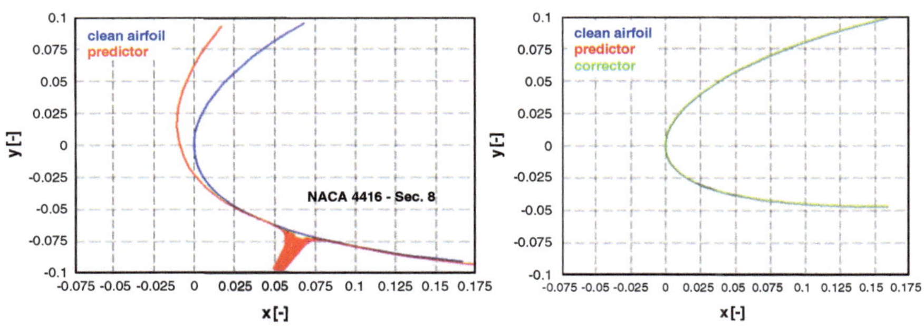

Fig. 4.55 Section 8 results (see Table 4.6): the trajectories of the water droplets on the clean profile (*left*) and the ice accretion shape (*right*)

detected. This is also verified by the coefficient of impingement very limited in these sections. It is known in fact from the figures that show the trajectories limit of the water particles on the profile, as the area of impact is very limited.

The heat transfer coefficient (expressed in kW/m^2) increases from root section to tip, mainly because of the effect of the relative velocity.

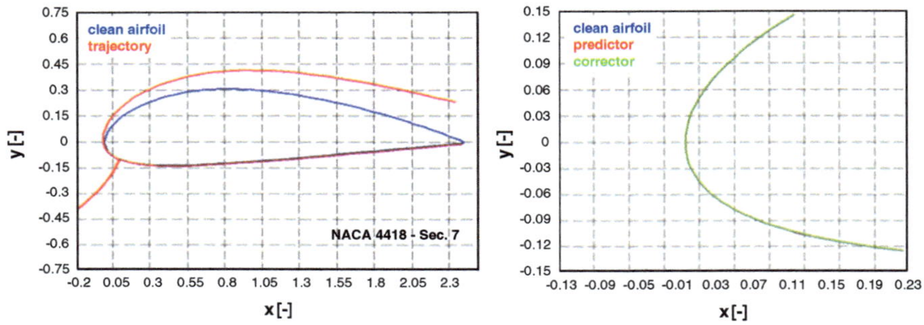

Fig. 4.56 Section 7 results (see Table 4.6): the trajectories of the water droplets on the clean profile (*left*) and the ice accretion shape (*right*)

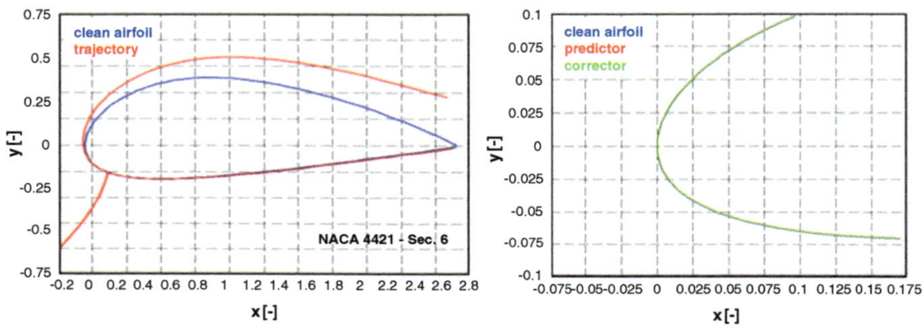

Fig. 4.57 Section 6 results (see Table 4.6): the trajectories of the water droplets on the clean profile (*left*) and the ice accretion shape (*right*)

4.9 Thermo-Fluid-Dynamic Processes at the Ice Surface

Since the book is devoted to the analysis of un-iced, heated surfaces, the approach followed to analyse the icing process was quite simplified, although complete. As the problem was described by neglecting dynamic processes at the phase interface, it is worth to add some hints for more sophisticated approaches to describe the ice accretion phenomenon. Various models are becoming available to analyse the multiphase flow and thermodynamics associated with the body ice accretion process. Two phenomena are briefly discussed here that are relevant in the recent literature: the micro physics of the surface and the runback water dynamics in glaze icing process.

4.9.1 The Micro Physics of the Surface

The development of the film of water on the surface generally undergoes complex dynamic interactions. The fluid motions adjacent to a solid-liquid interface may have relevant effects on the solid as it freezes from the fluid state.

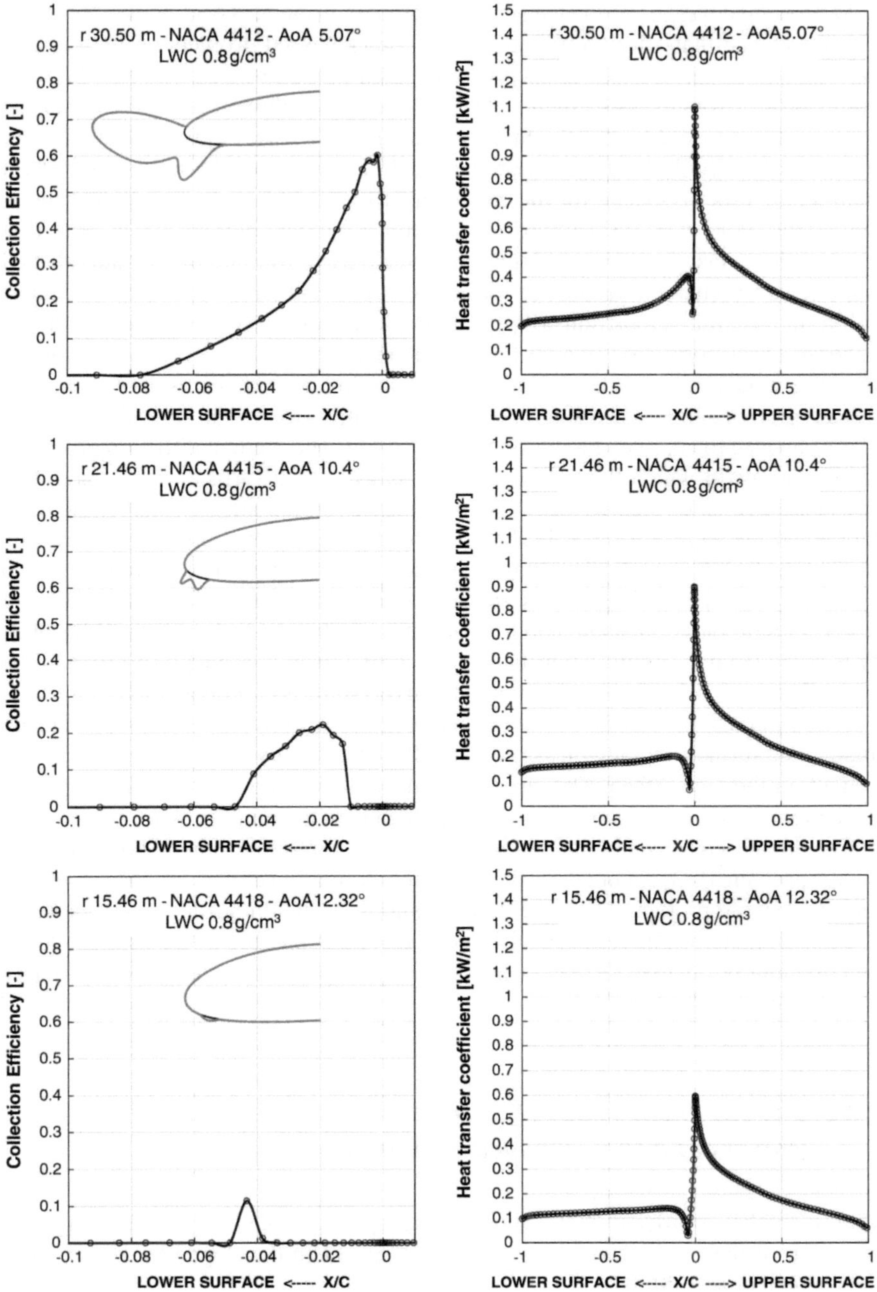

Fig. 4.58 Section 7 (*top row*): x/c upper = 0.033, x/c lower = 0.0487, iced area = 0.0002 m^2. Section 9 (*center row*): x/c upper = 0.0101, x/c lower = 0.046, iced area = 0.0012 m^2. Section 14 (*bottom row*): x/c upper = – 0.021, x/c lower = 0.076, iced area = 0.02 m^2

Microscopic movies of the ice accretion process made at NASA Lewis Research Centre indicate that after the initial flow of the coalesced droplets on the surface, the liquid does not flow but is caught and frozen in the grooves between the individual surface roughness elements. The process needs three-dimensional analysis of the icing surface. Most of the mathematical models used in previous studies are more macroscopic in nature because the roughness elements do not directly affect the freezing process except for the enhancing effect on the convective heat transfer coefficient. Tsao et al. [32] improved the high Reynolds number triple-deck theory, in order to take into consideration small-scale wave-like and/or irregular-shaped roughness on both the water and ice surfaces via an aero-hydro-thermodynamic interaction. In turn, this interaction will update the local convective heat transfer rate, in response to ice formation, which further enhances the local ice roughness formation. The model assumes that when the solidification front of a crystal is advancing into a region of supercooled liquid, its shape is subject to morphological instability, which can lead to complex modes of growth pattern. The thermal diffusion promotes configurations in which the growing solid has as large a surface area as possible. The latent heat is dissipated more rapidly in such configurations. Finally, the morphological instability of the solidification front is limited by capillary forces. The great variety of observed morphologies is due to the competition between these two forces, i.e. the supercooling and the surface tension. The typical scales of irregularity or corrugation on an unstable solid-liquid interface for the morphological instability is about 0.1 μm–1 mm. After this approach, the local phase equilibrium and the Gibbs-Thomson relation is used to model curved solid-liquid interfaces.

When glaze icing conditions are considered in the simulations, film rupture by small-scale surface ice roughness is obtained. Consequently, water beads are readily formed on the ice surface. This approach allows bypassing the need for empirical evaluations based on the surface wetness parameter shown in Fig. 4.42.

The often used conventional Stefan problem (the one-dimensional Stefan or phase change problem) [33] cannot describe the pattern formation of the kind seen in nature, since the Stefan problem lacks the dimensional information needed to set the scale of a crystal pattern. The results of this method are comparable to the Tsao model as the curvature of the body is small (the freezing point depression is appreciable only for very large curvature).

4.9.2 The Runback Water Dynamics in General Icing Process and the Extended Messinger Model

More recently, in order to better model the Stephan problem of phase change, an extension of the original Messinger approach has been developed by Myers [34], and Oezgen et al. [35].

The original Messinger model was developed to describe *the conditions that govern the equilibrium temperature of an insulated, unheated surface exposed to icing* [1]. Because the model is based on equilibrium temperatures, the transient evolution of the ice accretion cannot be modelled. It means that, when rime to glaze ice transition occurs, the freezing fraction steps suddenly from unity (rime condition) to another constant value, less than unity, and retains this new value till the next time step. The Stefan problem solution in contrast would predict that the freezing fraction will actually decrease monotonically from the unitary value of initial rime condition to its final equilibrium value. This results in an under-prediction of the ice thickness.

The hypothesis that the control volume assumes instantaneously everywhere the same temperature neglects transiently the heat conduction process through the layers of air, water, ice and body. In glaze conditions the isothermal assumption may still be considered acceptable for the air/water layers due to the mixing process, but this will not be the case for the water-ice and ice layer. According to the isothermal assumption no heat can be removed via conduction from the ice/water interface, therefore only the latent heat term survives in the energy equation during ice formation. This condition has been removed over the years by a number of authors [36–38], leading, as expected, to an increase of ice mass formation due to the additional cooling effect caused by the ice conduction. A comprehensive model will be outlined here, presenting the energy equations for the ice and water layers. This model is based on the cited models of Myers and Oezgen and Canibekcan and provides a more physical information about the temperature distribution inside the control volume and improves the heat transfer estimates. Although the Myers model can handle the 3D physical domain, here the heat transfer is assumed to be 1D in the control volume considered. Both rime and glaze ice growth can be simulated. The physical scheme is shown in Fig. 4.59.

The equations governing the phenomena are the one-dimensional Fourier equations for water, and ice, respectively, the mass conservation equation and the Stefan or phase change condition.

$$\frac{\delta T_w}{\delta t} = \frac{k_w}{\rho_w c_{p,w}} \frac{\delta^2 T_w}{\delta z^2} \tag{4.107}$$

$$\frac{\delta T_{ice}}{\delta t} = \frac{k_{ice}}{\rho_{ice} c_{ice}} \frac{\delta^2 T_{ice}}{\delta z^2} \tag{4.108}$$

$$\rho_{ice}\frac{\delta t_{ice}}{\delta t} + \rho_w \frac{\delta t_w}{\delta t} = \beta\ LWC\ W_\infty + \dot{m}_{w,in} - \dot{m}_{ev,subl} \tag{4.109}$$

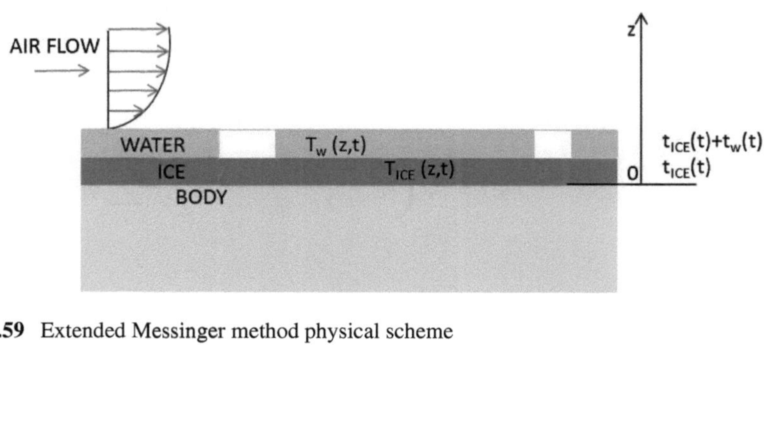

Fig. 4.59 Extended Messinger method physical scheme

$$\rho_{ice}\Delta h_f \frac{\delta t_{ice}}{\delta t} = k_{ice}\frac{\delta T_{ice}}{\delta z} - k_w \frac{\delta T_w}{\delta z} \tag{4.110}$$

where the coordinate z is normal to the surface. ρ_{ice} can take different values, for rime being $\rho_{ice,r}$ and for glaze $\rho_{ice,g}$ according to the local icing process. $\dot{m}_{ev,subl}$ indicates the occurrence of evaporation (glaze) or sublimation (rime) at the surface.

In order to determine the ice and water thicknesses together with the temperature distribution at each layer, appropriate boundary and initial conditions must be specified. These are based on the following assumptions:

- Ice is in perfect contact with the body surface and assumes instantaneously the same temperature
$$T_{ice}(0,t) = T_w(0,t)$$
- The temperature is continuous at the ice/water boundary and is equal to the freezing temperature:
$$T_{ice}(0,t) = T_s = T_{ref}$$
- At wetting onset (t = 0):
$$t_{ice} = t_w = 0$$
- At the air/water (glaze ice) or air/ice (rime ice) interface, the energy conservation Eq. (4.92) is equated to the conduction heat flux exchanged by the ice and water layer.

Therefore, still neglecting the radiant contributions two different conditions are analysed:

for rime ice at $z = t_{ice}$:

$$\begin{aligned} k_{ice}\frac{\delta T_{ice}}{\delta z} &= \dot{m}_{w,in}\, c_w\left(T_{w,in} - T_{ref}\right) - \dot{m}_{w,out}\, c_{w,l}\left(T_{w,out} - T_{ref}\right) \\ &+ \beta\, LWC\;\; W_{w,imp}\left[c_w\left(T_{w,imp} - T_{ref}\right) + \frac{W_{w,imp}^2}{2}\right] \\ &- K_A\left[\frac{h_c}{\rho_{air}\, c_{p,air}\, L^{2/3}}\frac{mm_w}{R}\left(\frac{e_{v,s}^{sat}}{T_s} - \frac{Rh\, e_{v,\infty}^{sat}\, p_e}{T_e\, p_\infty}\right)\right]\Delta h_{ev}^{T=T_s} \\ &- \dot{m}_{ice}\left[c_{w,ice}\left(T_{ice} - T_{ref}\right) + \Delta h_f^{T=T_s}\right] \\ &- h_c\left(T_w - T_\infty\right) - h_c\left(T_e + \frac{r\, w_e^2}{2\, c_{p,air}} - T_\infty\right) + \sum_i^n \dot{q}_r \end{aligned} \tag{4.111}$$

or

$$k_{ice}\frac{\delta T_{ice}}{\delta z} = \sum_i^n \dot{q}_r \tag{4.112}$$

for glaze ice at $z = t_{ice} + t_w$:

$$
\begin{aligned}
k_w \frac{\delta T_w}{\delta z} = {} & \dot{m}_{w,in}\, c_{w,l} \left(T_{w,in} - T_{ref}\right) - \dot{m}_{w,out}\, c_{w,l} \left(T_{w,out} - T_{ref}\right) \\
& + \beta\, LWC \;\; W_{w,imp} \left[c_w \left(T_{w,imp} - T_{ref}\right) + \frac{W_{w,imp}^2}{2} \right] \\
& - K_A \left[\frac{h_c}{\rho_{air}\, c_{p,air}\, L^{2/3}} \frac{mm_w}{R} \left(\frac{e_{v,s}^{sat}}{T_s} - \frac{Rh\, e_{v,\infty}^{sat}\, p_e}{T_e\, p_\infty} \right) \right] \Delta h_{ev}^{T=T_s} \\
& - \dot{m}_{ice} \left[c_{ice} \left(T_{ice} - T_{ref}\right) \right] \\
& - h_c \left(T_w - T_\infty\right) - h_c \left(T_e + \frac{r\, W_e^2}{2\, c_{p,air}} - T_\infty \right) + \sum_i^m \dot{q}_g
\end{aligned}
\tag{4.113}
$$

or

$$
k_w \frac{\delta T_w}{\delta z} = \sum_i^m \dot{q}_g \tag{4.114}
$$

4.9.2.1 Rime Ice Growth

This model allows to obtain a more specialised equation to evaluate the ice thickness compared to the simple expression of Eq. (4.103). Rime ice thickness can be obtained directly from the mass conservation Eq. (4.109) as water droplets freeze immediately on impact ($t_w = 0$):

$$
t_{ice,r}(t) = \frac{\beta \cdot LWC \cdot W_{w,imp} + \dot{m}_{w,in} - \dot{m}_{ev,subl}}{\rho_{ice,r}} \cdot t \tag{4.115}
$$

It can be demonstrated that when the ice thickness t_{ice} is below the critical value t_{ice}^* given by:

$$
t_{ice}^* = \frac{k_{ice}}{\beta \cdot LWC \cdot W_\infty\, c_{ice}}
$$

the temperature distribution is governed by:

$$
\frac{\delta^2 T_{ice}}{\delta z^2} = 0 \tag{4.116}
$$

The model represents the quasi-steady solution, a more general simplification valid when the layer is thin. The physical mean in this case is that the typical timescale for ice growth is less than that of conduction through ice, therefore the temperature can adapt during accretion.

As example, Table 4.7 indicates typical values of t_{ice}^* for the Tjæreborg wind turbine tip sections.

Table 4.7 Computational data for t^*_{ice}, t^{**}_{ice}, and t^{**}_w

Parameter	Section 10	Section 13
k_{ice} (W/mK)	2.18	2.18
c_{ice} (J/kgK)	2,050	2,050
c_w (J/kgK)	4,218	4,218
β_0 (–)	0.351	0.528
LWC (kg/m^3)	0.002	0.002
ϕ (–)	0.001	0.001
W_∞ (m/s)	58.25	70.45
c (mm)	1,500	960
t^*_{ice} (mm)	26.1	14.29
k_s/c (–)	0.0173	0.0148
t^{**}_{ice} (mm)	25.980	14.28
t^{**}_w (mm)	0.0068	0.0036

Integrating the above equation twice, one obtains the temperature distribution in the rime ice layer:

$$T_{ice}(z) = T_s + \frac{\sum_i^n \dot{q}_r}{k_{ice}} \cdot z \tag{4.117}$$

4.9.2.2 Glaze Ice Growth

By analogous considerations, if the ice layer thickness is less than t^{**}_{ice}:

$$t^{**}_{ice} = \frac{k_{ice}}{(1-\phi)\,\beta \cdot LWC \cdot W_\infty c_{ice}}$$

and

$$t^{**}_w = \frac{k_w}{\phi\beta \cdot LWC \cdot W_\infty c_w}$$

as indicated in Table 4.7. ϕ is the water fraction remaining liquid. The water layer has to remain sufficiently thin (about 3 mm in the tip section of the simulation of Table 4.7) compared to the other layers (ice and body) to allow its temperature to adjust.

The Fourier equation can be simplified as follows:

$$\frac{\delta^2 T_{ice}}{\delta z^2} = 0$$

and

$$\frac{\delta^2 T_w}{\delta z^2} = 0$$

Two times integration of these equations with the help of the boundary condition discussed above, gives the ice layer temperature distribution:

$$T_{ice}(z) = \frac{T_{ref} - T_s}{t_{ice}} z + T_s$$

and the water temperature layer distribution:

$$T_w(z) = T_{ref} + \frac{\sum_i^m \dot{q}_g}{k_w} \cdot z - T_{ice}$$

Integration of the mass conservation gives the water height:

$$t_w = \frac{\beta \; LWC \;\; W_{w,imp} + \dot{m}_{w,in} - \dot{m}_{ev}}{\rho_w} (t - t_{onset}) - \frac{\rho_{onset}}{\rho_w} (t_{ice} - t_{onset}) \quad (4.118)$$

The pedex *onset* defines the glaze ice onset conditions.

By substituting Eq. (4.118) into Eq. (4.110), a first order differential equation is obtained:

$$\rho_{ice,g} \, \delta h_f \, \frac{\delta t_{ice}}{\delta t} = \frac{k_{ice} \left(T_{ref} - T_s \right)}{t_{ice}} - \sum_i^m \dot{q}_g \cdot z \quad (4.119)$$

Continuity equation for glaze to rime ice applies:

$$\frac{\delta t_{ice,r}}{\delta t} = \frac{\delta t_{ice,g}}{\delta t}$$

at $t_{ice} = t_{onset}$.

By the aid of Eqs. (4.119) and (4.115) one gets finally the thickness at which glaze appears:

$$t_{ice,\,g} = \frac{k_{ice} \left(T_{ref} - T_s \right)}{\left(\beta \; LWC \;\; W_\infty + \dot{m}_{w,in} - \dot{m}_{ev,subl} \right) \Delta h_f + \sum_i^m \dot{q}_g} \quad (4.120)$$

4.9.2.3 The Modified Freezing Fraction

The modified freezing fraction, compared to the original Messinger definition for the given a panel is defined as the ratio of the amount of water that solidifies to the amount of water that impinges on the control volume plus the water that enters the panel as runback water, or, for rime and glaze respectively:

$$f_r = \frac{\rho_{ice,r} t_{ice,r}}{\left(\beta \, LWC \;\; W_\infty + \dot{m}_{w,in} \right) t} \quad (4.121)$$

$$f_g = \frac{\rho_{ice,r} t_{ice,g} + \rho_{ice,g} \left(t_{ice} - t_{ice,onset} \right)}{\left(\beta \, LWC \;\; W_\infty + \dot{m}_{w,in} \right) t} \quad (4.122)$$

The runback flow rate leafing the panel is therefore:

$$\dot{m}_{w,out} = (1 - f)\left(\beta\, LWC\ \ W_{\infty} + \dot{m}_{w,in}\right) - \dot{m}_{ev} \tag{4.123}$$

For both pressure and suction surfaces, the unfrozen water moves into the successive panel:

$$\dot{m}_{w,in} = \dot{m}_{w,out}$$

References

1. Messinger BL (1953) Equilibrium temperature of an unheated icing surface as a function of air speed. J Aeronaut Sci 20(1):29–42
2. MacArthur CD (1983) Numerical simulation of airfoils ice accretion. AIAA paper 83:0112
3. Wright WB (2002) User manual for the NASA green ice accretion code LEWICE-version 2.2.2. NASA Langley Research Center, NASA/CR-2002-211793
4. Wright WB (1995) Users manual for the improved NASA Lewis ice accretion code LEWICE 1.6. NASA Langley Research Center, NASA CR198355
5. Hedde T, Guffond D (1995) ONERA Three-dimensional icing model. AIAA J 33(6):1038–1045
6. Tran P, Brahimi MT, Paraschivoiu I, Pueyo A, Tezok F (1995) Ice accretion on aircraft wings with thermodynamic effects. AIAA J 32(2):444–446
7. Baruzzi GS, Habashi WG, Guvremont G, Hafez MM (1995) A second order finite element method for the solution of the transonic Euler and Navier-Stokes equations. Int J Numer Methods Fluids 20:671–693
8. Bourgault Y, Habashi WG, Dompierre J, Baruzzi GS (1999) A finite element method study of Eulerian droplets impingement models. Int J Numer Methods Fluids 29:429–449
9. Beaugendre H, Morency F, Habashi WG (2003) FENSAP-ICEs three-dimensional in-flight ice accretion module. J Aircr 40(2):239–247
10. Croce G, Beaugendre H, Habashi WG (2002) CHT3D: FENSAP-ICE conjugate heat transfer computations with droplet impingement and runback effects. AIAA paper 2002-0386
11. Lozowski EP, Stallabrass JR, Hearty PF (1983) The icing of an unheated, nonrotating cylinder. Part I: a simulation model. J Clim Appl Meteorol 22:2053–2074
12. Finstad KJ, Lozowski EP, Makkonen L (1988) On the median volume diameter approximation for droplet collision efficiency. J Atmos Sci 45(24):4008–4012
13. Langmuir I, Blodgett KB (1945) A mathematical investigation of water droplet trajectories. General Electric, RL 225 Ad 64354
14. Dorsch RG, Brun RJ, Gregg JL (1954) Impingement of water droplets on an ellipsoid with fineness ratio 5 in axisymmetric flow. National Advisory Committee for Aeronautics, Technical note NACA-TN-3099
15. Hacker PT, Brun RJ, Boyd B (1953) Impingement of droplets in 90° elbows with potential flow. National Advisory Committee for Aeronautics, Technical note 2999
16. Langmuir I (1946) Collected works of Irving Langmuir. Pergamon Press 10:348–393
17. Beard KV, Pruppacher HR (1969) A determination of the terminal velocity and drag of small water droplets by means of a wind tunnel. J Atmos Sci 26:1066–1072
18. Wang PK, Pruppacher HR (1977) An experimental determination of the efficiency with which aerosol particles are collected by water drops in subsaturated air. J Atmos Sci 34:1664–1669
19. Anderson DN (2004) Manual of scaling methods. Ohio Aerospace Institute, Brook Park, Ohio, Technical report, NASA/CR-2004-212875
20. Finstad KJ, Lozowski EP, Gates E (1988) A computational investigation on water particle droplet trajectories. J Atmos Ocean Technol 5:160–170

21. Wright WB, Potapczuk M (1996) Computational simulation of large droplet icing. FAA International conference on inflight icing, DOT/FAA/AR96/81
22. Wright WB (1995), Capabilities of LEWICE 1.6 and comparison with experimental data. AHS international icing symposium, Montreal, Canada
23. Hamed A (1981) Particle dynamics of inlet flow fields with swirling vanes. AIAA paper 81-0001, 19th AIAA aerospace sciences meeting. St. Louis (January 1981)
24. Farag KA, Bragg MB (1997) Three dimensional droplet trajectory code for propellers of arbitrary geometry. In: Proceeding of 36th AIAA Aerospace Sciences Meeting and Exhibit
25. Gelder FT, Lewis JP (1951) Comparison of heat transfer from airfoil in natural and icing conditions. NASA Langley Research Center, NACA TN 2480
26. Al-Khalil KM, Keith TG, De Witt KJ (1993) New concept in Runback water modeling for anti-iced aircraft surfaces. J Aircr 30(1):41–49
27. Bourgault Y, Beaugendre H, Habashi WG (2000) Development of a shallow-water icing model in FENSAP-ICE. J Aircr 37(4):640–646
28. Schmidt E, Wenner K (1943) Heat transfer over the circumference of a heated cylinder in transverse flow. NASA Langley Research Center, NACA TM 1050
29. Martinelli RC, Guibert AG, Morin EH, Boelter LMK (1943) An investigation of aircraft heaters, VIII—a simplified method for the calculation of the unit thermal conductance over wing. NASA Langley Research Center, NACA ARR (WR W-14)
30. Smith AG, Spalding DB (1958) Heat transfer in a laminar boundary layer with constant fluid proprieties and constant wall temperature. J R Aeronaut Soc 62:60–64
31. Incropera F, DeWitt D (1996) Fundamentals of heat and mass transfer, 5th ed. Wiley, New York
32. Tsao JC, Rothmayer AP (2002) Application of triple-deck theory to the prediction of glaze ice roughness formation on an airfoil leading edge. Comput Fluids 31(8):977–1014
33. Myers TG, Hammond DW (1999) Ice and water film growth from incoming supercooled droplets. Int J Heat Mass Transfer 31(42):2233–2242
34. Myers TG (2001) Extension to the Messinger model for aircraft icing. AIAA J 39(2):211–218
35. Oezgen S, Canibek M (2008) Ice accretion simulation on multi-element airfoils using extended Messinger model. J Heat Mass Transfer. doi:10.1007/s00231-008-0430-4
36. Huang JR, Keith TG Jr, De Witt KJ (1993) Efficient finite element method for aircraft de-icing problems. J Aircr 30(5):695–704
37. Bourgault Y, Beaugendre H, Habashi WG (2000) Development of a shallow-water icing model in FENSAP-ICE. J Aircr 37(4):640–646
38. De Witt KJ, Baliga G (1982) Numerical simulation of one-dimensional heat transfer in composite bodies with phase change, NASA CR-165607

Chapter 5
Ice Prevention Systems (IPS)

Abstract The chapter after proposing a procedure of ice prevention system assessments, compares the different IPS concepts that are presented and systematically compared. The advantages and disadvantages of current wind turbine IPS are then discussed. Emerging technologies are reviewed. They are the pneumatic de-icing system (already in use in aerodynamic field), microwave, low adhesion coating materials, the intermittent (cyclic) hot gas heating, the regenerative ice prevention system and finally the regenerative heating. Some simple calculations have been made to set up a comparison of the capabilities of such systems. From this discussion, a proposal of the energetic efficiency of an IPS is presented together with a synthetic model for estimating the anti-icing power and energy requirement. A worked example explains practically the theory. The chapters include the detailed calculation of the design of a hot air thermal anti-icing ice prevention system, developed on the basis of the knowledge developed from the previous chapters. It describes how the blade can be geometrically discretised, the thermo-aerodynamic model and the conjugate heat transfer model. Results are given and the simplification discussed.

5.1 Introduction

The selection and design of anti-icing systems for wind turbines shall be based on the reliable evaluation of the heat fluxes that the blades exchange with the environment during icing conditions and the duration of the phenomenon. As icing is a common occurrence for aviation, and marine also, the design methods for wind turbines can take profit from the past and current experiences. Some solutions can be adopted for its special needs, although basic differences make the technology transfer not obvious.

Today only a few dedicated numerical codes exist for analysis of wind turbine ice prevention system design, namely LEWICE ANTICE (from version 2.0) [1], adapted to predict anti-icing power requirements for wind turbines and TURBICE code [2] that simulates ice accretion on wind turbine blades developed by the Technical Research Centre of Finland (VTT). A special feature of the FENSAP ICE [3] also became recently available. The TREWICE code [4] developed by the University

L. Battisti, *Wind Turbines in Cold Climates*, Green Energy and Technology,
DOI 10.1007/978-3-319-05191-8_5

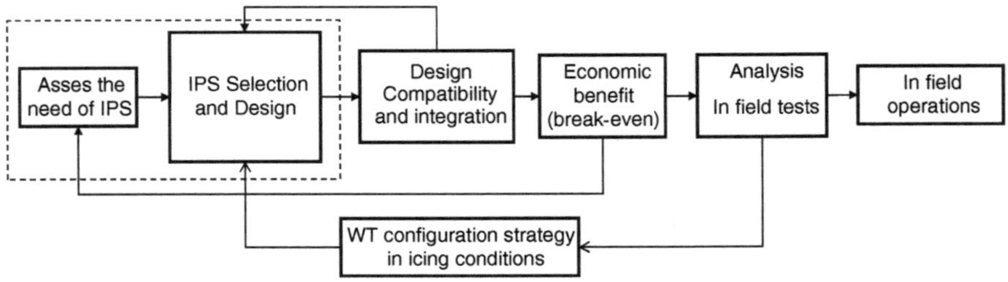

Fig. 5.1 Basic scheme for IPS design assessment

of Trento for research purpose allows either thermal anti-icing or de-icing analysis. TREWICE can model also hot air anti-icing systems either in continuous or in intermittent heating modes.

The selection, analysis and design process of anti-ice thermal ice prevention systems for wind turbines involves an iterative procedure, which scheme is given in Fig. 5.1.

The first design stage is to assess the need of the ice prevention system. This step requires reliable information on icing severity of the site and effect of ice on turbine production and loads. An aeroelastic code should assess whether the ice prevention system is actually required and its economic sustainability. IPSs affect both investment and operating costs in a different manner depending upon the site characteristics and the selected IPS typology.

Usually the system has to be integrated in an existing design philosophy rather than being a retrofitted part of a turbine designed for conventional sites. Field tests should complete the design procedure and fix the operating rules of turbine and IPS.

The outlined procedure is actually challenging because of a:

1. lack of meteorological and site parameters.
2. lack of data (both thermal and mechanical) on blade material behaviour at very low (−20 to 0 °C) and very high (70 to 110 °C) temperature for continuous service.
3. lack of tools for the design of ice preventions systems. Some numerical tools developed for aeronautical field are not actually tailored for use in wind turbines, because of very different climatic conditions encountered (LWC, MVD, etc.), the availability of such data, the presence of ice detection systems, the relative wind speed of the moving parts, on-board available power for anti-icing, maximum demand during specific operations, materials used, effect on performance, rotating parts, typical icing time, requirement of certification rules. Reliable numerical tools are quite difficult to be built up and validated. They require very structured and heavy computational modules, some of them are not ready available for private companies or research centres.
4. need of additional efforts (both for design and certification) to integrate the IPS in the current wind turbine design.

The suggested requirement in designing the blade ice prevention equipments is that the blade must be maintained sufficiently free of ice accretion in order to enable

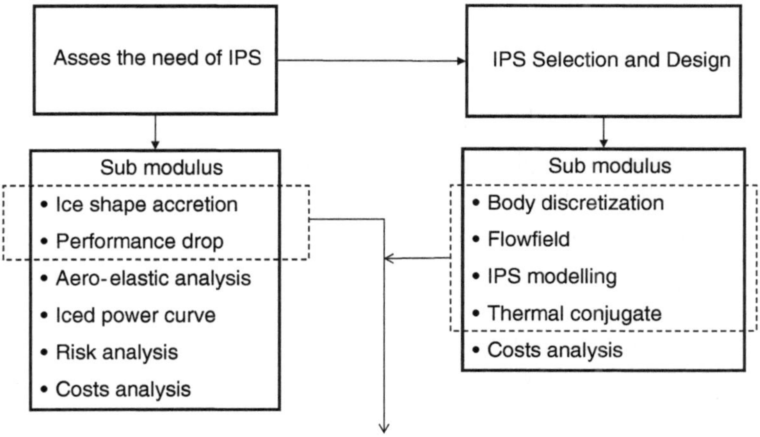

Fig. 5.2 Simplified path of the design step (contoured by the *dashed line* of Fig. 5.1)

continued, safe and economic operation in icing conditions. While in aeronautic field the recommended procedure is to design for the partial or complete evaporation of all impinging water in the most severe conditions likely to be encountered in a condition of continuous icing, some tolerance for ice formation could be admitted under certain conditions in wind turbines. A heated blade designed on this basis will remain clear of ice under most situations with only a relatively small amount of runback forming.

A standard ice accretion code alone is not sufficient for designing a thermal anti-ice IPS. Usually, the goal of ice accretions codes is to predict the local ice mass growing on the surface, while the anti-ice prediction codes are designed to assess the heat flows to be provided at the surface to ensure a given temperature to be maintained at the external surfaces. Ice accretion codes can be modified by adding an external heat source in the energy balance, which can possibly keep the blade external temperature above freezing. Adding external heat source means introducing a conjugate heat transfer model. Usually the conjugate heat transfer analysis encompasses the analysis of the external flow field, the blade wall conduction process and the inner flow field analysis (when blade warm in-duct air circulation is employed).

The adoption of steady-state conditions for anti-icing analysis introduces some simplification in the calculation path compared to unsteady processes involved in ice accretion. The simplified path of the design step, which is contoured by the dashed line of Fig. 5.1, is shown in Fig. 5.2.

5.2 A Procedure of Ice Prevention System Assessments

As previously stated, the ice prevention systems (IPSs) should ensure continued, safe and economic operation in icing conditions. Continued operation in icing conditions means no or reduced unavailability due to:

- ice accretion on blades,
- ice detection failure,
- IPS failure.

Safe operation in icing conditions means no or limited:

- potential damages to people or properties due to ice,
- potential risk for maintenance people.

Economic operation in icing conditions means minimal:

- energy yield loss due to icing,
- additional unavailability due to ice accretion,
- extra-investment costs for IPS,
- running costs for IPS (power level, energy requirement and maintenance),
- insurance costs.

Since IPSs of wind turbines cause additional investment and operational costs, some criteria are needed for the assessment of the benefits of such systems.

Figure 5.3 shows a possible integrated design path for ice prevention systems applied to the rotor as an explanatory situation.

On the basis of environmental data and technical guidance issues, manufacturers' design requirements and objective, the procedure first assigns the boundary conditions for a series of numerical simulations and experimental campaigns. Ice accretion analysis is necessary to provide the expected contaminated blade profile shape and the induced performance loss and loads modification. In this step, the aeroelastic code simulations evaluate the ice accretion on the blades and the consequent surface and mass loads. A performance deterioration analysis allows the energy output to be evaluated (see DLM matrix of Chap. 2). Shed-ice trajectory and structural damage tolerance analysis is also an output of the integrated procedure. A technical level decision is now made on the need of ice prevention systems. For light icing, where negligible structural and performance penalties and an acceptable risk for people and goods is expected, no ice prevention system is required and only limitations can be prescribed for icing operations. If the use of ice prevention systems is necessary, instead, a preliminary selection of the technology is made on the basis of engineering experience.

Four types of systems are indicated in Fig. 5.3: thermal, mechanical, ice-phobic surfaces and hybrid ones. At this stage, simplified methods can be used to evaluate the anti-icing installed power and its energy consumption, and a break-even analysis is carried out to assess the economic viability of the system. In the case the system ends as non-economically worth, only limitations are prescribed for operation in icing conditions. If, for instance, the selection of continuous anti-ice technology is confirmed as preliminarily viable, Fig. 5.4 indicates the full calculation path for system executive design, verification and certification. Numerical and in-field tests on models are necessary to assess the technical availability and reliability of the ice prevention systems. This step ends with the analysis of the component's redesign. If a major redesign is needed, a new economic break-even is required; otherwise, only technical approval path is followed.

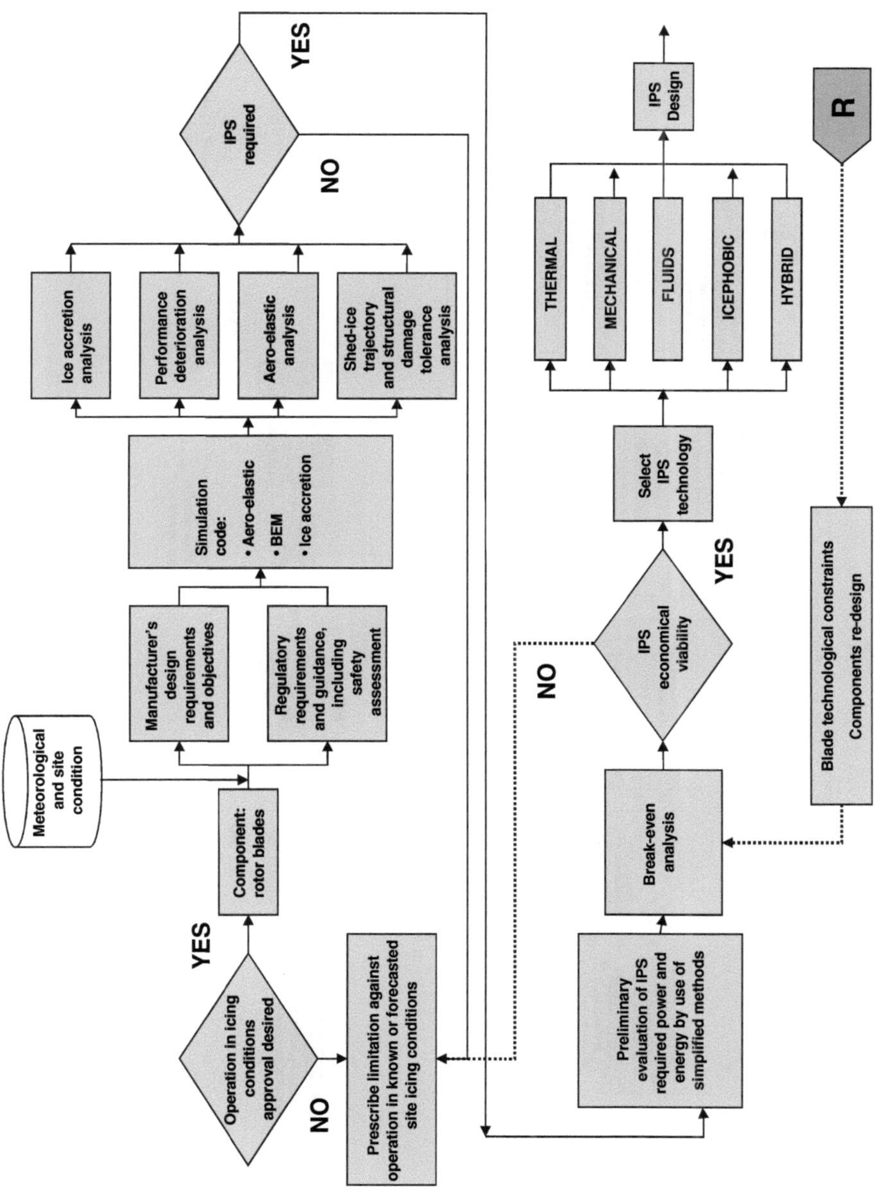

Fig. 5.3 Ice prevention systems integrated design

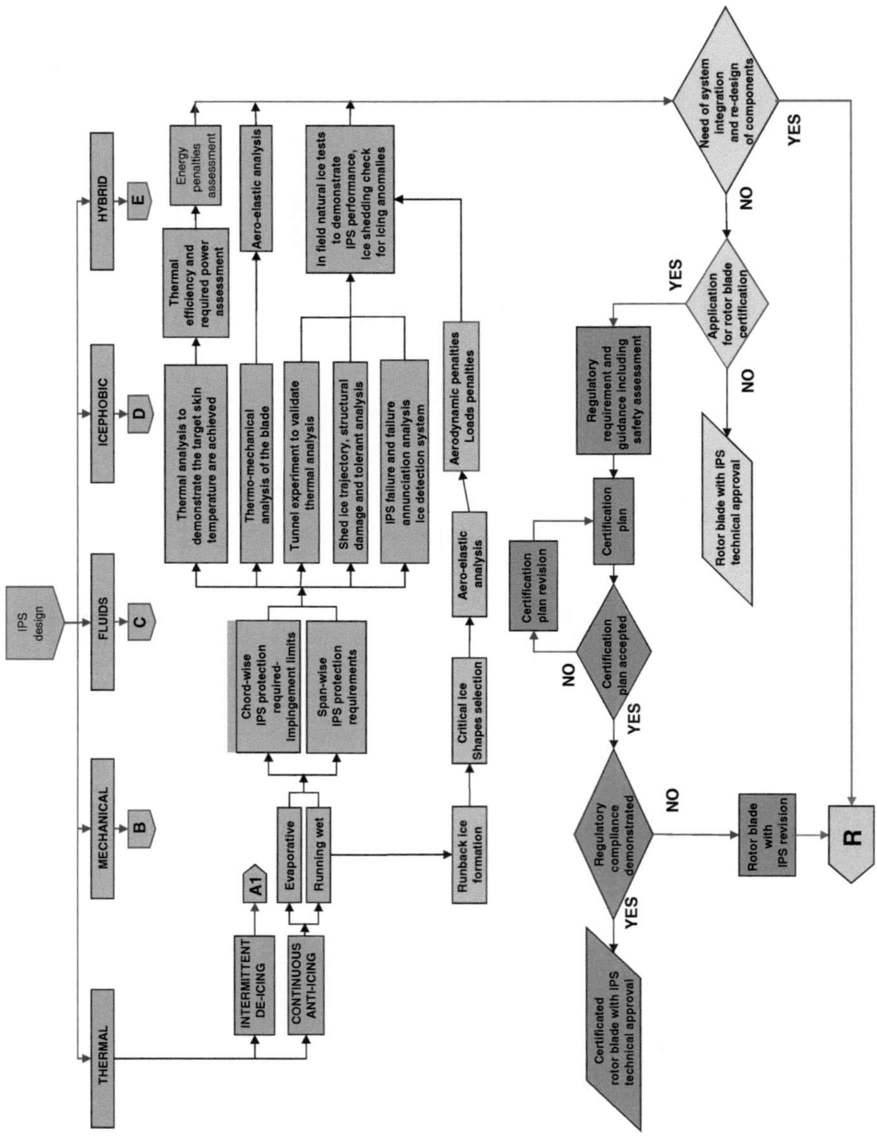

Fig. 5.4 Ice prevention system executive design, verification and certification

5.3 IPS Concepts Comparison and Discussion

Bibliographic search in open literature and various databanks on anti-icing and de-icing applications identified more than 40 techniques at different stages of development in different fields such as air, marine and rail transport, electrical networks, telecommunications, etc. All such systems can be grouped according to classification criteria.

5.4 IPS Classification

Three types of classification can be used for IPS:

1. the first is very intuitive and is based on the principle of operation adopted as thermal, mechanical or other;
2. the second is based on the duration of the applied means and is analysed by the definition of the intermittency factor (discussed after).
3. the third is based on the energy requirement (dividing IPSs in active and passive systems);

In the following discussion, the first classification will be used to discuss the available technology options. The energy requirement will be associated to each principle of operation and discussed in a paragraph dedicated to energetic efficiency. Finally, a separate paragraph will expand the concept of intermittency.

5.4.1 IPS Classification Based on the Principles of Operation

IPS can be differentiated on the basis of the principle of operation used as thermal, mechanical or others, the latter grouping all systems not included into the first two.

5.4.2 IPS Mechanical Methods

In *IPS mechanical methods*, mechanical shocks are induced to break the ice accretion. According to this operation they are better recognised as de-icing systems. The main methods are forced vibration, electromagnetic impulses, ultrasounds and surface deformation methods such as electro and magneto-strained envelopes, changes in the alloy memory phase and inflatable boots. Also, manual polishing of the blade skin is adopted in wind energy. Blade flexibility is also one of the factors helping in ice break-up and can be enumerated within this category.

Aeronautic field makes wide use of mechanical systems, especially for small aircraft. Moveable leading edge skin by means of both mechanical and electromagnetic devices are in use along with leading edge ice scrapers. After the ice forms, the skin can be stretched and the ice will crack and flake off. The skin material can be a latex-based elastic or a silicone-based elastic. The latter category of system is of different implementation for wind turbines.

5.4.2.1 IPS Thermal Methods

In principle, in *IPS thermal methods*, a suitable heat source can be used either to raise the water droplet temperature before they struck the body, or by warming up a portion of the surface encompassing the water impingement region (plus an additional safety area). The first method will be discussed in the paragraph dedicated to the emerging technologies. In case of the second method, the heat needed to prevent ice can be provided using two heating strategies, i.e. by a distributed heat source generation (electrical pads) or a concentrated heat source generation (hot air). In the former case, the heat is generated at the desired location by means of electrical power dissipation (Joule effect), while in the latter the heat is produced in one location and then distributed by means of a suitable carrier (generally air, but other intermediate heat carriers can also be used). Microwave can be included in this class.

Heat can be used to prevent freezing of the impinging water droplets either by evaporating or maintaining them at a liquid state (running wet option). In this case, we refer to an anti-icing system. In evaporating practice, sufficient heat is supplied to evaporate all water droplets impinging upon the heated surface, while in running wet practice only a minimum heat is provided to prevent freezing on the heated surface. Beyond the heated surface of a running wet system, the water can freeze, resulting in runback ice. For this reason, running wet systems must be used carefully so as not to permit build-up of runback ice in critical locations.

When heat is used to rebound or melt an existing ice cap, one refers to de-icing systems.

In thermal ice prevention systems, the heat source can be supplied either by extract it from the plant's own electrical system, or by a separate on-board heat generation unit. An interesting opportunity in wind turbines is to seek for regenerative energy sources as the heat released from electrical devices or heat exchangers, as shown in the Fig. 5.5.

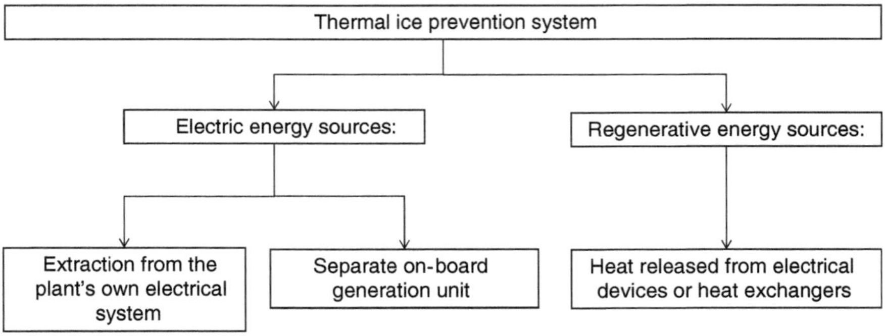

Fig. 5.5 Means to generate the heat source

5.4.3 Other IPSs

This classification encompasses all means that are not purely mechanical or thermal, as:

- solar radiation absorbent coating;
- hydrophobic and ice-phobic coating, viscous products and greases.

Coatings techniques have the advantage to be passive, i.e. they do not require energy for operation.

Special coatings of hydrophobic and ice-phobic material shall reduce the shear forces between the ice and the blade's surface. Expected advantages of coating the whole surface of rotor blades should be relatively low costing, as there are no moving mechanical parts, and no special lightning protection are required.

On the other side, these systems do not always hinder ice formation nor remove it and can only help by limiting ice effects, and have complex in-field maintenance. Some possible aerodynamic performance issues and risk of wear can be of concern. Moreover some coatings can be toxic or corrosive.

By painting the leading edge blades black, as shown in Fig. 5.6, the blades can heat up during daylight and the ice melts earlier than with white painted ones. The blade surface behaves as a black body, boosting absorption of incoming radiation. While this will cause heating during sun light, the same mechanism will increase the outgoing radiation with the unwanted effect of cooling the blades during clear nights.

In summertime, the high temperature achieved by the blade's surface may affect the material properties of the glass fibre-reinforced plastics (GRP) which are sensible to high temperatures, but this conclusion is quite controversial.

Black blades appear to add relatively low solar gain in windy areas, in neither summer nor winter. Moreover, they are not active during the night and less efficient

Fig. 5.6 Vestas V47–660 (*left*) and Bonus 150 kW (*centre*) with black-painted fluorourethane (StaClean) coating. Searsburg turbines (*right*) with *black paint* applied to the blades: note the layer of ice along the blade leading edge (Color figure online)

in cloudy days. As icing problems are prominent during the wintertime on mountains and at subarctic locations where the sun is scarce, this method is expected to provide marginal effects. Tests on a 900 kW NEG-micron wind turbine near Äppelbo in Sweden ended with records of long stops due to ice accretion [5].

A disadvantage of both systems is that they tend to generate jeopardised ice accumulation with consequent possible unsymmetrical ice accretion that leads to unbalance. Ice throw during operation can be a further concern.

5.5 IPS Classification Based on Duration of the Applied Means

For energetic analysis purpose, the most interesting way to classify thermal IPS is by the time duration of the applied mechanical or heat means to de-ice. For this aim the intermittent factor τ is introduced as the ratio of the duration of the heat application to the surface to the total time of the icing event, which will be the sum of the time duration of heat supply and the time duration where no heat is supplied. Equation (5.1) shows this ratio:

$$\tau = \frac{T_{heat-on}}{T_{heat-on} + T_{heat-off}} = \frac{T_{heat-on}}{T} \tag{5.1}$$

On this basis, de-icing systems are characterised by an intermittency factor less than unity, which indicates cyclic operation, while anti-icing systems operate at intermittency factor equal to unity, indicating continuous operation. In de-icing practice (which is widely used in the aeronautic field), sufficient heat is rapidly applied to the ice-surface interface in order to melt the bonding layer of ice; aerodynamic or centrifugal forces then remove the bulk of the ice. These two operating options determine quite different power and energy requirements, and different loads on the blades, thus leading a different cost impact on the whole system.

5.6 IPS Classification Based on the Energy Required

In this class of systems, coatings are not the only energy consuming systems, while all others will have different energy requirements for their operation depending on the efficiency of the systems as will be discussed later on.

5.7 Wind Turbine IPS in Use

Among active systems, several companies claimed for commercially available systems: Kelly Aerospace, VTT (KAT), Enercon, EcoTEMP, Ice CODE/Goodrich, Vests, Siemens, LM Glasfiber, etc. Actually most of them seem to be in progress and under development. Hardly any data are unfortunately freely available up now neither on development phase nor in some cases satisfactorily describing the technology actually used. The following paragraphs are thus built up on the basis of information gathered in congresses, and personal discussion with some operators in exhibitions.

5.7.1 Rotor Blade Electric Heating

Electrothermal systems use electrical foils either embedded into the blade wall or glued on the blade skin. The system supplies the blade's external surface the heat required to keep it at a prescribed temperature. Ideally, if any panel could be singularly heated, this strategy would lead to an isothermal surface and hence to the minimum heating supply. Some unavoidable heat wall back-side losses have to be considered however.

Electrically heated foils at the LE (heating wires or carbon fibres) were initially developed at VTT (Technical Research Centre of Finland) in 1992. In 1993, a 220 kW wind turbine was erected on the top of Pyhätunturi Fell. The turbine was the first one provided with the blade heating system and it has been used as a test bed for the development of the blade heating systems. Several modifications have been tested winter after winter.

The blade heating system was operated using plastic-based heating foils on the blade surface. The first commercial electric blade heating systems were delivered in 1996 by JE-System by Kemijoki Arctic Technology Oy [6]. In Fig. 5.7, the details of the application are shown indicating the heated (ice free) part of the LE extending to the aft portion of the blade.

The electrical pads are manufactured and laminated on blade surface with post-surface finishing. The typical installed thermal power was about 15 kW/blade for a 600 kW rated power wind turbine. The electrical supply and control system include supply transformer and cable, safety switches, lightning protection, control unit and ice detector, control unit and turbine control interface.

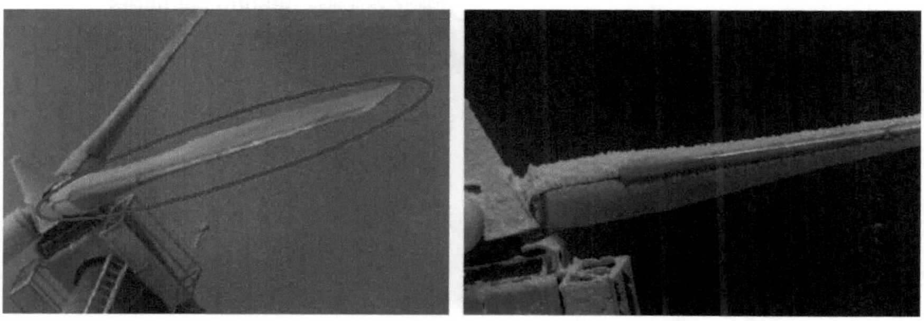

Fig. 5.7 Details of the JE-System by Kemijoki Arctic Technology Oy [6]

Experimental in-field tests indicated energy consumption ranging from 1 to 4 % of AEP. The system is quite effective for harsh conditions and works well as a de-icing system too. After the original concept, different regions of the blade can be separately heated to gain maximum efficiency of the system, but electronic problems and difficulties to monitor the temperature for each heated panel suggested switching to a more simple control system. There are three some unsolved problems related to such a system: one is related to the material of the heating elements, metal or carbon fibre, which can attract lightning strokes. Another reason is related to the structural problems caused by the position of the heating elements at the LE. As depicted in Fig. 5.8, the gravity forces acting as the rotor turns cause high deterministic loads on the blade's structure. High strains in the GRP load carrying girder will cause even higher strain in the wires of fibres, respectively, of the heating elements. This is especially true if the heating elements are carbon fibre made. Their Young's modulus

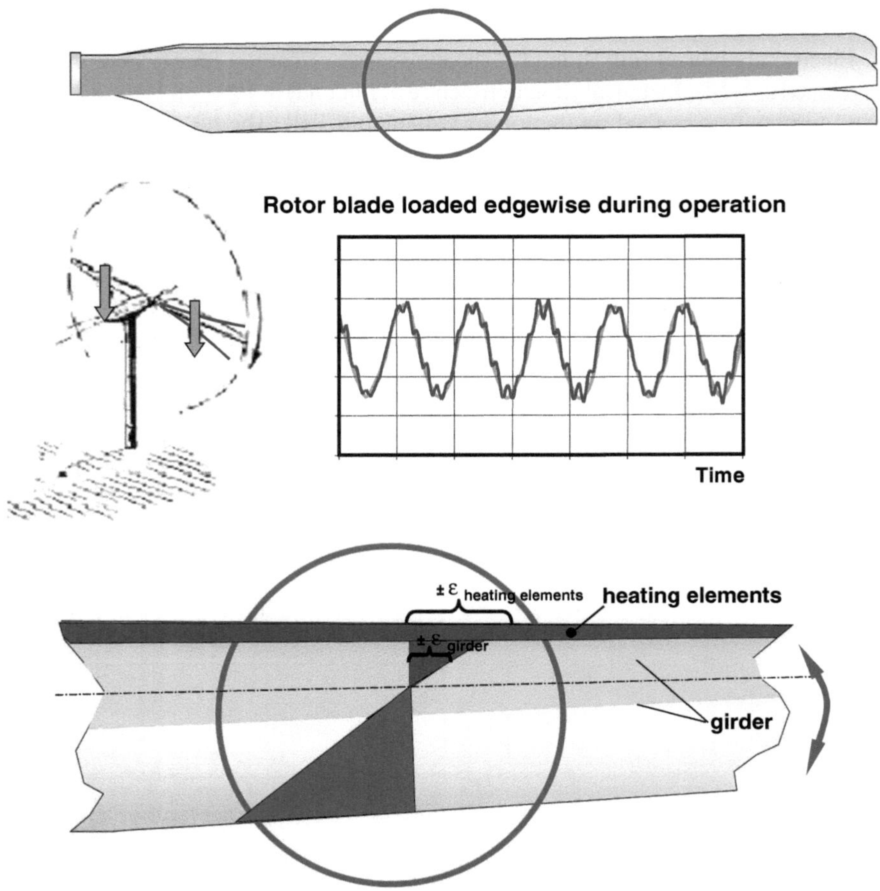

Fig. 5.8 Details of the structural behaviour of the JE-System by Kemijoki Arctic Technology Oy [7]

is much higher compared to glass fibres and the consequences are that the loads are carried by the heating fibres preferably. A third class of problems is due to the runback water freezing the aft of the leading edge area. While the system is very apt to keep the leading edge clean, the water tends to freeze the aft of forming ridges, which, as shown in the Chap. 3, has the greatest impact on depleting aerodynamics.

LM Glasfiber carried out some tests on heating foils for a LM 35 0P blade in field. Designed for low sensitivity to lightening, the foils take the outer half part of the blade and the power installed is about 17 kW per blade [8].

Recently, Kelly aerospace, see Fig. 5.9, presented the solution of a continuous heat-conducting outer layer called *Tedlar*, embedded with adhesives and protective erosion shields. A zoned heater system is utilised in a variety of lengths, widths and thicknesses to match available power. The impingement area, or leading edge, is kept warm continually for melting impinging ice, or simply, 'runs wet'. During a de-ice cycle, the voltage is increased raising the temperature of this shed zone, determining the detachment of the ice bond and its shedding away via aerodynamic force. Once power is removed from the heater, the shedding zones immediately freeze and continue to collect ice until the next de-ice cycle. This cycle is claimed to have 1 °C/s temperature rise to 10 °C/s temperature rise depending on power supplied. The power is cycled around the wind turbine to maintain symmetrical/balanced de-icing characteristics. Each zone takes 20–40 s to de-ice depending on blade temperature and the total de-ice time takes 714 min.

As a final remark it is evident that the electrical power supplied to electrothermal heating systems cannot simply be obtained by connecting a power cable from the grid to the de-icer element on the blade; it is obvious that any such direct connection is impossible because of the rotor spinning. This is a very common problem, however, and many solutions exist. LM Wind Power (former LM Glasfiber) was probably able to supply the 60–80 kW of power for the blade heaters with an off-the-shelf rotary power transfer device, most likely a slip ring.

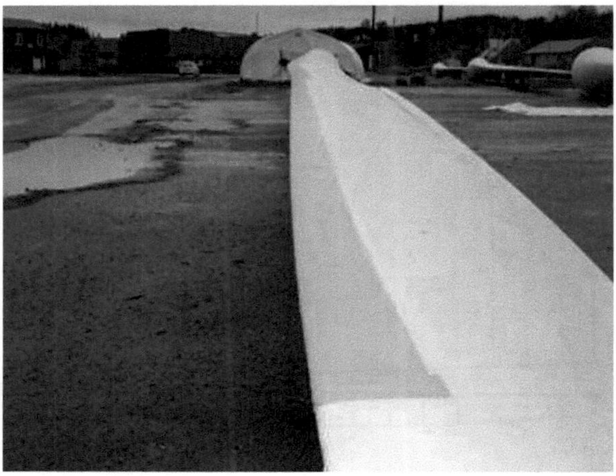

Fig. 5.9 Picture of Kelly aerospace electrothermal de-icer [9]

5.7.2 Hot Air In-duct Circulation Systems

In-duct circulating air systems for anti-icing and de-icing of aircraft components, especially airfoils, were originally developed in the 1950s and now represent the most often adopted systems for aircrafts wings. They were soon though for use in wind turbines, too: in fact, the first description of such an application for wind turbine rotors dates back to a German patent of 1949 [10], shown in Fig. 5.10.

Here an in-duct circulating air system in the blade was described, wherein the air was heated by way of the waste heat from the electrical generator. The system did not come into commercial use until the 1980s, by which time more rational concepts had been developed for blades equipped with both de-icing and anti-icing systems [11].

The use of hot air in wind turbines comes from the great simplicity and reliability of the system.

There are basically two possible configurations of in-duct circulating air anti-ice system, i.e. the open-circuit and the closed-circuit systems, both obtained by arranging suitable partitioning walls inside the blade shell. Those configurations are schematically illustrated in Figs. 5.11 and 5.12. In the open-circuit system, the hot air enters at the blade root and flows towards the tip in a channel built in the leading edge area, where the anti-ice heat demand is greatest. The partitioning wall may possibly coincide with the blade spar. The air is expelled at the blade tip.

The air static pressure changes along the channel and is locally determined by a balance among the centrifugal forces, pressure losses and heat losses through the blade wall. A naturally aspirated flow can possibly be established in the channel for small fast rotating turbines, depending on the inlet air pressure and temperature, the

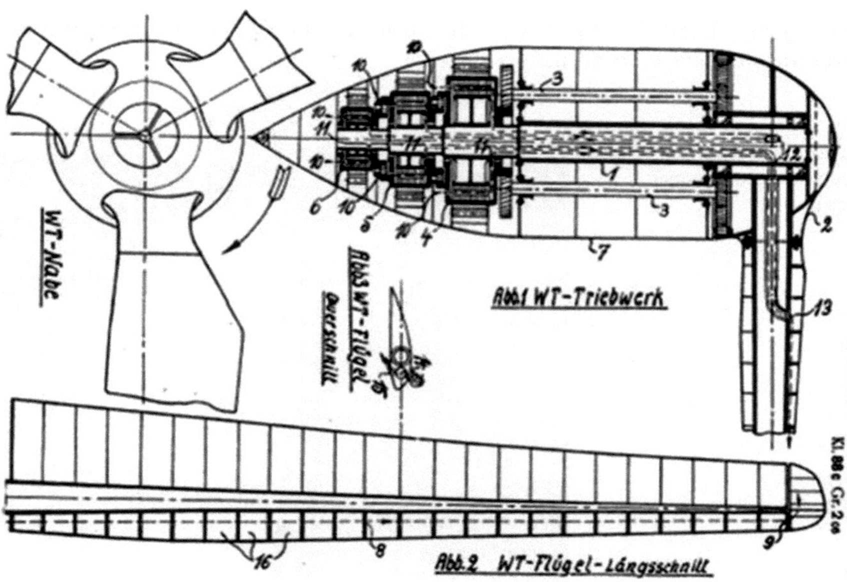

Fig. 5.10 First description of such an application for wind turbine rotors dates back to 1949 [10]

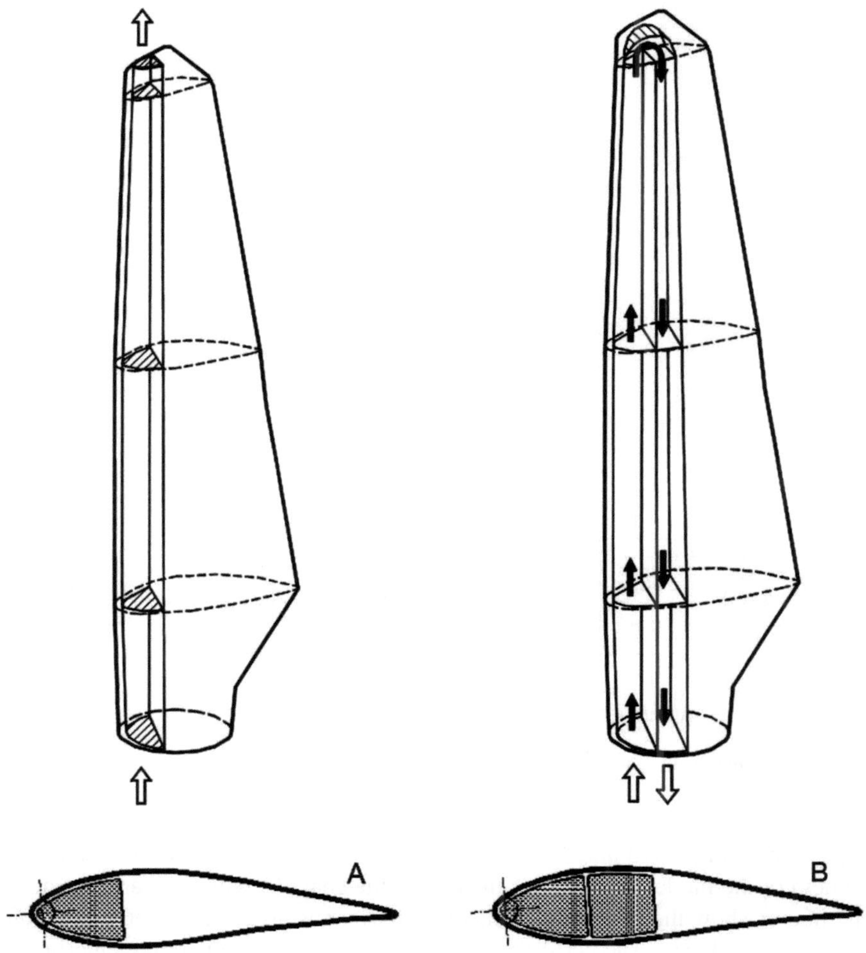

Fig. 5.11 Schematic of the open-circuit and closed-circuit systems [12]

rotor turning speed and the shape of the circuit (channel section, aspect ratio and blade length). For given ambient and operating conditions, the thermal performance of the system is thus determined by the air mass flow and the inlet air pressure and temperature [13].

In the closed-circuit system, two (or more) channels are created within the blade shell to enable the air to flow back from the tip towards the blade flange through return channels. The exhausted air, back from the tip, is then reheated and cyclically routed into the rotor blade. This solution needs a fan to establish the flow within the circuit, but it also provides ice protection on the blade aft portion [13, 14]. Where glaze icing is a concern, this system prevents runback freezing. Its thermal performance is better than the open channel one, despite its greater complexity and the need for

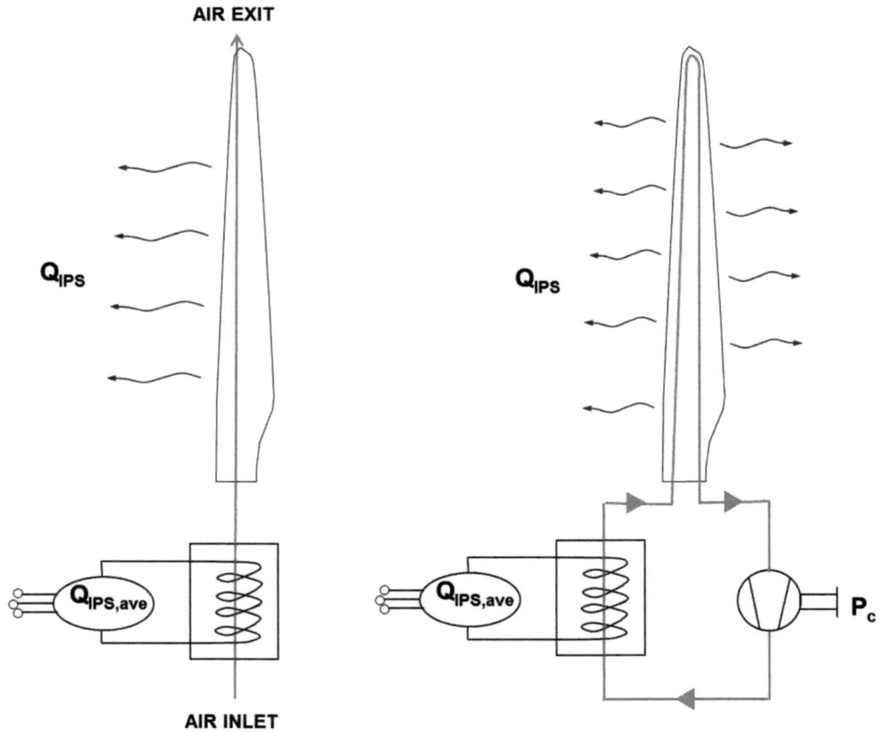

Fig. 5.12 Arrangement of the open-circuit and closed-circuit systems [12]

higher mechanical power to promote the air circulation. This is evident from the inspection of Figs. 5.13 and 5.14 which resembles several simulations (made with the help of the model described at the end of the chapter) with both arrangements. The figures show the operating limits of both systems, as the ambient temperature and wind speed vary. The performance depends on the air temperature entering the blade root. The results indicate the important limitation of the open-circuit system that would need very high hot air temperatures (read enthalpies) to operate even at moderate freezing temperatures. Instead, with realistic air blade inlet temperature compatible with material resistance, the closed-circuit allows extending theoretically the operational range of the turbine till −4 °C.

Only a few works are available in the published literature on this topic, and they almost always refer to aeronautical systems (see for all [15]), which have important differences compared to wind turbines because of the different materials and airfoil manufacturing techniques adopted. Even fewer papers are available on the related experimental issues, and they refer only to the use of electrical pads for anti-icing instead of internal air circulation [1]. The distribution of the air within the blade shell makes the control of the surface temperature a challenging task. Figure 5.15 shows the computed distribution of the wall external temperature for a generic blade

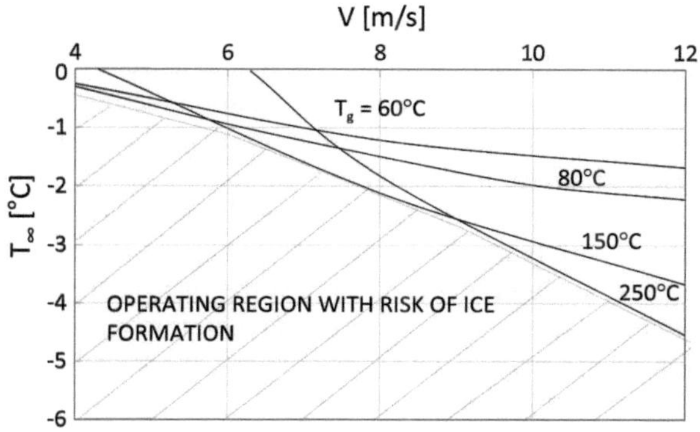

Fig. 5.13 Operational performance envelope of the open-circuit arrangement

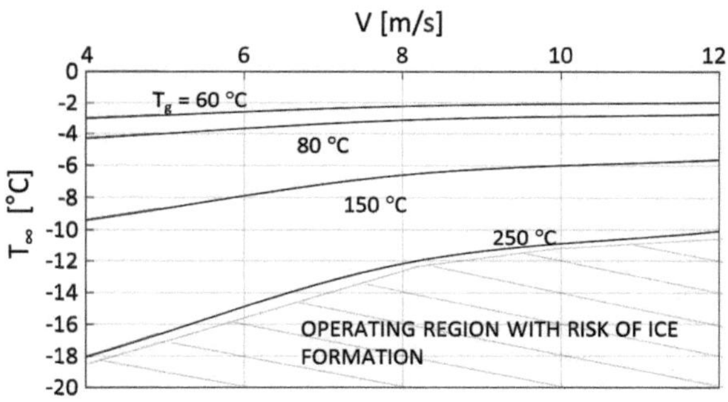

Fig. 5.14 Operational performance envelope of the closed-circuit arrangement

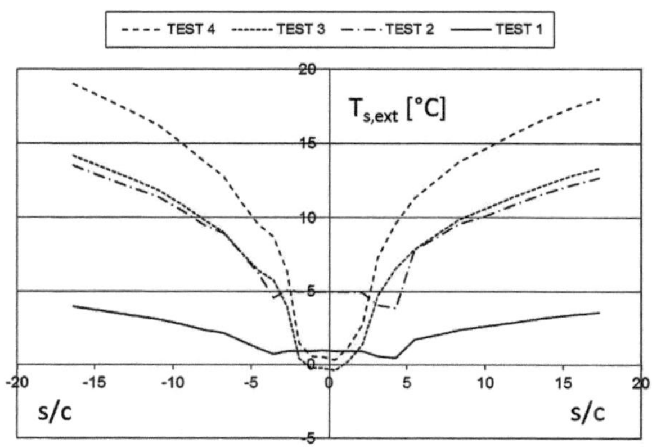

Fig. 5.15 Blade surface external temperature of a tip section: test 1 electrothermal pads, test 2-3-4 hot air internal flow increasing blade root air temperature

tip section at 0° AoA [12]. Test 1 refers to electrical heated foils while Tests 2-3-4 indicate hot air heating with increasing air root temperature.

The requirement of a minimum wall temperature at the blade LE leads to large temperature gradients in the aft part, moving the design goal away from the isothermal wall, with the consequence of much higher heat fluxes compared to electric heated foils. The claimed advantages of the hot air thermal system are that the LE area and in general the blade's aerodynamics are not affected by the IPS system and there are no negative effects on the lightning protection system. Disadvantages are due to the fact that the GRP material is a good insulator and therefore during high wind speeds or during rotation of the rotor at low temperatures, the heat losses due to forced convection will require very high heating power.

Although most of the aeronautic anti-icing systems are designed to evaporate a large part of the water caught by wings, such a strategy is not suitable for wind turbines due to the prohibitive amount of required heat (more than 20 kW/m^2).

The concept is currently adopted by Enercon Gmbh, as shown in Fig. 5.16. Enercon provides two different operating principles for the rotor blade ice prevention system, one for operating turbine (anti-icing) and the other for stand still one (de-icing). Each rotor blade has an individual de-icer system. A fan heater is installed at each blade root, feeding an electrical heat exchanger which heats up the air to a maximum of 72 °C. The inner blade shell creates a closed-circuit channel allowing the exhaust air returning to the root to be reheated, with the thermal efficiency benefits already mentioned.

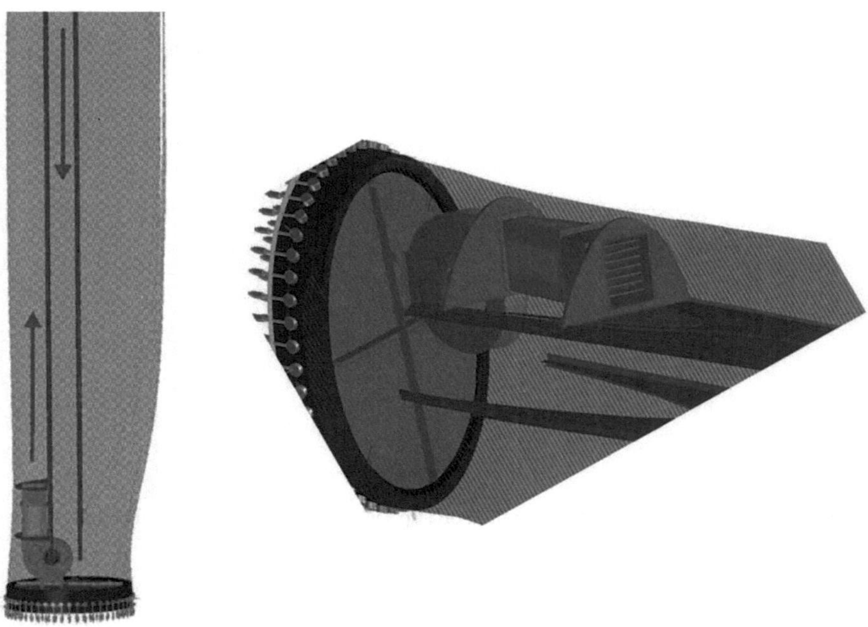

Fig. 5.16 Details of the in-duct air circulation of Enercon [16]

The power consumption per blade ranges from 12.2 kW for the about 20 m blade length of the E44 900 kW model to 23.8 kW for the about 40 m blade length of the E82 2,000–3,000 kW turbine.

The blade de-icing system can be operated either automatically and manually. In the automatic mode when ice is detected, the WT is shut down and the system is activated. The turbine restarts automatically after a defined heating period. De-icing operation with the turbine at standstill reduces the risk of ice throw. When de-icing is used during turbine operations, the system is claimed to detect ice and melt it at a very early development stage. If very harsh conditions are detected and ice still grows despite de-icing system activation, the turbine will be stopped and standstill de-icing will be used. An empirically determined tolerance range to ice is applied to the specific turbine. This is based on simulations and tests developed in several years. It is reported in [17] that this system is only activated once the turbine is at standstill. The rotor blade heating system does not prevent ice build-up, but considerably reduces the time required for the ice to melt. Turbines equipped with rotor blade de-icing restart after the heating period, which is usually several hours. If it is required that the turbine remains shut down once ice has been detected, the control system parameters can be set accordingly. In this case, the turbine can only be restarted manually.

In Fig. 5.17, a picture of a blade with de-icing on (left) and off (right) is shown.

The rotor blade heating is activated at (measured) external temperatures equal to or less than +2 °C, provided the control system is informed that ice has formed on the rotor blades. The power curve is contextually measured.

Figure 5.18 shows effects of a heating system used to provided warm-air from electrical heating transistors placed along the inner side of the leading edge of E70 turbines installed in the Wind Farm Moschkogellocated at 1,600 m a.s.l. in the Austrian Alps [18].

The system was replaced in summer 2008 and equipped with an amended heating system based on warm-air circulation inside the blade. The total heating power of that new system was about 70 kW per turbine. The detection of ice is made through

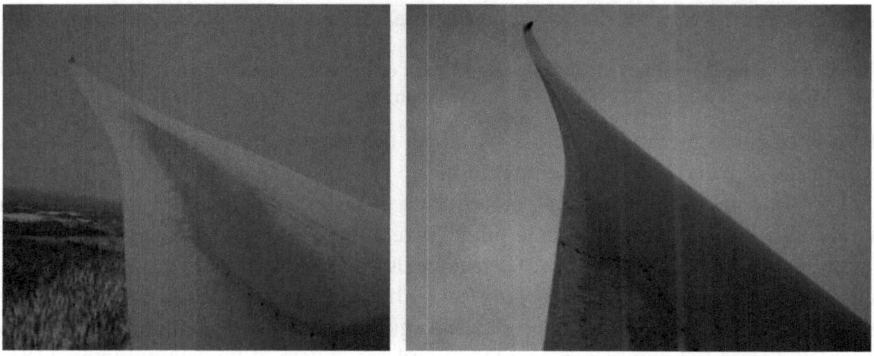

Fig. 5.17 Picture of a Enercon E82 blade with de-icing on (*left*) and off (*right*) [17]

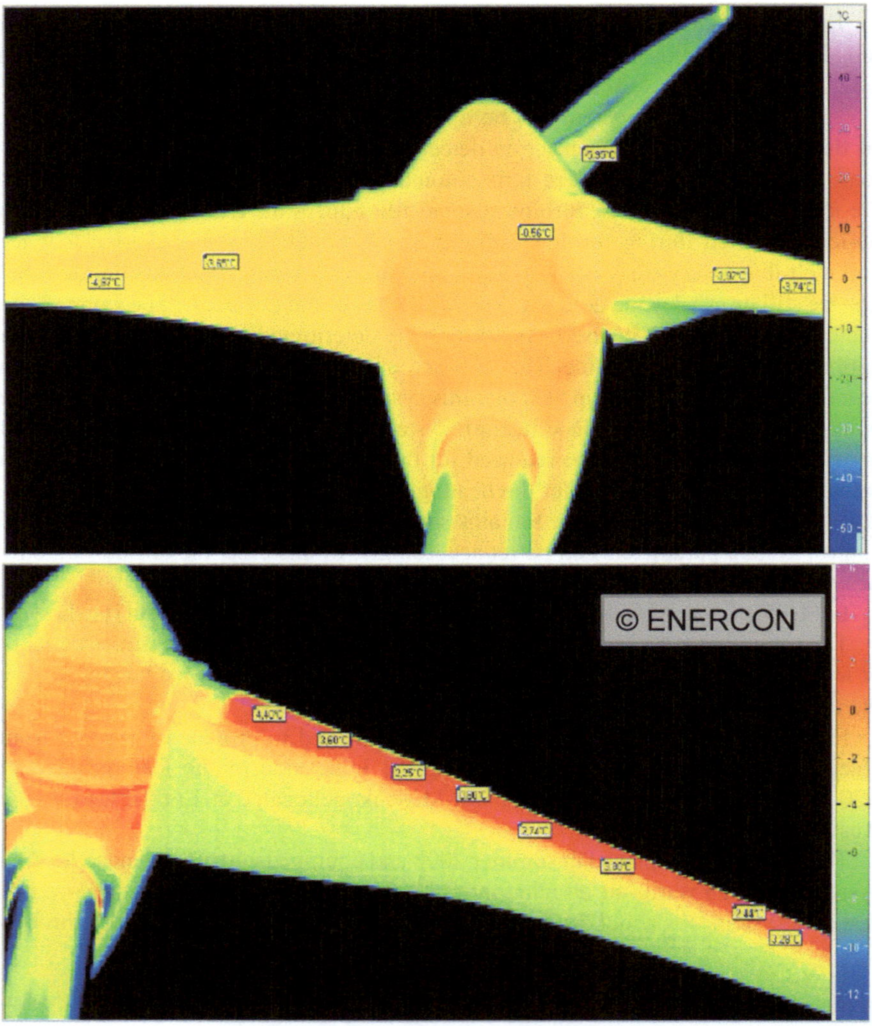

Fig. 5.18 Infrared picture of the turbine temperature of the Enercon turbine installed in the Wind Farm Moschkogel located at 1,600 m a.s.l. in the Austrian Alps [18]

the power curve method, which is based on the sensitivity of rotor blade profiles against ice-induced blade shape alteration.

The drop in WT operating performance is used to detect ice build-up (via wind/rotational speed/power/blade pitch correlation). A disadvantage of the power curve method is that it is not able to detect ice during standstill of the rotor. The calculation of production losses due to icing (and other turbine errors) is based on comparison of the actual 10-min energy production against the theoretical energy production calculated from the power curve and the wind speed data of the anemometers [18] .

Despite its simplicity, any system employing in-duct hot gas, presents the following weakness:

- high heat flux is needed due the high wall thermal resistance (unlike aeronautic field) and
- air promoting circulation systems are required.

Due to the low thermal conductive material used for blade shells ($\lambda_m < 1$ W/mK), and the thick walls adopted (form about 10 mm at the tip to about 70 mm at the root), the thermal problem is dominated form the heat resistance through the wall. Only a marginal benefit is therefore obtained (a part of the very tip regions) through enhanced internal air circulation (as used in aircraft wings or propellers blades). As a consequence, very high temperatures are required at the blade inner wall, which results in unacceptable working conditions for common GRP, preventing the system for a very efficient use at low external temperature (see Figs. 5.13 and 5.14). The use of more thermal conductive blade would meliorate the problem. This goal could be accomplished by means of one or more of the following strategies:

(a) drastic reduction of thickness;
(b) use of polymeric matrix charge variation;
(c) use of thermal bridges;
(d) optimised allocation of the thermal power.

The last option is very evident in the picture of Fig. 5.19 of the technical detail of an Enercon blade [19], where the piping of the hot air directing the thermal power flow towards the outer blade portions is shown.

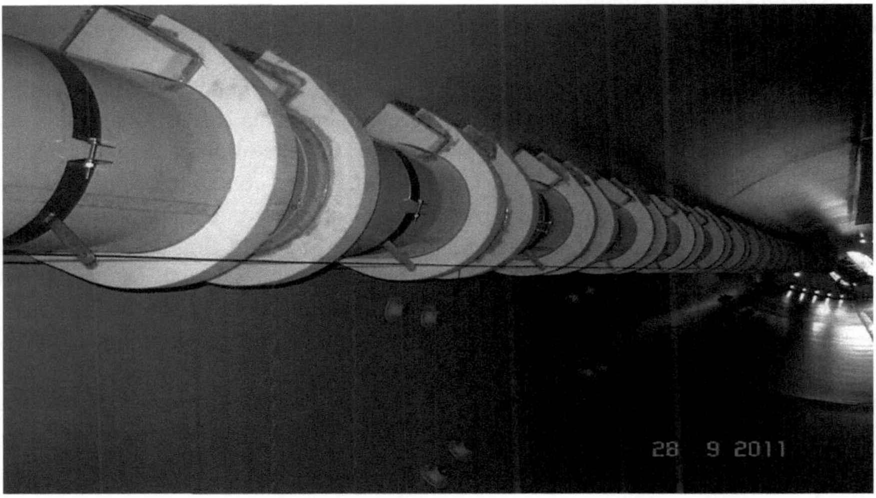

Fig. 5.19 Picture of a Enercon showing the piping of the hot air directing the power source towards the outer blade portions [19]

To conclude, thermal hot air circulation-based anti-icing systems, as will become evident in the simulation presented in the next paragraph, show the following characteristics:

- very high power density for anti-icing (1–10 kW/m^2);
- safe operations are provided in only moderate icing conditions. The system becomes ineffective for ambient temperatures below let us say −3 °C;
- applicable only for a small portion of blade surface to prevent excessive power requirement;
- when ice detection systems fails, or some delay in icing onset occurs, the available power for de-icing becomes inadequate to remove the ice formed.

5.8 The Design of an In-duct Hot Air Anti-icing System

As discussed, in the most simple architecture of in-duct hot air anti-icing systems, as illustrated in Figs. 5.11 and 5.12, the heat is supplied by a stream of hot air flowing within the blade shell and is discharged at the tip. The air is heated up by a dedicated electrical heater, or in alternative, by collecting the heat released from the electrical generator. As the air enters the blade root, it is conveyed within the blade shell, warming up the blade wall by a convection/conduction mechanism.

Due to the characteristics of the system considered, i.e. a wind turbine blade, some general assumptions can be made:

- the climate and meteorological conditions are kept constant during wind turbine operations in icing conditions.
- the heating strategy considers not ice tolerant operations, so the wall is maintained wet.
- the problem of determining the extension of the blade wetted surface is solved by a particle trajectory analysis carried out by a lagrangian approach. Since anti-icing operation is not based on a ice-tolerant strategy, and the impingement rate on the profile is not changing in time (the surface remains not contaminated by ice), the collision efficiency is computed only once and considered invariant with time.
- due to the thick insulating wall of the blade, the heat conduction through the wall is essentially 1D. This feature has been confirmed by a 2D analysis carried out with ANSIS and is shown in the following sections.
- The flow is modelled as a 2D stationary flow past the blade and the inner flow is 1D and the air motion is a fully-developed flow regime.
- The warm-air evolving in the blade shell is following a polytropic process and the polytropic exponent is computed segment by segment.

The thermodynamic conditions evolve from blade station to blade stations, by March ing in space from blade root to tips sections, matching sector-by-sector internal and external fluid dynamic and thermal boundary conditions. The resulting model is a quasi 3D stationary thermofluid dynamic model. The solution gives the external surface temperature along the blade for assigned ambient conditions and turbine operating conditions, and the anti-ice thermal power to be supplied to each blade.

The solution of the whole model is accomplished by the following steps:

1. definition of the blade geometry (geometry module),
2. determination of the thermal and flow field past the blade (thermofluid dynamic module),
3. determination of the conjugate heat transfer (thermal module).

5.8.1 The Geometry Module

As schematically shown in Fig. 5.20, each rotor blade is divided into a series of segments in the radial direction (from i = 1 to i = N). Each segment is in turn divided into a series of chordwise panels (from j = 1 to j = M). One or more partitions can be generated inside the shell so as to delimit the hot air internal pattern according to the internal layout scheme.

5.8.2 The Thermofluid Dynamic Module

The external flow field is modelled using a 2D approach. A panel code builds up the flow field past the body to have pressures and velocity distributions past the

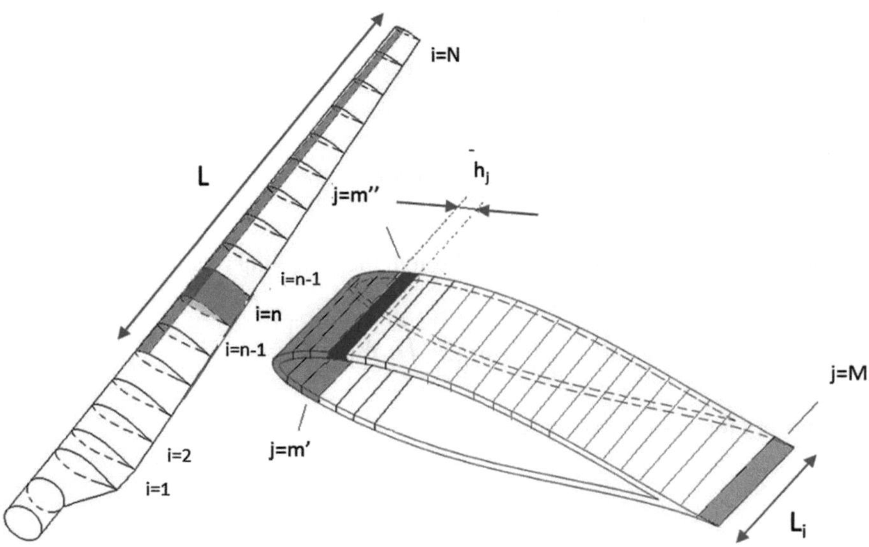

Fig. 5.20 Blade geometry discretization

profiles at various sections and different incidence angles. Integral boundary method is used from the data generated by the panel code to obtain the pressure $p_e^{i,j}$ of each panel/segment spanwise/chordwise position at the edge of the boundary layer. The velocity v_e and temperature T_e at the edge of the boundary layer are thus computed by the usual thermodynamic relationships, already introduced in Chap. 4:

$$p_e = p_\infty + \frac{1}{2}\rho_\infty v_\infty^2 \left[1 - \left(\frac{v_e}{W_\infty}\right)^2\right]$$

$$T_e = T_\infty^0 \left(\frac{p_e}{p_\infty^0}\right)^{\left(\frac{\gamma-1}{\gamma}\right)}$$

$$T_\infty^0 = T_\infty \left(\frac{p_\infty^0}{p_\infty}\right)^{\left(\frac{\gamma-1}{\gamma}\right)}$$

$$p_\infty^0 = p_\infty + \frac{1}{2}\rho_\infty W_\infty^2$$

The transition point from laminar to turbulent boundary layer is inferred by the application of the integral boundary layer approach taken from Kays and Crawford [20]. It occurs at:

$$Re_\theta > 1.74 \left(1 + \frac{22{,}400}{Re_s}\right) Re_s^{0.46}$$

where Re_θ and Re_s are respectively referred to the momentum thickness and to the quantity evaluated at the coordinate s measured from the stagnation point:

$$Re_\theta = \frac{W_e\,\theta}{\mu_{air}}$$

$$Re_s = \frac{W_e\,s}{\mu_{air}}$$

The momentum thickness is evaluated through the following relationship:

$$\theta = \left[\left(0.45\frac{\mu_{air}}{W_e^6}\int_s^0 W_e^6 ds\right)\right]^{\frac{1}{2}}$$

The recovery temperature is computed as:

$$T_{rec} = T_e + r\frac{W_e^2}{2c_p}$$

5.8.3 The Conjugate Heat Transfer Module

This module determines the radial temperature and pressure evolution of the internal air stream flowing within the blade shell. The internal flow is assumed to have a uniform radial velocity distribution at any station, and the fully-developed within the channel. This hypothesis is supported by the consideration that there is a relatively long entering part of the blade (up to about 10–20 % of the blade length) where the flow can develop, such a part, being aerodynamically not profiled, is not taken into account in the computation. If the radial segments are sufficiently short, a constant polytropic exponent can be used for the pressure-volume process. The polytropic relationship allows calculating the local segment pressure from the temperature data.

Three control volumes are employed to model the conjugate heat transfer problem, i.e. volumes I, II and III, as shown in Fig. 5.21. For each of them, the mass and energy conservation principles are written under the hypothesis of a steady-state and fully-developed flow.

5.8.3.1 Volume I

For each i-station, and for each j-panel, the mass conservation states that:

$$\dot{m}_{w,imp} + \dot{m}_{w,in} - \dot{m}_{w,out} - \dot{m}_{ev} = 0 \tag{5.2}$$

The mass flow of standing water due to surface tension effects, and the mass flow of shedding water are here neglected. The energy conservation equation is given by:

$$\begin{aligned}[\dot{m}_{w,in}c_w\left(T_{w,in} - T_{ref}\right) - \dot{m}_{w,out}c_w\left(T_{w,out} - T_{ref}\right) \\ + \dot{m}_{w,imp}c_w\left(T_{w,imp} - T_{ref}\right) + \dot{m}_{w,imp}\frac{W^2_{w,imp}}{2} - K_A\dot{m}_{ev}h_{ev}^{T=T_{s,ext}} \\ - h_{c,ext}\left(T_{s,ext} - T_{rec}\right)]A_{ext} + \frac{1}{R}\left(T_{s,int} - T_{s,ext}\right)A_m = 0\end{aligned} \tag{5.3}$$

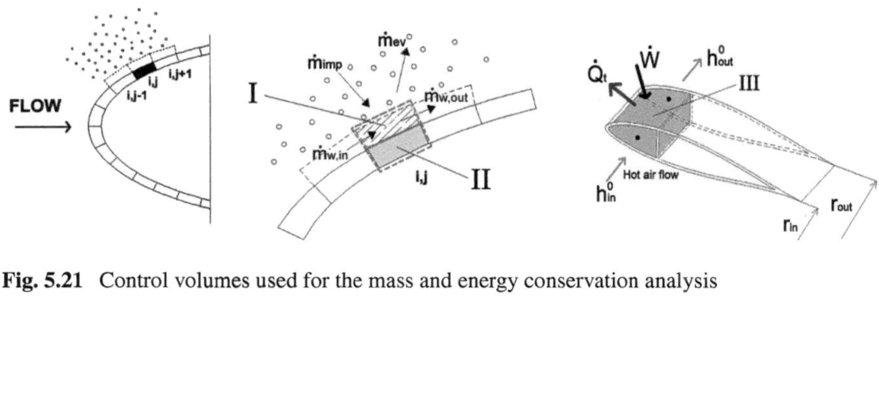

Fig. 5.21 Control volumes used for the mass and energy conservation analysis

As already discussed in Chap. 4, the model can neglect radiant heat contributions (due to long-wave and short-wave radiation) because of their discontinuous presence and negligible magnitude compared to the other contributions. For the sake of brevity, Eq. (5.3) is summarised by the following heat balance equation:

$$\Delta \dot{Q}_{ice} + \frac{1}{R}\left(T_{s,int} - T_{s,ext}\right) A_m = 0 \tag{5.4}$$

5.8.3.2 Volume II

The energy conservation for volume II yields:

$$h_{c,int}\left(\bar{T}_{a,m} - T_{s,int}\right) A_{int} = \Delta \dot{Q}_{ice} \tag{5.5}$$

5.8.3.3 Volume III

The mass conservation for volume III yields:

$$\dot{m}_{a,in} = \dot{m}_{a,out} = \dot{m}_a \tag{5.6}$$

The energy conservation equation applied to volume III is written as:

$$\dot{m}_a\left(h^0_{in} - h^0_{out}\right) - \Delta \dot{Q}_{ice} = 0 \tag{5.7}$$

By splitting the total enthalpy difference, we have:

$$h^0_{out} - h^0_{in} = u_{out} - u_{in} + \frac{u^2_{r,out} - u^2_{r,in}}{2} + \frac{u^2_{t,out} - u^2_{t,in}}{2} + \frac{p_{out}}{\rho_{out}} - \frac{p_{in}}{\rho_{in}}$$

In incompressible flows, the term $u_{out} - u_{in}$ corresponds to the friction losses ΔW_d and can be computed by:

$$\Delta W_d = \frac{L}{D_h} f \frac{\left(\frac{W_{out} + W_{in}}{2}\right)^2}{2}$$

f is the friction factor, given by:

$$f = a + b\left(Re_{D_h}\right)^c$$

being a, b and c constants.

The evolution of pressure inside the channel segments is given by relating temperature and pressure by the polytropic process. The relationship is given by:

$$p_a T_a^{\frac{m}{m-1}} = \text{const.} \tag{5.8}$$

and the polytropic exponent m is calculated through:

$$m = \frac{c_p - c_m}{c_v - c_m} = \frac{c_p - \left(\frac{\Delta \dot{W}_d}{dT} - \frac{\Delta \dot{Q}_{ice}}{dT}\right)}{\frac{c_p}{k} - \left(\frac{\Delta \dot{W}_d}{dT} - \frac{\Delta \dot{Q}_{ice}}{dT}\right)} \tag{5.9}$$

The following operational constraints are set for the solution of the problem:

- for the open-circuit (one way)

$$(T_{s,ext}^{i,j} - T_{min}) > 0$$
$$(T_{max} - T_{s,int}^{i,j}) > 0$$
$$p_{out}^N - p_\infty = 0$$

- while for the closed-circuit it holds:

$$(T_{s,ext}^{i,j} - T_{min}) > 0$$
$$(T_{max} - T_{s,int}^{i,j}) > 0$$
$$0.1 - Ma > 0$$

T_{min} represents the minimum external surface temperature preventing ice formation, while T_{max} is the maximum allowable material working temperature for continuous operations.

The problem is solved iteratively by imposing a guess temperature $T_{in}^{1,j}$. For each panel the local value of the heat transfer coefficient is thus calculated. The external convective heat transfer coefficient is evaluated through integral boundary layer calculation model [20, 21] introduced in Chap. 4.

The internal convective heat transfer coefficient is evaluated by means of the Petukhov, Kirillov and Popov [22] relationship.

$$h_{c,int} = \frac{k_{air}}{D_h} \frac{\frac{f}{8} Re_{D_h} Pr}{1.07 + 12.7 \left(\frac{f}{8}\right)^{\frac{1}{2}} (Pr^{\frac{2}{3}} - 1)} \tag{5.10}$$

where f is given by:

$$f = \left(1.82\, log_{10} Re_{D_h} - 1.64\right)^{-2} \tag{5.11}$$

Re_{D_h} is the Reynolds number evaluated with the hydraulic diameter given by:

$$D_h = \frac{4A_P}{P}$$

being A_P the surface of the area of the section and P the wetted contour.

A local equivalent thermal resistance $R^{i,j}$ was assumed for the composite blade wall assembly and the contact resistances. The equivalent transverse thermal conductivity for hollow fibre-reinforced composites was estimated through the relationship given by Halpin-Tsai [23].

5.8.4 The Rate of Intercepted Water

The local impingement has been computed through a particle trajectory analysis as described in Chap. 4. This model superimposes the flow field solution around the profile, obtained by a panel flow solver which is followed by the reconstruction of the streamlines to the results of water particle motion.

This model neglects the effect of the migration of the water film towards outboard sections due to the centrifugal effect to which the film is subjected to.

5.8.5 Design Results

The Tjæreborg wind turbine [24] blade platform has been adopted with hypothesis to realise an open channel inside the hollow blade shell by inserting internally a partitioning rib slightly aft the leading edge. The rotor operational features and anti-icing working conditions are listed in Tables 5.1 and 5.2, respectively. Operating conditions at the wind turbine rated speed were chosen.

Table 5.1 Input data for the simulations (ambient and turbine)

Parameter	Value
p_∞ (Pa)	99,000
T_∞ (°C)	−2.0
LWC (g/m^3)	0.2
MVD (μm)	20.0
Humidity (−)	0.99
Wind velocity (m/s)	13.79
Rotational speed (rpm)	22.15

Table 5.2 Input data for the simulations (open-circuit system)

Parameter	Value
$T_{a,max}$ (°C)	130
$\dot{m}_a$ (kg/s)	1
t_{min} (mm)	10
t_{max} (mm)	50
λ_{mat} (W/mK)	0.5
A_c (m^2)	0.025

Table 5.3 General blade data of Tjæreborg wind turbine [24]

Section	r/R (−)	Radius (m)	Profile type	c (m)	t (m)	t/c (%)	β_{twist} (°)	ϕ_{flow} (°)	AoA (°)	W (m/s)
0	0.00	0.00	Circular	1.80						
1	0.05	1.46	Circular	1.80						
2	0.09	2.75	Circular	1.80						
3	0.10	2.96	Circular	1.80						
4	0.21	6.46	NACA 4430	3.30	1.009	30.58	8.00	36.16	30.16	19.35
5	0.31	9.46	NACA 4424	3.00	0.723	24.10	7.00	26.23	21.23	25.32
6	0.41	12.46	NACA 4421	2.70	0.571	21.15	6.00	19.41	15.41	31.66
7	0.51	15.46	NACA 4418	2.40	0.449	18.71	5.00	15.32	12.32	38.19
8	0.61	18.46	NACA 4416	2.10	0.353	16.81	4.00	13.03	11.03	44.80
9	0.70	21.46	NACA 4415	1.80	0.278	15.44	3.00	11.40	10.4	51.50
10	0.80	24.46	NACA 4414	1.50	0.216	14.40	2.00	10.12	10.12	58.25
11	0.90	27.46	NACA 4413	1.20	0.16	13.33	1.00	8.87	9.87	65.04
12	0.95	28.96	NACA 4412	1.05	0.134	12.76	0.50	7.95	9.45	68.43
13	0.98	29.86	NACA 4412	0.96	0.122	12.74	0.20	6.82	8.62	70.45
14	1.00	30.50	NACA 4412	0.94	0.118	12.59	0.16	3.23	5.07	71.92

The geometry of the Tjæreborg wind turbine is given in Table 5.3 and Fig. 5.22. The inner channel has an average hydraulic diameter of 0.145 m and the channel area was set to 0.250 m^2. The evolution of the wall thickness in the LE region and the inner and outer surfaces areas are shown in Fig. 5.23. The evolution of static pressure p_a and temperature T_a in the inner channel is depicted in Fig. 5.24.

For the most representative sections, the impingement water is shown in Fig. 5.25. As can be noted, accordingly to the model presented in Chap. 4, the middle section (Section 7) is only marginally wetted compared to the outboard Sections (9–13).

Figure 5.26 shows the temperatures of the inside and outside wall in the impingement area for the section positioned at r = 15.460 m (Section 7), r = 21.460 m (Section 9) and r = 28.960 m (Section 13) together with the external heat transfer coefficient $h_{c,ext}$ and overall heat transfer coefficient U, defined as:

$$U = \frac{1}{\dfrac{1}{h_{c,ext}A_{ext}} + \dfrac{1}{2\lambda_{mat}}\ln\left(\dfrac{A_{int}}{A_{ext}}\right) + \dfrac{1}{h_{c,in}A_{int}}}$$

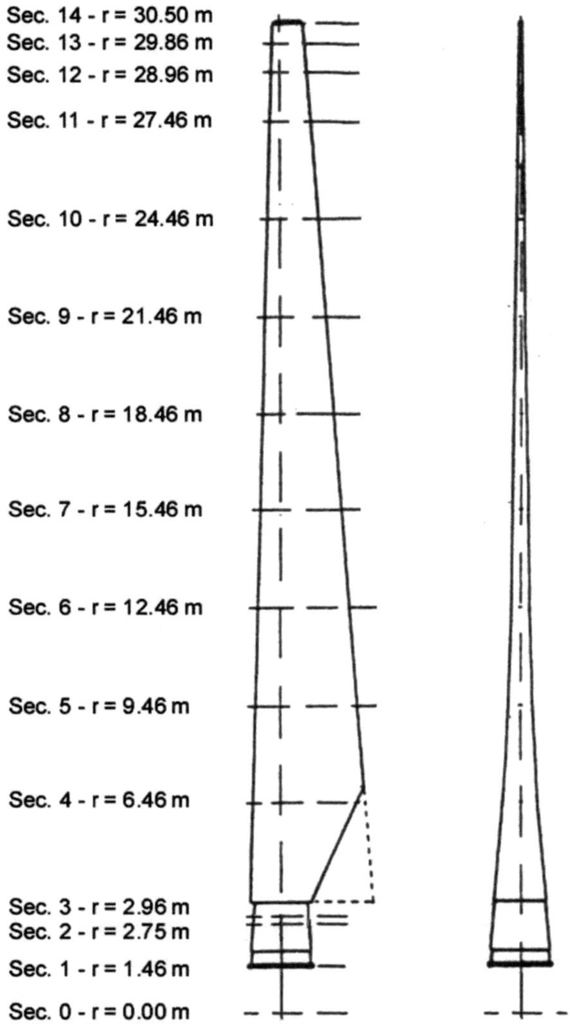

Fig. 5.22 Tjæreborg wind turbine blade planform

Figure 5.27 shows in detail the main heat fluxes contribution as functions of the nondimensional curvilinear coordinate s/c for the tip section (Section 13).

The sensible heat flux has a negligible contribution. Since the heat exchange is dominated by the heat convection, the minimum wall temperature is reached where the convective heat transfer coefficient is maximum (at the leading edge region), while the maximum is obtained at the stagnation point where heat convection is minimum and water mass transfer heat fluxes are negligible (see Fig. 5.26). The effect of the thinner blade profile and the higher relative flow velocity of the tip is to produce an higher amount of water collection and more intense convective heat transfer coefficient.

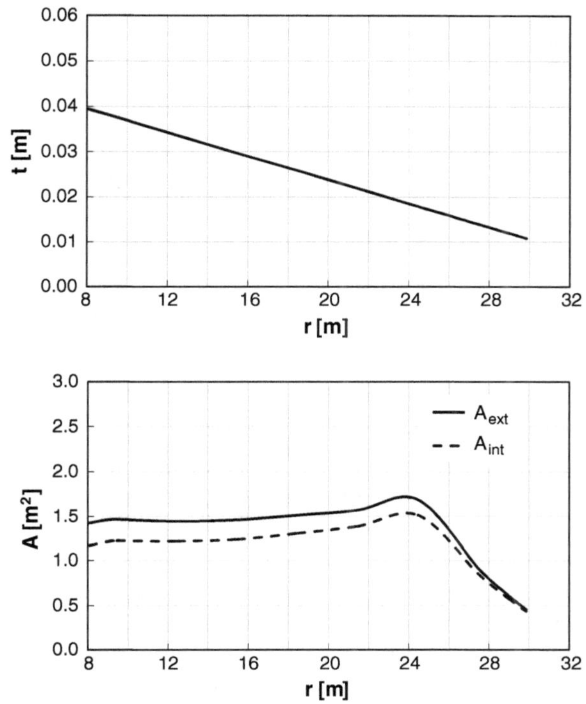

Fig. 5.23 Layout of the blade geometry

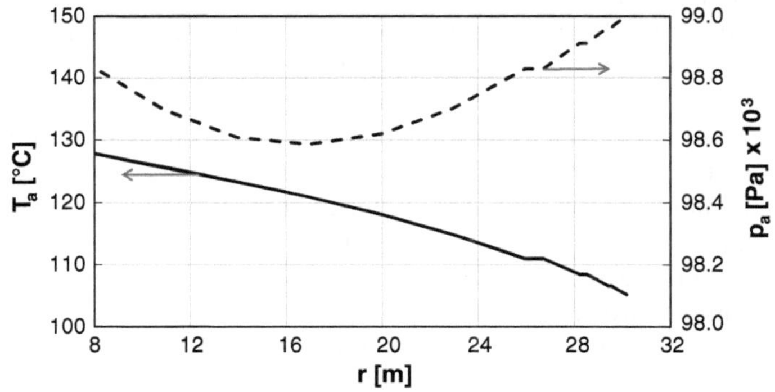

Fig. 5.24 Evolution of static pressure p_a end temperature T_a in the inner channel

Here, however, a complex temperature pattern develops on the suction surface since the mass heat transfer contribution becomes comparable with the other ones. The increased heat flux, compared to one of the inboard section, is responsible for the generalised wall temperature reduction (more relevant at the leading edge).

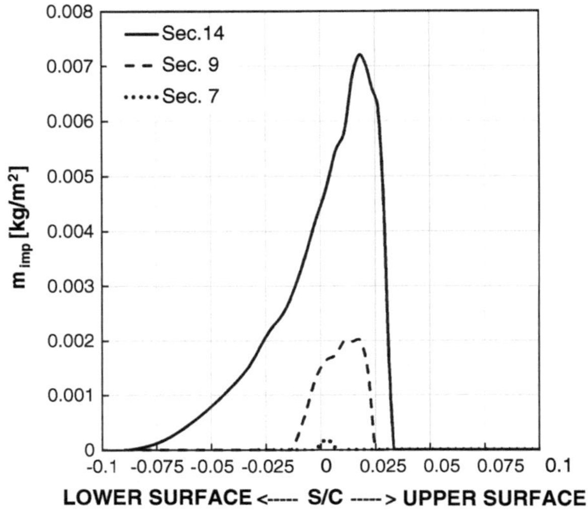

Fig. 5.25 Impingement mass per unit area for Sections 7, 9 and 13

Figure 5.28 shows the comparison of the net specific heat fluxes for sections r = 15.460 m (Section 7), r = 21.460 m (Section 9), and 28.960 m (Section 13).

The net heat flux requirement along the blade, i.e. the thermal power to be provided by the anti-ice system is shown, section by section, in Fig. 5.29. As expected, tips regions are dominated by more intense net heat fluxes. About 27.6 kW are required for one blade.

Figure 5.30 shows the blade external wall temperature averaged on the eight panels around the impingement region for each section. The wall external temperature shows a dropping trend from the root regions towards the outwards ones until the surface from dry becomes wetted. From this point on it falls toward a minimum close to the blade tip because of the increasing wall cooling effect. As mentioned, tip losses at the very tip section reduce such cooling and a small wall temperature recovery after the point of minimum is observed. The nonlinear increase of the average heat flux is also highlighted. With 1 kg of air ingested at the blade root, the system operational parameters are summarised in Table 5.4.

The blade minimum external temperature T_{min} results to be about one degree below freezing threshold. But this is a single one point on the contour of the tip section, and since lateral conduction is neglected, the average temperature computed over a minimum set of panels at leading edge is more informative about actual conditions as Fig. 5.30 shows. The blade acts as a very poor heat exchanger, the temperature drop is of only 24.8 °C and to avoid external wall temperature to drop below 0 °C, an inlet air temperature at the root section of about 130 °C is needed, which determines the upper operating temperature for composites material mechanical resistance (about 80 °C).

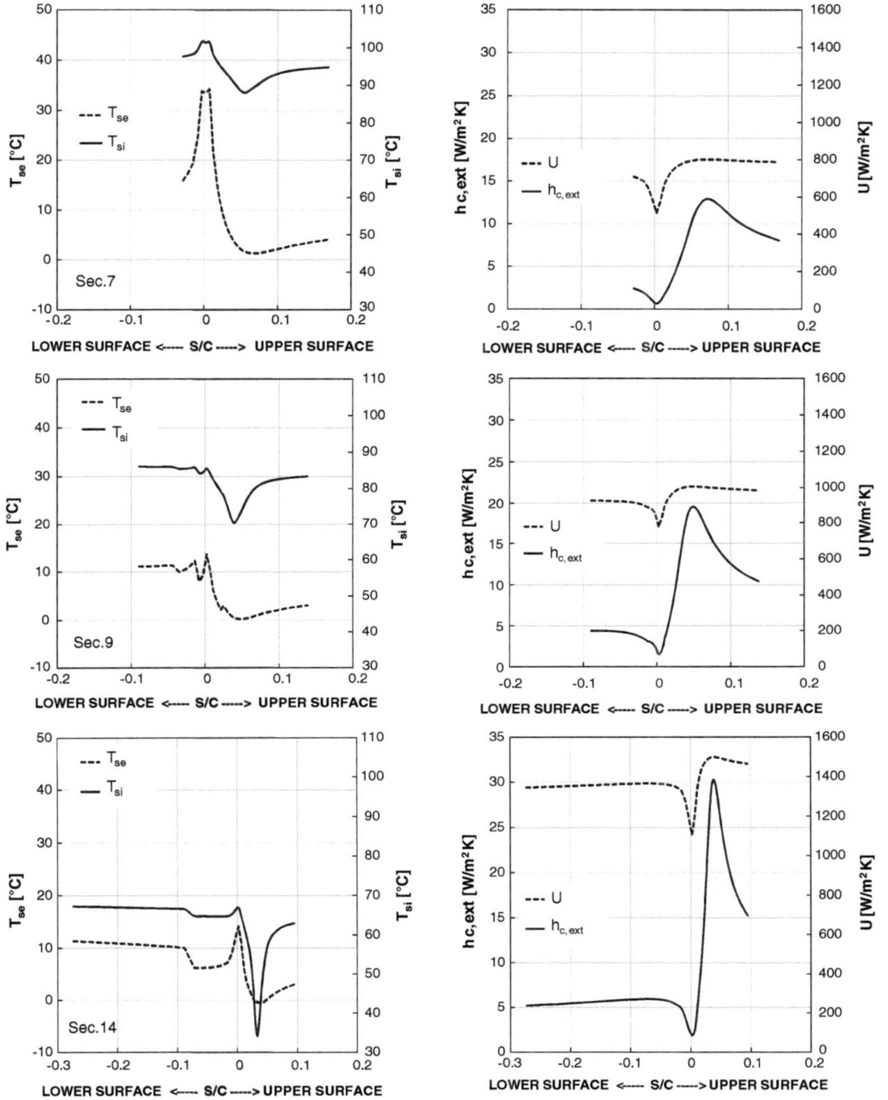

Fig. 5.26 Temperatures of the external and internal wall in the impingement area for the sections positioned at r = 15.460 m (Section 7), r = 21.460 m (Section 9) and r = 28.960 m (Section 13), external heat transfer coefficient $h_{c,ext}$ and overall heat transfer coefficient U

On the basis of this simulation a total thermal power of about 83 kW is required for the anti-icing of the investigated three-bladed rotor configuration. The mechanical power needed to circulate the hot air, if a global electric efficiency of 0.7 is considered for the system, is about 1.2 kW.

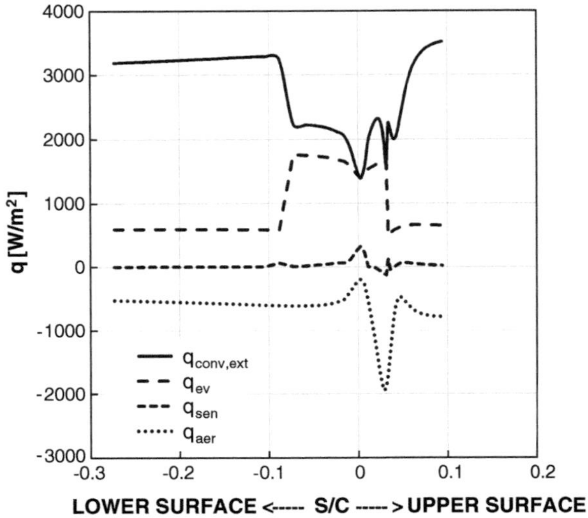

Fig. 5.27 Main specific heat fluxes for the tip section (Section 13) as function of the nondimensional curvilinear coordinate s/c

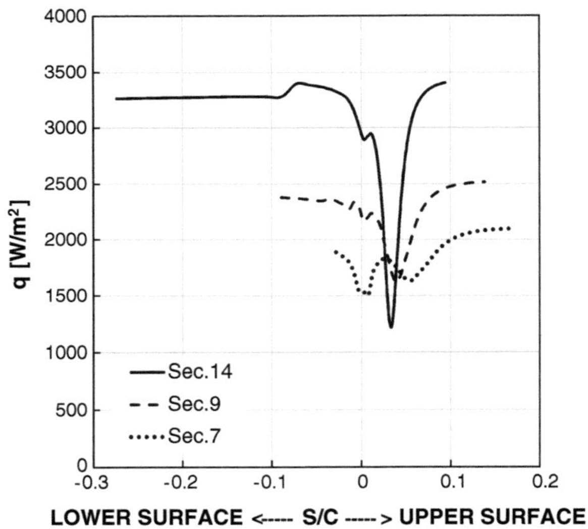

Fig. 5.28 Specific net heat fluxes for sections r = 15.460 m (Section 7), $r = 21.460$ m (Section 9) and r = 28.960 m (Section 13)

To check the accuracy of the assumption of 1D conduction into the blade wall, additional simulations were carried out with ANSYS® on 2D numerical domains that reproduce actual blades sections. An equivalent heat transfer coefficient was set as external boundary condition of the domain, according to:

$$h_{eq,ext}^{i,j} = \frac{\Delta \dot{Q}_{IPS}^{i,j}}{A_{ext}^{i,j}(T_{s,ext}^{i,j} - T_{\infty})}$$

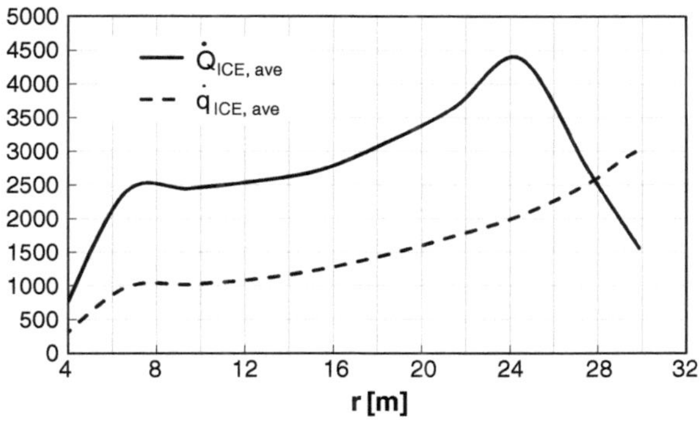

Fig. 5.29 Section averaged anti-icing specific heat flux and total thermal power

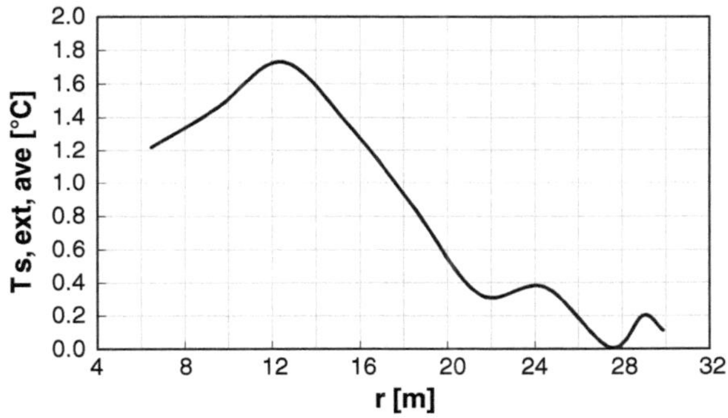

Fig. 5.30 Blade external wall temperature averaged on the eight panels around the impingement region

Table 5.4 Output data of the simulations

Parameter	Value
Number of stations	14
$\dot{Q}_{IPS,ave}$ (per blade)	27.56 (kW)
$\dot{Q}_{IPS,ave}$ (per 3 bladed rotor)	82.78 (kW)
Blade exchanger thermal efficiency	0.18
T_{min}	−0.713 °C
T_{max}	130.9 °C
T_{out}	105.2 °C
p_{out}	99,001.4 Pa
ΔT	24.8 °C
Δp	99,853 Pa

Here the local temperature and heat flux were computed through the 1D code. The internal heat transfer coefficient is the same as the one used for the 1D code. Figure 5.31 shows the 1D to 2D comparison of the external wall temperature distribution with nondimensional curvilinear coordinate. As expected, due to the low materials thermal conductivity a good agreement is found where small temperature gradients occur (at the leading edge region), while some discrepancies arise at the

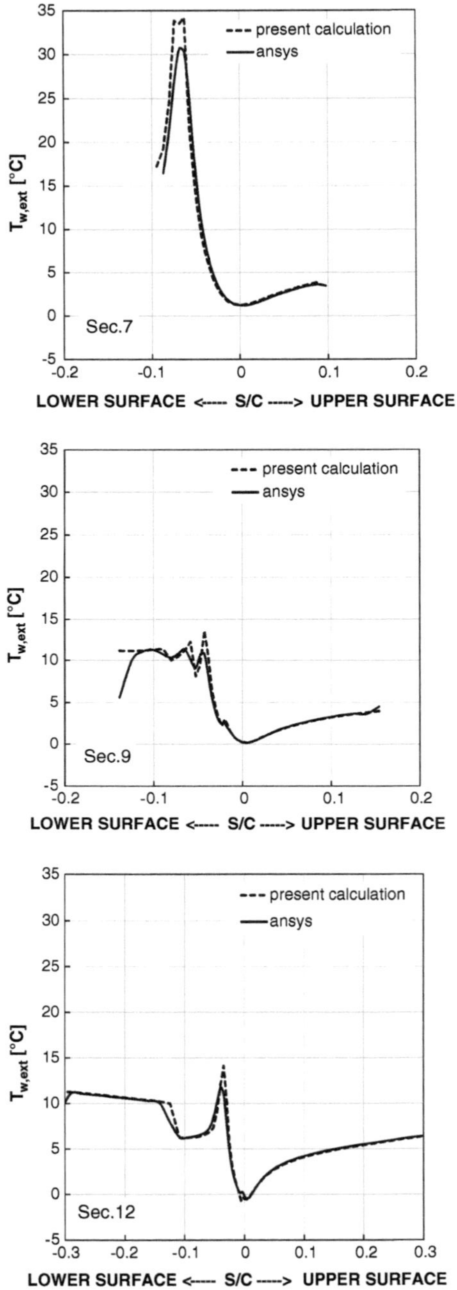

Fig. 5.31 Comparison of the wall external temperatures computed with the present model and with ANSYS® for the sections positioned at r = 15.460 m (Section 7), r = 21.460 m (Section 9), and r = 28.960 m (Section 13)

stagnation point where minor 2D conduction effects were revealed. Since these discrepancies only affect a few volumes around the stagnation point, the influence on the anti-icing specific heat flux evaluation is marginal: a maximum heat flux difference of about 2 % at the blade hub is revealed, where however poor heat exchange occurs (see Fig. 5.26).

5.9 The Energetic Efficiency of an IPS

The presence of an ancillary system as the IPS introduces an energetic penalty in AEP. During IPS operations, the energy harvest is:

$$E_{ice,\,ave} = \int_{T_{heat-on}} P\,dt = \int_{T_{heat-on}} \frac{1}{2}\rho\, V^3\, C_P(V)\, \eta_m\, \eta_{el}\, \eta_{aux}\, \eta_{IPS}\, dt \tag{5.12}$$

The period of operation of the IPS, $T_{heat-on}$, is given by Eq. (5.1). This duration may coincide or not with the icing duration. This depends upon the anti-icing strategy adopted, and the efficacy of the ice detection systems to correctly switch on and off the IPS.

The efficiency of an IPS can be defined as the ratio between the thermal power required from the IPS and the average electric power delivered in the icing period:

$$\eta_{IPS} = \frac{\dot{Q}_{IPS,\,ave}}{P_{ice,\,ave}} \tag{5.13}$$

For an electrothermal system, the actual average heating power of the IPS is given by the ratio of the ideal average heating power (as to keep the surface at the prescribed temperature) and the heating efficiency, which takes into consideration the heat losses into the wires and towards the blade back wall:

$$\dot{Q}_{IPS,\,ave} = \frac{\dot{Q}_{ice}}{\varepsilon_{IPS}} \tag{5.14}$$

As hot air systems are of concern, the IPS heating efficiency is computed by taking into consideration that the surface temperatures differ from the (minimum) required. The efficiency includes also the mechanical power needed to promote the hot air flow within the system and the actual average heating power of the IPS is written as:

$$\dot{Q}_{IPS,\,ave} = \frac{\dot{Q}_{ice}}{\varepsilon_{IPS}} + \overline{P}_m \tag{5.15}$$

Table 5.5 states that the maximum efficiency is achieved with heating foils, while for hot air IPS a considerable amount of heat is wasted because of the practical inability to obtain the theoretical minimum surface temperature for an optimum heat distribution.

Table 5.5 Average thermal efficiencies of IPS

IPS principle	ε_{IPS}
Heating foils	0.9
Hot air, closed channel	0.6
Hot air, open channel	0.3

5.10 A Simplified Approach for Estimating the Anti-icing Power Requirement

This paragraph, on the basis of the model described in the preceding chapter, introduces a rapid method for the preliminary estimation of the anti-icing power and average energy requirement.

As stated previously, anti-icing thermal power depends on several variables, which can be conveniently grouped as: meteorological, wind turbine characteristics and operation, and ice prevention system characteristics variables. The following functional relationship can be written:

$$\begin{aligned} \dot{Q}_{IPS,ave} &= f\left(\underbrace{V,\ T,\ LWC,\ MVD,}_{meteorological}\ \underbrace{Z,\ \text{blade planform},\ \text{control},\ \lambda_{mat},}_{turbine}\ \underbrace{T_{s,\,min},\ A_{heat},\ \eta_{IPS}}_{ice\ prevention\ system}\right) \\ E_{IPS,ave} &= g\left(\dot{Q}_{IPS,ave},\ t_{ice}\right) \end{aligned} \tag{5.16}$$

For any set of values of the variables expressed by Eq. (5.16) conductive to actual icing events, the anti-ice thermal power can be calculated by means of suitable models [2, 4], which involve basically the steps listed in Fig. 5.5. With reference to the body discretization scheme of Fig. 5.20, the thermal power required to protect the ice-free surface of the rotor is expressed as:

$$\dot{Q}_{ice} = Z \sum_{i=1}^{N} \dot{q}_i\, L_i\, H_i \tag{5.17}$$

being Z the number of blades, L_i the spanwise length and H_i the chordwise length of the heated panel, and

$$\dot{q}_i = \frac{\sum_{j=M'}^{M''} \dot{q}_{ice,j}}{\sum_{j=M'}^{M''} A_j} \tag{5.18}$$

$\dot{q}_{ice,j}$ is the specific heat flux supplied to the blade wall at the single panel.

The anti-ice power requirement is proportional to the heated area. The extension of the heated part either in spanwise and chordwise direction should be set by consideration involving both the length of the impingement (deriving from the water impingement calculation) and a safety coefficient (k_s) based on engineering experience (e.e.). The heated chordwise extension is therefore given by:

$$H_i = \underbrace{k_s}_{e.e.} \left(\underbrace{s_u + s_l}_{droplet\ traj.\ calc.} \right) \tag{5.19}$$

The starting section of the spanwise length depends on the risk connected to icing of the root section.

In Eq. (5.18) the term $\dot{q}_{ice,j}$ is the chordwise specific heat flux exchanged by the j-panel of the i-station of the blade, while the summation gives the spanwise heat flux of the N stations. $\dot{q}_{ice,j}$ is obtained by solving for each j-panel the mass and energy balance given by (see Chap. 4 on icing process for further details).

Determining $\dot{Q}_{ice}$ from $\dot{q}_i$ is rather complex and tedious, since each contribution has to be computed and integrated over the surface. Some simplifications can be obtained from information derived from a sensitivity analysis aimed to highlight the variables which mainly influence the anti-icing heat flux to be supplied to the leading edge area. For this analysis two significant turbine parameters (i.e. the imposed surface temperature $T_{s,ext}$ and the distance r/R of the section of blade from the rotational axis), and six environmental parameters (i.e. air temperature, site altitude, wind speed, liquid water content, average droplet diameter and relative humidity) have been chosen. As will be evident after, the section of the blade located at the distance of about $r/R = 70\,\%$ represents in average the specific heat flux $\dot{q}_{ice}$ of the whole blade. Therefore referring to the a blade which characteristics are reported in Table 4.2 and to the ranges relative to a "baseline condition" as in Table 5.6, a reasonable variability for the parameters has been explored on the heat fluxes contributes. This variability is indicated, together with the baseline value, at the foot of each graph.

The results of the analysis are shown in Fig. 5.32, where the heat fluxed have been grouped in cooling and heating ones and the legend at the figure bottom helps the interpretation of each case.

Table 5.6 Range of the variables used for the sensitivity analysis and baselines conditions

Parameter	Set (−)	Baseline	Set (++)
T_s (°C)	1	2	3
T_∞ (°C)	−1	−3	−5
V (m/s)	9	13.79	17
R_h (m/s)	0.81	0.9	0.99
LWC (g/m^3)	0.10	0.15	0.20
MVD (μm^3)	10	0.30	50
z (m)	0	1,000	2,000
R (m)	15	30.5	45

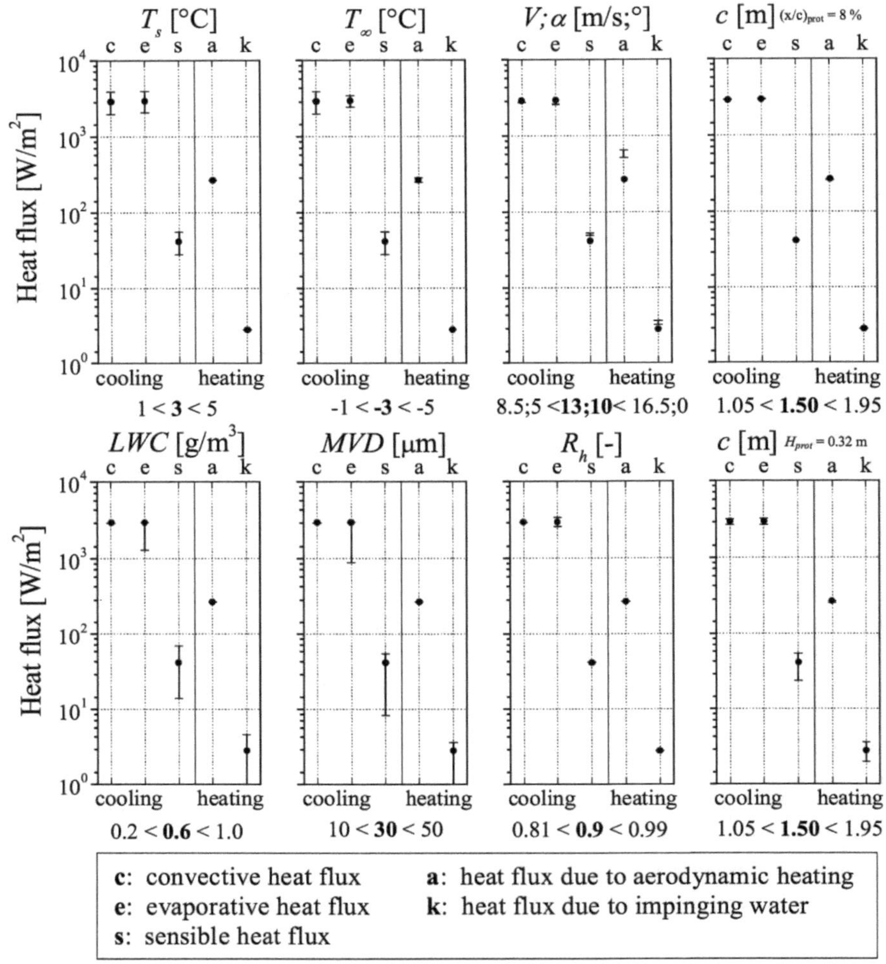

Fig. 5.32 Sensitivity analysis of relevant heat fluxes at the blade LE

The relevant comments to Fig. 5.32 are:

- there is a great variability in the magnitude of the single heat flux that the blade leading edge exchanges with the outside environment;
- the higher heat fluxes are those of convection and evaporation;
- the sensible heat flux for high values of the parameters LWC and MVD can reach far from negligible values, whereas the impact heat flux can be completely disregarded;

- the two parameters that affect much the magnitude of all cooling heat fluxes and consequently also the value of the anti-icing heat flux requirement are the imposed surface temperature and the temperature of free stream air;
- the effect of wind speed is negligible above the rated value, while under this value (if a full-variable machine is of concern) it contributes to diminish both the cooling heat fluxes and the heating heat fluxes;
- the effect of the site altitude were found to be practically negligible (small variations in the heat fluxes depend on the convective heat transfer drop due to Reynolds number).

The graphs of Figs. 5.33 and 5.34, with respect to the "baseline condition", show in a more suitable manner the effect of the imposed external surface temperature, and free stream air temperature, respectively, on the heat fluxes contributes.

Moving from this analysis, a simplified version of Eq. (4.104) can be deduced, that considers the more relevant contributions:

$$\dot{q}_{ice,j=1} = \dot{q}_{conv} + \dot{q}_{ev} + \dot{q}_{sens} - \dot{q}_{kin} - \dot{q}_{aer} \tag{5.20}$$

where:

$$\dot{q}_{conv} = h_c \left(T_{s,ext} - T_\infty\right)$$

$$\dot{q}_{ev} = \frac{0.622\, h_c\, 2.5 \times 10^6}{c_{p,air}\, p^0_\infty L^{2/3}} 27.03 \left(T_{s,ext} - T_\infty\right)$$

$$\dot{q}_{sens} = \beta_{3D,0}\, LWC\, (\Omega\, r)\, c_w \left(T_{s,ext} - T_\infty\right)$$

$$\dot{q}_{kin} = \beta_{3D,0}\, LWC\, (\Omega\, r)\, \frac{W^2}{2}$$

$$\dot{q}_{aer} = h_c \frac{r\, W^2_\infty}{2 c_{p,air}}$$

Therefore Eq. (5.21)

$$\dot{q}_{ice,j=1} = \dot{q}_{conv}\, (1 + \alpha_{ev} + \alpha_{sens} + \alpha_{aer}) \tag{5.21}$$

where the coefficients α_{ev}, α_{sens}, and α_{aer} are taken from the bottom Figs. 5.33 and 5.34.

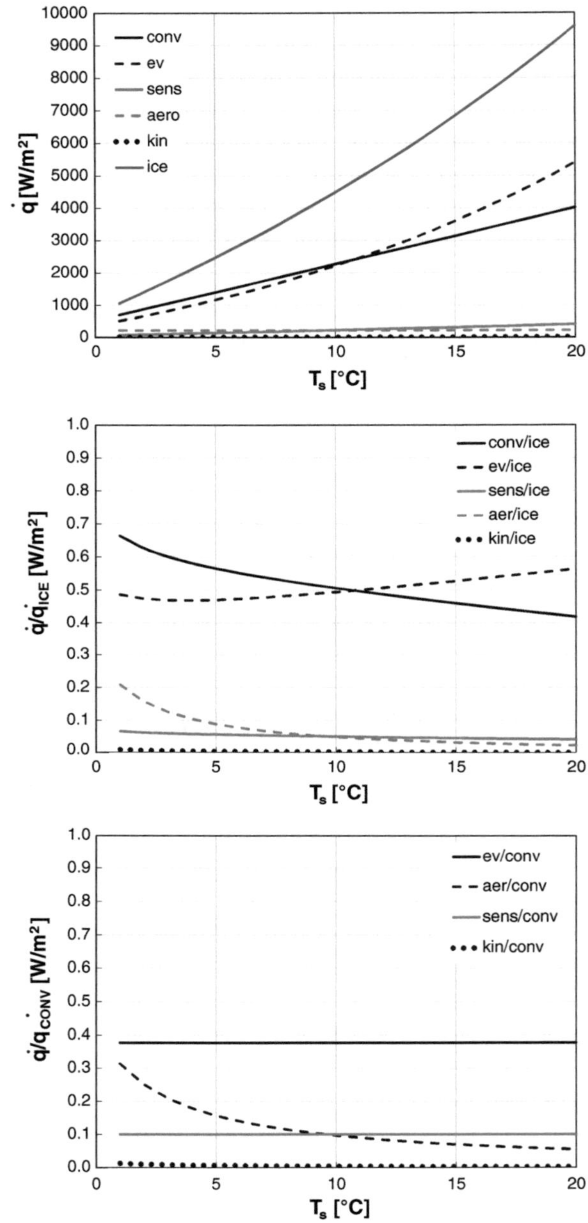

Fig. 5.33 Variation of the specific LE heat fluxes (*upper*) with the imposed external surface temperature ($T_\infty = -3\,°C$), normalised to the net heat flux at the LE $\dot{q}_{ice,j=1}$ (*middle*) and normalised to the LE convective heat flux at the LE $\dot{q}_{conv}$ (*bottom*)

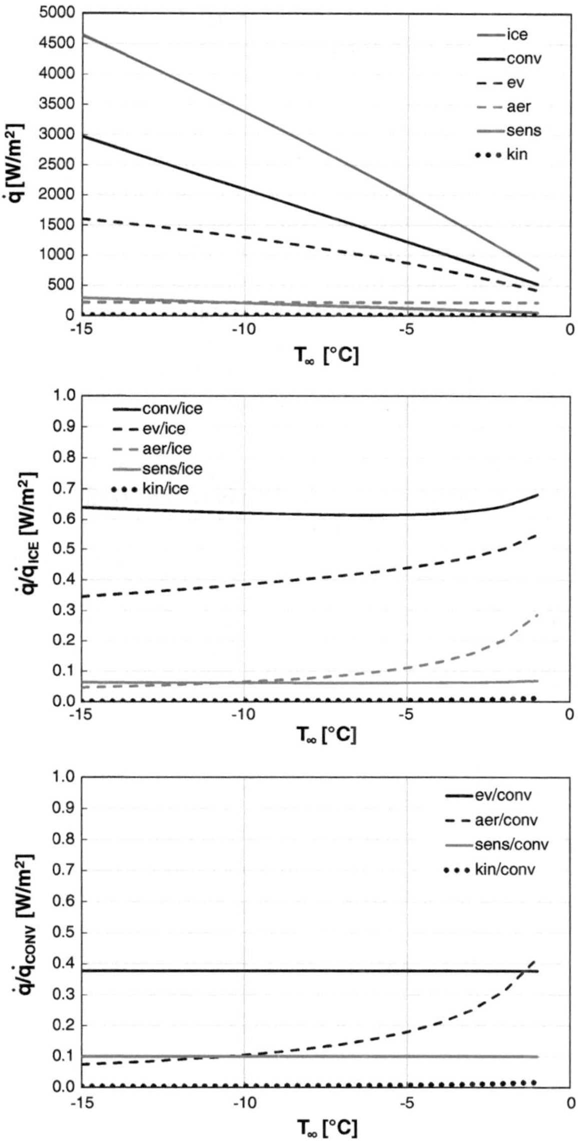

Fig. 5.34 Variation of the specific LE heat fluxes (*upper*) with the ambient temperature, normalised to the net heat flux at the LE $\dot{q}_{ice,j=1}$ (*middle*) and normalised to the LE convective heat flux at the LE $\dot{q}_{conv}$(*bottom*), $T_s = +2\,°C$

5.10.1 *Assessment of the Anti-icing Heat Requirement of Different Types of Turbines*

The simple model presented indicates that due to the increasing relative air velocity along the blade radius, the anti-icing heat flux is expected to increase from the root to the tip of the blade, with an approximately linear law as results in Fig. 5.35 which show the variation of the specific anti-icing heat flux $\dot{q}_{ice}(W/m^2)$ along the blade radius. The average is located approximately at 70 % section of the blade

With the help of the simplified model, it is worthwhile to investigate the variability of the thermal anti-icing power in relation to the type (1, 2 or 3 blade) and size of the turbine. A first comparison has been made between a two-bladed and a three-bladed wind turbines, rated at the same power output of 1,000 kW. The main characteristics of those two machines are shown in Table 5.7 and are taken from manufacturers free literature data.

Furthermore, with the goal to extend the comparison with a one-bladed turbine, three wind turbines having a rated power output between 300 and 350 kW where considered, by having the same rotor diameter and number of blades equal, respectively, to one, two and three. The main characteristics of these three turbines are listed in Table 5.8. Also for these turbines the data has been taken from manufacturer's free available data. Tables 5.7 and 5.8 indicate also the extension of the heated spanwise length along the blade radius L, and the collective width of the heated zone H, used for the simulations.

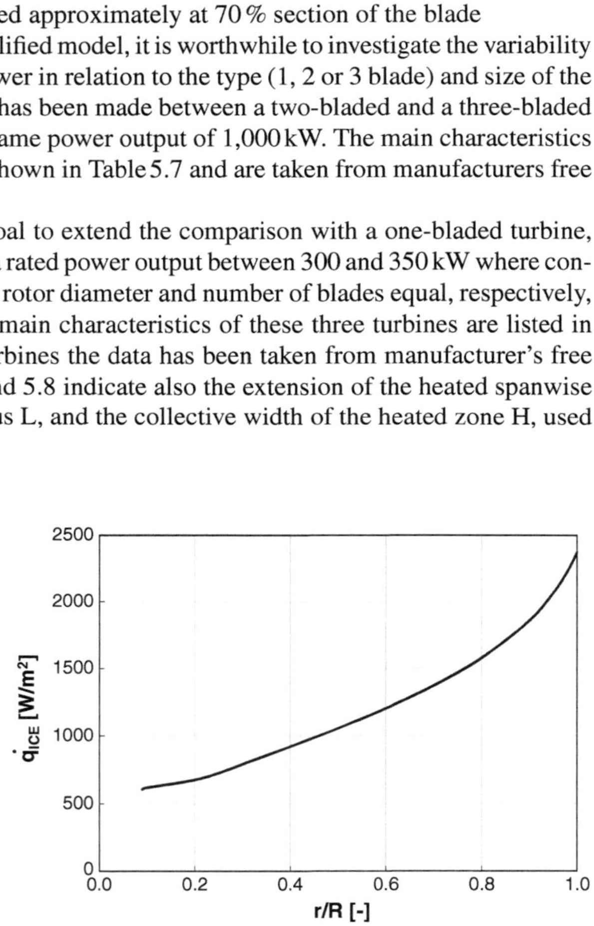

Fig. 5.35 Distribution of the net heat flux $\dot{q}_{ice}$ (W/m^2) as a function of the dimensionless blade radius

Table 5.7 Variables used for a comparison of the thermal anti-icing power between two-bladed and three-bladed 1 MW size turbines

Turbine	P_R (kW)	D (m)	Z (−)	Ω (rpm)	W (m/s)	Airfoil type	Length of the protected area L (m)	Height of the protected area H (m)
m1	1,000	54	3	22	15	NACA 63xxx	20.2	0.3
m2	1,000	54	2	22	15	NACA 63xxx	20.2	0.3

Table 5.8 Data used for a comparison of the thermal anti-icing power between one-bladed, two-bladed and three-bladed turbines of medium size

Turbine	P_R (kW)	D (m)	Z (−)	Ω (rpm)	W (m/s)	Airfoil type	Length of the protected area L (m)	Height of the protected area H (m)
m3	300	33.4	3	31	14	NACA 44xx	12.5	0.3
m4	320	33	2	41.2	13	NACA 44xx	12.5	0.3
m5	350	33	1	57	12	NACA 44xx	12.5	0.3

Table 5.9 Data used for a comparison of the thermal anti-icing power between one-bladed, two-bladed and three-bladed turbines of medium size

T_s (°C)	T_∞ (°C)	p_∞ (Pa)	Rh (−)	LWC (g/m^3)	MVD (m)
+2	−3	89,870	0.99	0.1	20

The anti-icing thermal power requirement $\dot{Q}_{ice}$ [W] for the considered wind turbine have been calculated by using Eqs. (5.17) and (5.21), with respect to the tabled rated operational condition. The site variables assumed in the simulations are those shown in Table 5.9.

The external surface temperature is set to be at +2 °C with the task to realise a running wet condition. Figures 5.36 and 5.37 show that the higher optimal rotational speed (at rated power) of the two-bladed turbine leads to a direct increase in the cooling thermal fluxes. At the same time, however, there is also an increase in the heat flux due to aerodynamic heating, so there is actually no significant difference as number of blades varies in terms of the specific heat flux. However, the anti-icing thermal power requirement $\dot{Q}_{ice}$, which depends on the number of blades, nonetheless decreases considerably from a three-blade to a single-blade turbine, in a manner that is nearly proportional to the number of blades.

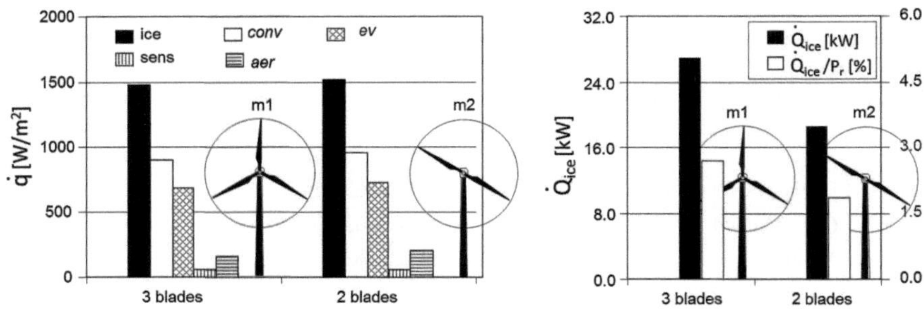

Fig. 5.36 $\dot{q}_{ice}$ (kW/m^2) for the blade 70 % section, anti-icing thermal power requirement for the wind rotor $\dot{Q}_{ice}$ (kW) and ratio between rotor anti-icing thermal power and rated turbine's power (Table 5.7)

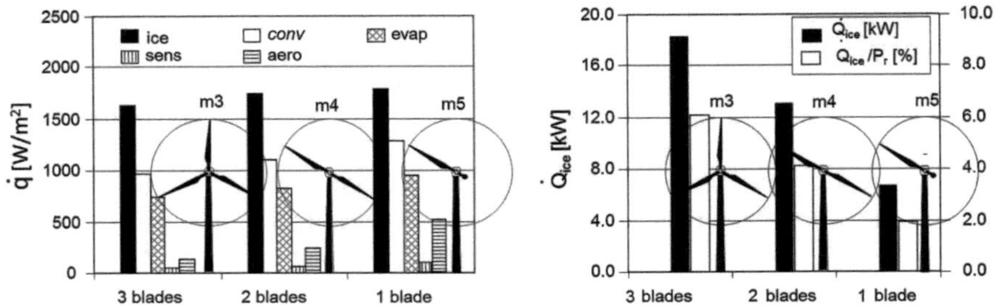

Fig. 5.37 $\dot{q}_{ice}$ (kW/m^2) for the blade 70 % section, anti-icing thermal power requirement for the wind rotor $\dot{Q}_{ice}$ (kW) and ratio between anti-icing power requirement and rated turbine's power (Table 5.8)

Again by considering the Figs. 5.36 and 5.37, we can note that when the specific anti-icing heat flux $\dot{q}_{ice}$ of the two three-bladed machine of different size is compared, this parameter increases as the machines size drops; this is true as well for the two-bladed machines. When turbines with the same rated power are considered, the reduction of the number of blades leads to a decrease of the ratio of rotor anti-icing thermal power to rated turbine power. If a different rated power for the same number of blade is considered, the larger turbines are less penalised in terms of total anti-icing requirement compared to smaller ones. Please note that the figures are not taking into consideration the IPS efficiency, and the actual anti-icing thermal fluxes and the ratios to the rated power shall be corrected by the values indicated in Table 5.5.

This analysis has been extended to six three-bladed turbines of different rated output, whose main characteristics are listed in Table 5.10. The data are taken from the open manufacturer literature. For the six turbines, the same width of the heated zone is assumed, according to the fact that as the turbine size decreases, decreases as well the chord dimension while the water collected by the airfoil increases. The heated spanwise length have been chosen equal to the profiled length of the blade.

Table 5.10 Turbines data used for a comparison of the anti-icing thermal power with respect to rotor's size

Turbine	P_R (kW)	D (m)	Z (−)	Ω (rpm)	W (m/s)	Airfoil type	Length of the protected area L (m)	Height of the protected area H (m)
M1	300	33.4	3	31	14	NACA 63xxx	12.5	0.3
M2	600	44	3	27	15	NACA 63xxx	15.5	0.3
M3	1,000	54	3	22	15	NACA 63xxx	20.2	0.3
M4	1,300	62	3	19	15	NACA 63xxx	23.2	0.3
M5	2,000	76	3	17	15	NACA 63xxx	28.5	0.3
M6	2,300	82.4	3	17	15	NACA 63xxx	30.9	0.3

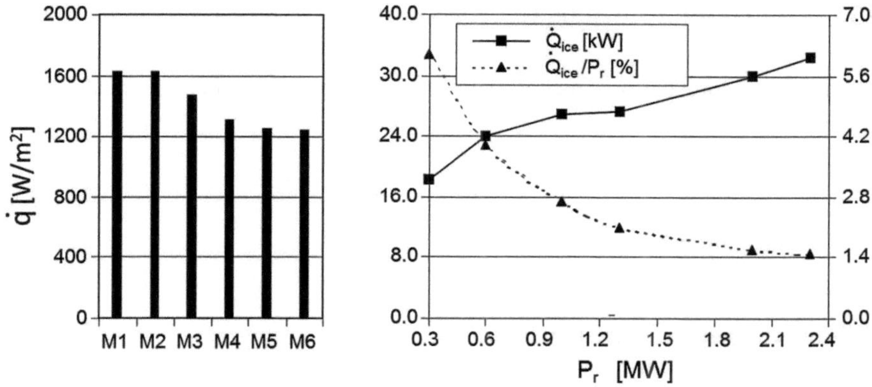

Fig. 5.38 $\dot{q}_{ice}$ (kW/m^2) for the blade 70 % section, anti-icing thermal power requirement for the wind rotor $\dot{Q}_{ice}$ (kW) and ratio between anti-icing power requirement and rated turbine's power $\dot{Q}_{ice}/P_r$ (Table 5.10)

According to Fig. 5.38, as the rotor size increases, the specific heat fluxes at 70 % section tend to become asymptotic due to the progressive balance of cooling an heating contribution.

As for anti-icing power, the power produced increases with increasing machine sizes, but so does the length of the blade and consequently also the surface to heat since the hypothesis of constant heated surface area per unit blade length). As shown in Fig. 5.38, however, if the ratio of anti-icing power to rated output is analysed for the turbines considered, this ratio appears to be unfavourable for the small size turbines.

As the above results in terms of anti-icing power distribution have been developed regardless the ice prevention system technology adopted, the use of the results for the design of the latter must be carefully evaluated. The IPS thermal efficiency η_{IPS} plays a basic role in defining the actual power requirement. This means that the results in terms of anti-icing power distribution for different turbines type and size, can be with a marginal error, directly extended only to electrothermal IPS, since the efficiency can be considered not far from unity, as Table 5.5 shows. On the other hand, if the ice prevention system is based on hot air circulation, the extension of the results is restricted to the comparison between turbines of the same size. No straightforward conclusions can be drawn from the comparison for turbines of different size because of the different blade geometrical feature and the different off-design operations. For this situation, a detailed analysis should be dedicated. Three-blade turbines of MW size, depending on environmental conditions, require an anti-icing power for a running wet system of 2–8 % of the machine's rated output, and this output is confirmed by the data gathered from commercial IPS systems.

5.11 Emerging Solutions for IPSs

Among the ice prevention concepts, some new solutions have been proposed in the last years:

1. mechanical
   - pneumatic de-icer system,
   - electroimpulsive/expulsive devices;
2. thermal
   - electromagnetic radiation heating,
   - intermittent heating,
   - regenerative heating,
   - effusive heating;
3. low ice-adhesion surfaces.

5.11.1 Mechanical

5.11.1.1 Pneumatic De-icing System

Small airplanes often use mechanical de-icing systems by means of so-called inflatable rubber boot on the LE of the wings and control surfaces although such systems lead to aerodynamic disturbance and additional noise. Goodrich developed the technology depicted in Fig. 5.39. In the normal non-inflated state, the tubes adhere to the airfoil surface to which the de-icer is bonded. After the build-up of a controlled layer of ice, the de-icers are inflated with compressed air. The inflation cycle lasts for a few seconds to achieve optimal ice shed and prevent additional ice formation on the inflated surface.

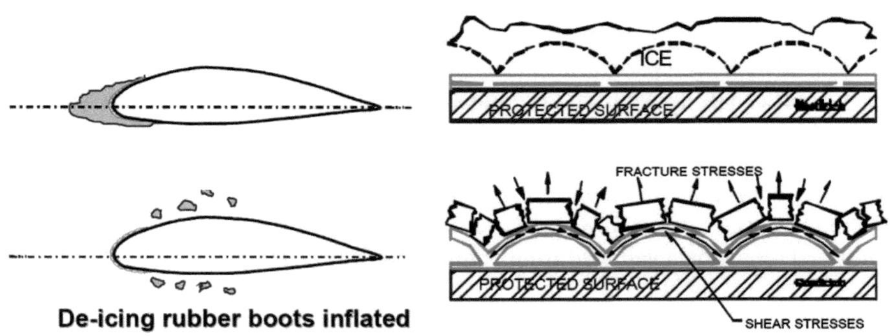

Fig. 5.39 Working principle of inflatable rubber boots [25]

Fig. 5.40 Pneumatic de-icing system, picture of the experimental rig [25]

The ice cracked is thus naturally removed through centrifugal and aerodynamic forces. The de-icer is then allowed to deflate by evacuating the air into atmosphere, and after by applying vacuum the system ensures that there is no lifting of the rubber boots on the suction side of the airfoil [25]. The system is quoted to have very low energy consumption [26]. Goodrich tested first this technology on a simulated wind turbine rotor blade 1.5 MW laboratory facility, building up a 6 m by 1m de-icer as shown in Fig. 5.40. Tests indicated the need to work at high pressures for wind turbine applications, and the system proved to satisfactory remove ice glaze at temperatures above −10 °C and residual ice at temperatures between −10 and −20 °C.

During in-field operation, additional benefits are gained from the fact that residual ice is removed due to the helping effect of blade vibration and centrifugal forces.

A part of the high risk of additional unavailability due to the increased complexity of such IPS, the effect of the roughness of the boots on the aerodynamic performance and noise emission is not clear yet. Also the wear and repair issues seem to pose serious limitation to the system use in field. Ice shed is also a concern.

5.11.1.2 Electro Impulsive/expulsive Devices

This quite new de-icing system has not yet been tested on wind turbines, but is mentioned here for its potential. The method uses very rapid electromagnetically induced trains of vibration pulses that flex a metal abrasion shield and crack the ice [27]. This is currently accomplished by a spiral coil placed near the surface of the blade. When current is applied to the coil, the magnetic field created between the coil and the blade wall determines a sudden displacement of the surface and the shed of the ice. The method has been recently certified for use on Raytheons Premier I business jet. It is used by Hydro-Quebec for transmission lines and Goodrich is currently developing this method for aeronautical applications. The system is efficient, environmentally friendly, has low energy consumption and causes no interference with Hertz transmission. Application procedure, lightening issues, and maintenance on WT blades should, however, be also carefully evaluated.

5.11.2 Thermal

5.11.2.1 Electromagnetic Radiation Heating

Electromagnetic heating can be accomplished either by infrared (700 μm–1 m wavelength) and microwave radiation (1 mm–1 m wavelength).

Direct and indirect heating can be in principle be used. Direct heating means transmitting the microwave energy directly through the air or through a dielectric surface waveguide, toward the super-cooled water droplets approaching the body.

The power absorbed in the liquid water or ice is given by:

$$P_{ab} = I_0 \left(1 - e^{\alpha \cdot t}\right)$$

where I_0 is the intensity of the magnetic field, α is the absorption coefficient per length, and t is the characteristic thickness of the medium (vapour, water, ice). The absorption coefficient of water is a function of the water phase (vapour, liquid, solid), and the radiation frequency, and it is therefore characterised by spectra for the different phases. In general, the absorption behaviour at infrared and microwave frequency can change considerably.

If the water is in the liquid state, the power required per unit surface to rise the water temperature is:

$$\dot{q} = LWC \cdot W \cdot c_w \left(T_w - T\right) = LWC \cdot W \cdot c_w \Delta T$$

The unitary surface is the far upstream area crossed by the droplets travelling at speed W. The power per unit surface required to heat the water by infrared or microwave radiation of one Kelvin degree is therefore:

$$\dot{q}_{ab} = LWC \cdot W \cdot c_w \left(1 - e^{\alpha \cdot d}\right)$$

being d the droplet diameter. By assuming LWC $= 0.002\,\mathrm{kg/m^3}$, W $= 60\,\mathrm{m/s}$, $c_w = 4.182\,\mathrm{J/kgK}$, d $= 20\,\mu\mathrm{m}$, and the liquid water absorption coefficients according to the relative wavelength, the required power per unit surface to rise the temperature of the incoming water at different wavelength emission by one Kelvin degree is shown in Fig. 5.41.

The figure indicates the high specific power requirement of the microwave higher band range.

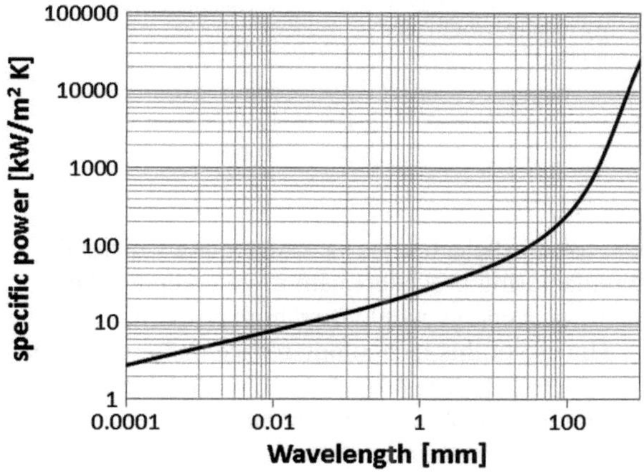

Fig. 5.41 Required power per unit surface to rise the temperature of the incoming water at different wavelength emission by one Kelvin degree

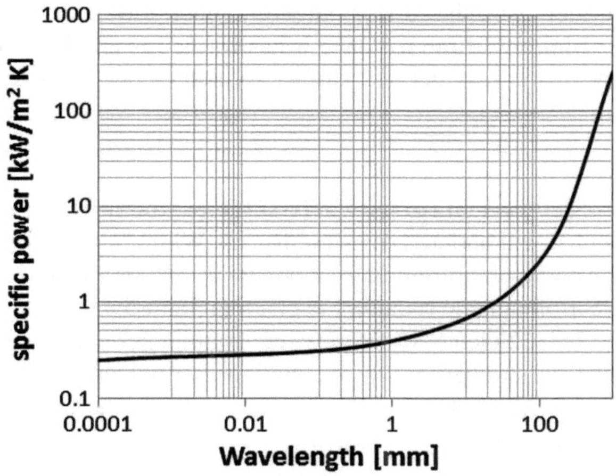

Fig. 5.42 Required power per unit surface to increase the temperature of a layer of 0.1 mm of water at different wavelength emissions by one Kelvin degree

The water can be also heated up on the surface. In this case, the required power per unit surface to heat a layer of $t_w = 0.1$ mm of water up to one Kelvin degree on the surface is shown in Fig. 5.42, β_∞ is assumed to be 0.5:

$$\dot{q}_{ab} = \beta_\infty LWC \cdot W \cdot c_w \left(1 - e^{\alpha \cdot d}\right)$$

Even if also here high specific power requirement of the microwave higher band is required, the absolute power appear to be more reasonable. This power has to be continuously supplied in order to keep the water layer in a liquid state.

The poor water energy absorption makes microwaves inefficient to heat water or melt ice. An high amount of energy is wasted by transmitting microwaves through the air, as they are emitted stereoscopically. Because of the reflection of microwave energy from the surface to be de-iced, a further reduction in efficiency loss is expected.

Table 5.11 Electromagnetic energy requirement to melt 1 mm of ice

Wavelength (mm)	Energy (MWh)
1,000	83.54
100	0.88
1	0.13
0.0001	0.08

The use of dielectric surface waveguides make the system more efficient, but the manufacturing procedures are more complex. Therefore, the direct transmission of microwave energy would need high power fluxes, making this solution not practical.

The use of this technology for ice melting is actually under investigation by LM Glasfibers [8]. A test rig for a LM 19.1 blade has been set up with a microwave generator of 6 kW at 2.54 GHz and a power emission of less than 10 mW/m^2. The power unit proved to have, also during prototyping phase, a low impact on blade costs, and no lightening issues, but the limited energy absorption in glaze ice obtained needs further test to be carried out with higher power.

Melting of the ice once it has formed is another option of this technology. The specific energy per unit surface to melt the ice layer is:

$$E_{ice} = \Delta h_f \rho_{ice} t_{ice}$$

Melting 1 mm of ice will be sufficient for the whole ice layer to be shed away from centrifugal and aerodynamic forces. With $\Delta h_f = 334\,\mathrm{J/kg}$ and $\rho_{ice} = 900\,\mathrm{kg/m^3}$, the energy requirement is about 300 kJ/m^2 or about 83 kWh. The power provided by electromagnetic radiation is given in Table 5.11.

The efficiency of microwaves is quite poor. This suggests the use of the indirect heating strategy, by an irradiating thermal propagation tube, which has a highly absorptive coating on the interior surface. The microwave energy is emitted from one or more microwave generators. The microwave energy is converted into thermal energy, which is conducted through a low thermal resistance spar to the blade skin surface, which may be integral with the leading edge of a blade section. As the blades are made of high thermal resistance materials, thermal propagation through the wall thickness is quite ineffective.

Increasing the frequency to millimetre waves indicates that infrared waves can be very efficient in heating water and melting ice [28].

5.11.2.2 Intermittent (Cyclic) Hot Gas Heating

Preceding paragraphs indicate that the energetic penalties for turbine using continuous heating to protect blades against icing are large and in some cases may be prohibitive. An advantageous method [29, 30] for icing protection in wind turbines can be accomplished by cyclical de-icing. By this principle, some ice is permitted to form on the surfaces and then is removed periodically during relatively short, intensive application of heat.

A water film between the surface and the ice is promoted by an internal heat application and this permits removal of the ice by centrifugal forces. Because the heating is pulsed, heat is supplied successively to relatively small surface areas, and a constant heat load is thus maintained on a heat source. The total heat input for cyclical de-icing, therefore, can be greatly reduced compared to continuous heating.

The use of cyclical de-icing using electric power as heat source has the inherent disadvantage of large system weight, susceptibility to failure by damage of the heating circuits, high cost of maintenance and fire hazard when the system fails because of heater burn-out. Economical considerations also lead to the conclusion that the most economical icing protection system with respect to the installed cost (and weight for high elevation installations) consists on a system utilising hot gas from a convenient heat source. If the large heating requirement associated with continuous hot gas heating can be reduced by use of a hot gas cyclical de-icing system while the devices relatively low installed weight is maintained, the wind turbine performance penalties could be decreased form those incurred with the continuous heating system.

Pulse Electrothermal De-icing (PETD), has been proposed by using local electrical pads. PETD has been implemented to de-ice a large variety of surfaces from power lines and bridges to freezers and ice makers. It utilises the electrothermal heating of a high-power but low-energy pulse to remove ice from any surface. But by analysing the power level needed for a MW size turbine, it follows that blades that are nearly 40 m long with a blade surface area of some 10 sq. m, a huge amount of power is needed to achieve a high efficiency in a PETD system. With an average power density of 2.5 kW/m^2, an heated surface area of about m^2, see Table 5.10, a pulse power of about 25 kW per blade or 73 kW for the whole rotor would be needed. Since the turbine generates power typically at 590 VAC and 50 Hz, about 130 amps of current are needed for such power level. It appears that an induction-based rotary power transmission device of this capacity is a challenging task with current technology.

A hot gas cyclical based pulsed de-icing system has instead the inherent advantage of integral design with the wind turbine structure, low maintenance cost and elimination of the possible source of inboard fires.

Even though the average thermal power level required for the anti-icing and de-icing cases is comparable, the heating period ($t_{heat\text{-}on}$) for the de-icing practice is much shorter than for the anti-icing system ($t_{heat\text{-}on} + t_{heat\text{-}off}$). Contrary to aeronautical de-icing systems that employ ratios $\bar{P}_{de\text{-}icing}/P_{anti\text{-}icing}$ up to 10 [31, 32], lower values of this ratio, i.e. 0.5–1.5 can be typical for WT. This is caused by the different thermal conductivity of the materials used to manufacture the blade wall. The effect can be explained by the Fourier number:

$$\mathrm{Fo} = \frac{k_{mat}\, t}{\rho_{mat}\, c_{mat}\, t_{mat}}$$

which expresses the ratio between the thermal conduction and the thermal inertia of a body. Bodies with large Fourier numbers, i.e. with large conductivity k_{mat} and small thermal inertia $\rho_{mat}\, c_{mat}\, t_{mat}$ (such as airplane wings made of aluminium alloy), allow large heat fluxes to readily diffuse through the volume.

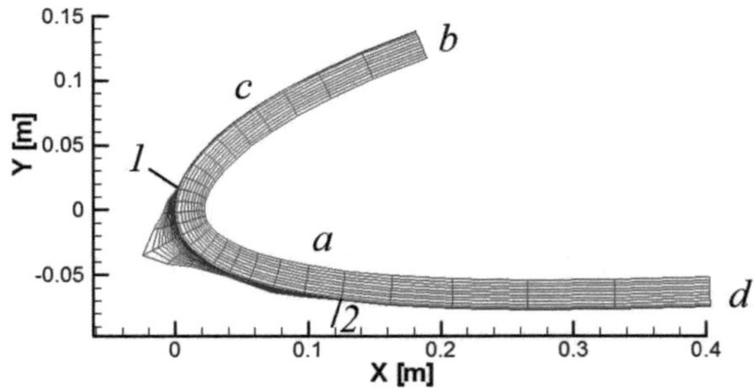

Fig. 5.43 Blade/ice discretization

On the other hand, large heat fluxes cannot diffuse quickly through thick-walled WT blades made of composites (low Fourier number), resulting in a large increase of the "inner-side" (warm surface) temperature, without any significant heat diffusion at the "outer-side" (cold-surface) in short times. Therefore, in the latter case, low thermal power densities should be preferred for long warming periods.

Numerical simulations [30] have confirmed the validity of this approach. The de-icing problem was tackled numerically by means of a 2D finite difference code to solve the energy conservation equation into a solid with a generic shape. In particular, the code allows computing the heat diffusion through a multilayer domain, one of whose materials being subjected to melting. Figure 5.43 shows the discretization scheme of a typical wind turbine blade section with a given ice accretion on the leading edge region.

The blue portion represents the blade wall, which is made out of a single layer of composite material with uniform thickness and thermal properties. The red portion of the domain shows a typical ice accretion at the leading edge region. The ice shapes and accretion times considered were provided by CIRA (personal courtesy, not published data) that uses a predictor-correct method to compute the ice growth, as a function of the flow filed around the blade and of the environmental conditions [33] (i.e. ambient temperature, MVD and LWC).

The equation for the 2D transient heat diffusion through a multilayer domain is considered in its conservative form:

$$\frac{\partial u_i}{\partial t} = \frac{\partial}{\partial x}\left(k_i \frac{\partial T_i}{\partial x}\right) + \frac{\partial}{\partial y}\left(k_i \frac{\partial T_i}{\partial y}\right) + S_i \tag{5.22}$$

where the temperature is written in terms of internal energy by means of a material's density and specific heat. For any wall layer, the following relationship between temperature and internal energy holds:

$$T_i = \frac{u_i\,\rho_i}{c_i} \tag{5.23}$$

A more complex relationship has to be considered for the ice layer, since it partially melts during warming. In particular, the latent heat of fusion has to be taken into account together with density and specific heat of ice and water (the pedex w and ice denote liquid water and iced water respectively):

$$T = \left\{T_w + \frac{\Delta T}{u_w - u_{ice}}\left(u_{ice,w} - u_{ice}\right)\right\} H\left(u_{ice,w} - u_{ice}\right) H\left(u_w - u_{ice,w}\right) \tag{5.24}$$
$$+ \left\{T_w + \frac{u_{ice,w} - u_{ice}}{c_{ice}\rho_{ice}}\right\} H\left(u_{ice} - u_{ice,w}\right) + \left\{T_w + \Delta T + \frac{u_{ice,w} - u_w}{c_w \rho_l}\right\} H\left(u_{ice,w} - u_w\right)$$

where H is the Heaviside step function. The internal energy relative to ice u_{ice} and water u_w at the melting point, are here defined as:

$$u_{ice} = \rho_{ice} c_{ice} T_w$$

$$u_w = u_{ice} + \frac{\rho_{ice} T_w + \rho_w (T_w + \Delta T)}{2} \Delta H_f$$

being ΔH_f the latent heat of fusion.

Therefore, the ice-water temperature–internal energy relationship is divided into three regimes: the liquid phase, the melting phase and the solid phase. Moreover, it is assumed that the ice melts over a small temperature range ($\Delta \mathrm{T} = 0.05\,^\circ\mathrm{C}$). The last approach allows the abrupt phase transition to be undertaken numerically without introducing unsteadiness in the heat flux computation.

The ADI [34] method with forward time and central space finite differences is adopted as a solving scheme. Since the blade heating is supplied by means of warm-air flowing inside the blade shell, heat convection boundary conditions are imposed on both the inner and the outer side of the blade wall ("a" and "c" in Fig. 5.43). The heat exchange at the inner side of the blade wall is computed as a function of uniform warm-air temperature $T_{a,in}$ and convection coefficient $h_{c,in}$:

$$\dot{q}_{conv,in} = h_{c,int}(T_{a,in} - T_{s,int}) \tag{5.25}$$

Similarly, the heat exchange at the outer side is calculated as a function of cold air temperature T_∞ and "overall heat exchange coefficient" coefficient h_{amb} (see [29] for further details):

$$\dot{q}_{out} = h_{amb}\left(T_{s,ext} - T_\infty\right) \tag{5.26}$$

h_{amb} is a function of the heat convection coefficient, which, in turn, is largely influenced by the air velocity distribution on the blade's surface and by the wall temperature; therefore it continuously changes during the ice formation process. However, since the ice accretion computation with time (and therefore the heat convection coefficient computation) is extremely demanding, the following simplified approach is adopted when evaluating h_{amb}:

- the ice-water layer thickness remains unchanged on the surface during the warming period, and is set equal to the maximum growing during the de-icing cycle period;
- the overall heat exchange coefficient remains unchanged during the ice growing time on the blade surface.

With regard to the latter assumption and in order to adopt a conservative approach, the overall heat exchange coefficient occurring during the most severe warming operation, for given environmental conditions, is employed. This condition occurs in anti-icing practice, and the ice formation is inhibited by keeping all the impinging water in the liquid state. In this situation, the relatively high temperatures of the surfaces determine the most relevant heat fluxes toward the ambient.

For the simulations presented in the following, the values of h_{amb} are evaluated assuming a minimum outer surface temperature of 0.5 °C.

Figure 5.44 shows the results of a simulation in terms of temperature distribution, while Table 5.12 lists the parameters used as boundary conditions.

De-icing simulations were carried out to compute the local thermal energy and power requirement for a section placed at 70 % of the blade span, operating at the rated condition. Even though the section considered is not the most significant in terms of icing potential, it presents some criticisms to the warming of the wall due to its considerable thickness (0.022 m). Table 5.13 shows the blade's wall and ice–water layer features.

As can be seen in Fig. 5.45, large values of the overall heat exchange coefficient (up to 900 W/(m^{2}K)) arise at the suction surface, mainly at the leading edge region, due to the strong air acceleration and to the amount of water locally collected by the profile.

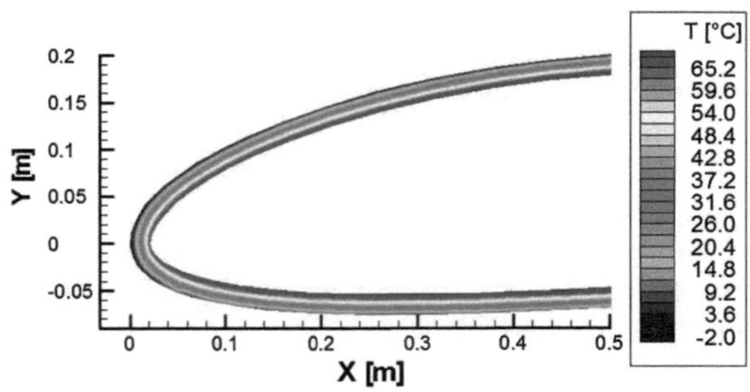

Fig. 5.44 Temperature distribution for a steady-state simulation

Table 5.12 Environmental and functional parameters for simulation shown in Fig. 5.44

Parameter	Input
T_∞	−2 °C
W	55.75 m/s
LWC	0.2 g/m^3
MVD	40 μm
$T_{a,in}$	117°C
$h_{c,in}$	77 W/(m^2 K)

Table 5.13 Multilayer domain features

Parameter	Blade wall	Ice layer	Water layer
Chord (m)		1.8	
Thickness (m)	0.022	0.03 (max)	–
Conductivity (W/(m K))	0.7	2.23	0.517
Spec. heat (J/(kg K))	1,000	2,064	4,280
Density (kg/m^3)	1,400	940	1.000

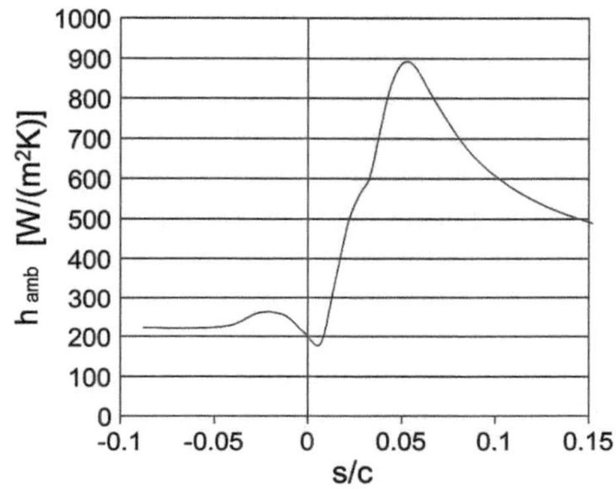

Fig. 5.45 Outer surface overall heat exchange coefficient distribution

This method is valid if the flow field is only moderately influenced by the ice accretion, as in the cases considered. For the present case, ice formations with a maximum thickness of 3 cm were considered that are developed on a blade profile with a 1.8 m long chord.

Since low conductive materials are used for the blades fabrication, the chordwise heat diffusion is negligible and adiabatic wall conditions are applied on the remaining boundaries of the physical domain (see Fig. 5.43 "b" and "d").

The code was validated comparing stationary (anti-icing operating mode) simulations obtained by means of the commercial code ANSYS on an ice-free blade wall with the present finite differences code. The results showed a good agreement both in terms of temperature and heat flux distributions. The percentage difference observed between the numerical solutions is lower than 1 % for the temperatures, and lower than 5 % for the heat fluxes.

The set of data and boundary conditions have been used for a comparative evaluation of the effectiveness of hot air anti-icing and de-icing systems. For given environmental conditions, the two ice prevention systems were compared in terms of system sustainability and thermal energy requirement. Additionally, two different ice shapes were investigated for the ambient temperatures of –3 and −6 °C. Different accretion times were, however, computed for each ice formation, as a consequence of the LWC value analysed from time to time. Ice accretion times of 144, 72 and 36 min were, respectively, considered for the three values of LWC.

Regarding the first issue, the maximum blade working temperature during warming was used as a feasibility threshold parameter. The practicability of the de-icing system was assessed by means of the "intermittent factor" τ. It is defined as the ratio between the warming time and the total cycle time:

$$\tau = \frac{T_{heat\text{-}on}}{T_{heat\text{-}on} + T_{heat\text{-}off}} = \frac{T_{heat\text{-}on}}{T} \tag{5.27}$$

The warming time $T_{heat\text{-}on}$, is dependent on the thickness of the water layer to be produced in order to provoke the ice layer shedding. A body/ice-water interlayer 1 mm thick was here considered to be sufficient for the ice to be swept away due to centrifugal and aerodynamic forces.

The total cycle time equals the time needed for the ice to grow on the blade's profile, assuming that the wall achieves the initial temperature at the end of each warming period. This is true only for long ice accretion times (i.e. for LWC $= 0.25\mathrm{g/m^3}$ in the simulations performed), while the final average wall temperature is higher than the initial for short ones. This assumption is therefore conservative for most cases.

The energy consumption ratio parameter ε_τ is introduced to compare de-icing and anti-icing systems, in terms of energy usage. It is defined as:

$$\varepsilon_\tau = \frac{\dot{q}_{de\text{-}icing,ave}\, t_{heat\text{-}on}}{\left(t_{heat\text{-}on} + t_{heat\text{-}off}\right)} = \frac{\dot{q}_{de\text{-}icing,ave}}{\dot{q}_{IPS,ave}}\tau = \xi\, \tau \tag{5.28}$$

where ξ is a magnification factor representing the fraction of extra power required for de-icing. $\dot{q}_{de\text{-}icing,ave}$ is the thermal power per linear metre (averaged with respect to $t_{heat\text{-}on}$ and heated area) submitted at the inner surface during the warming phase. $\dot{q}_{anti\text{-}icing,ave}$ is the thermal power per linear metre submitted at the inner surface during the anti-icing operation. The latter parameter was quoted about 2,600 W/m with regard to the ambient temperature of $-3\,^\circ$C, and about 5,300 W/m for the air temperature of $-6\,^\circ$C. The LWC has a minor effect on the thermal power.

The same through-flowing area, i.e. the same warm-air channel, was considered for the computation of both thermal powers, since they strongly depend on the extent of the surface considered (see the h_{amb} distribution in Fig. 5.45). In particular, a total wall chordwise length of 0.64 m was considered around the stagnation point (see Fig. 5.43). The length was evaluated as a suitable one for preventing ice formation due to direct impingement water and runback water during anti-icing operation [35].

Figure 5.46 shows the intermittent factor variation versus the warm-air temperature and LWC, for cold air temperature equal to $-3\,^\circ$C. The intermittent factor drops as the hot air temperature is increased due to the reduction of the time required for the ice to melt. The chart shows an asymptotic trend as the internal warm-air flow temperature increases. Figure 5.46 also shows that during de-icing operation, when warm-air temperature is higher than about 140 °C, the maximum inner surface temperature rises above 100 °C. This figure could affect the structural integrity

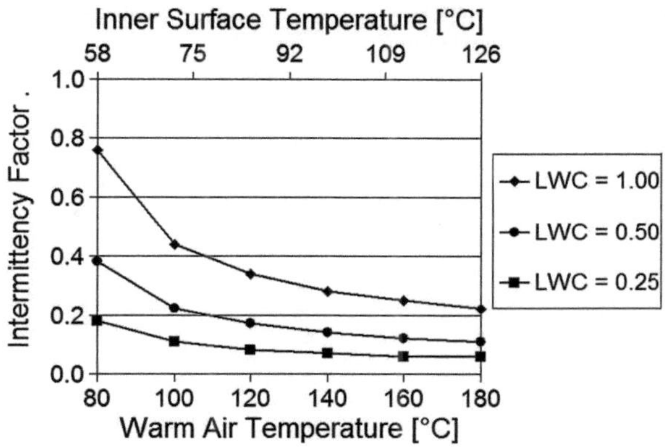

Fig. 5.46 Intermittent factor versus warm-air temperature at $T_\infty = -3\,^\circ\mathrm{C}$

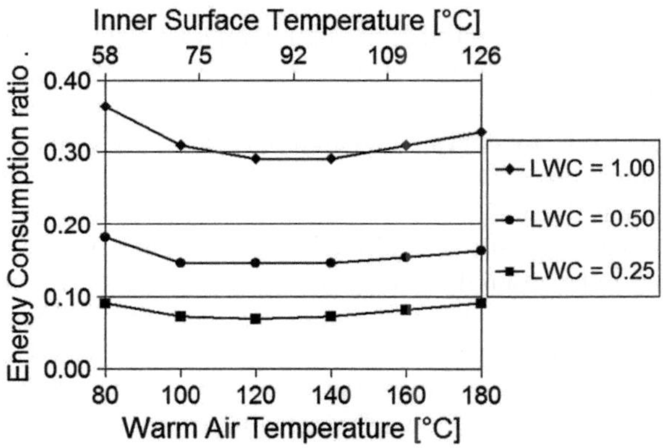

Fig. 5.47 Energy consumption ratio versus warm-air temperature at $T_\infty = -3\,^\circ\mathrm{C}$

of the blade, and has been assumed as the maximum working temperature of the component.

A minimum warm-air temperature between 70 and 80 °C is required for the de-icing practice; below that, the intermittent factor approaches rapidly the limit of 1.

Figure 5.47 compares the anti-icing and de-icing system in terms of energy consumption ratio for operation at ambient temperature of −3 °C. The chart shows that thermal energy requirements lower than 40 % are needed in all the cases considered, when de-icing is adopted instead of anti-icing. All the three curves present a minimum at about 120 °C, meaning that the best compromise between employed heating power and de-icing Heat-On time is here obtained. This operational condition has to be preferred, if it is compatible with the needed air heater power to be installed and the blade safe-working temperature.

The anti-icing operation adopted as a reference for the computation of $\varepsilon_t au$ would require a warm-air temperature of 180 °C to ensure an outer surface minimum temperature of 0.5 °C. Such a condition imposes a maximum wall temperature of about 130 °C. The simulations show therefore that for the investigated blade's section, the anti-icing operation is inhibited for ambient temperatures lower than about −2 °C.

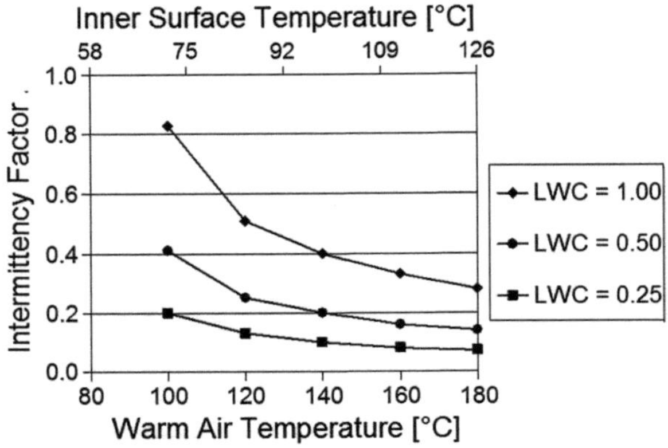

Fig. 5.48 Intermittent factor versus warm-air temperature at $T_\infty = -6\,^\circ\mathrm{C}$

Such data are therefore used for comparison purpose only, since such blade working temperatures are unrealistic for practical use.

Figure 5.48 shows the intermittent factor variation versus warm-air temperature and LWC, for cold air temperature equal to $-6\,^\circ\mathrm{C}$. It can be noted that the intermittent factor values evaluated for each given warm-air temperature are higher here compared to that of in Fig. 5.46 since an higher thermal power, and thus a longer time, are needed to melt the water layer. Moreover, a minimum working temperature of about $100\,^\circ\mathrm{C}$ has to be ensured for the proper working of the de-icing system.

When de-icing and anti-icing systems are compared in terms of energy consumption ratio at $T_\infty = -6\,^\circ\mathrm{C}$ (see Fig. 5.49), it can be noticed that the curves take values lower than that of $T_\infty = -3\,^\circ\mathrm{C}$. In this case, the de-icing practice becomes even more convenient with respect to the former case as a consequence of the large increase of $\dot{q}_{ave,anti\text{-}icing}$, which occurs as the ambient temperature drops. The minima of the curves move toward higher values of warm-air temperature, reaching the "safe-working temperature" limit. However, the curves are rather flat in the temperature range considered; therefore, operation far from the maximum temperature

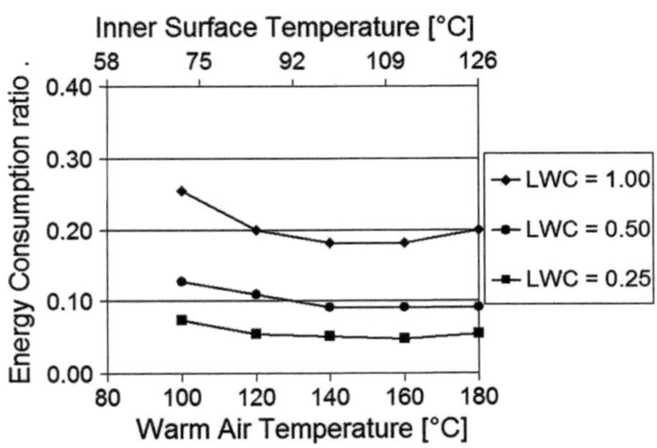

Fig. 5.49 Energy consumption ratio versus warm-air temperature at $T_\infty = -6\,^\circ\mathrm{C}$

limit could be taken into consideration without any significant increase of the energy consumption.

The simulations demonstrate the feasibility of the cyclic warm-air de-icing at moderately low ambient temperatures. Relatively low temperatures of the warm-air (80–120 °C) are suitable for the de-icing process, allowing lower temperatures (60–80 °C) of the blade's wall, if compared with the common anti-icing continuous practice. This has a positive impact on the air heater since, for the same amount of air mass flow warmed up, the de-icing approach allows the heater to work at a lower temperature, if compared to the anti-icing strategy. Therefore, heaters with lower nominal-power can be installed and the system operates for a limited time, up to 10 % of the icing event.

The short warming periods and the moderate thermal power densities employed dramatically reduce the amount of thermal energy required with respect to that needed for the anti-icing operation.

5.11.2.3 The Regenerative Ice Prevention System

The regenerative option concerns with the opportunity to use available on-board heating sources to supply thermal power (and energy) during anti-icing/de-icing operations. A fraction of the thermal power dissipated from the electrical generator and the electrical converter could be collected for this purpose. In Fig. 5.50 the conceptual scheme is shown.

With reference to Fig. 5.50, if the electrical generator is regarded as an heat exchanger, its overall thermal efficiency, expressed as ratios of enthalpies differences is given by:

$$\varepsilon_g = \frac{h_{g,2} - h_{g,1}}{h_{w,g} - h_{g,1}} \tag{5.29}$$

The reheat factor is defined as:

$$R = \frac{h_{g,2} - h_{g,1}}{h_{g,in} - h_{g,1}} \tag{5.30}$$

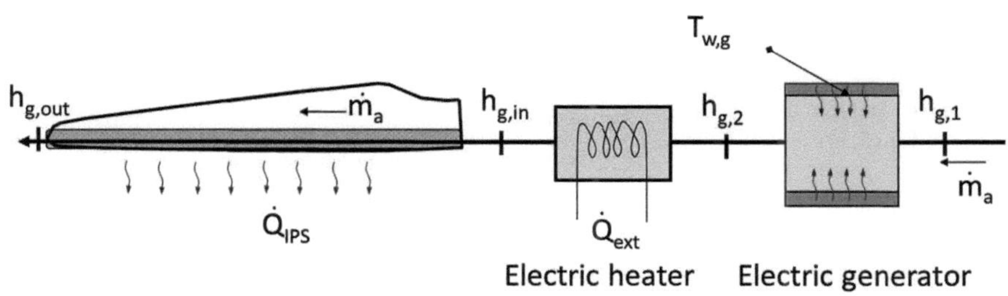

Fig. 5.50 Conceptual scheme of the regenerative ice prevention system

The enthalpy of the air at the exit of the electric generator $h_{g,2}$ is depending upon the geometry of the generator, the casing adopted to convey the warmed air passing through it and the Reynolds number realisable by this layout, all features determining the heat transfer coefficient. A $\varepsilon - NTU$ approach can be adopted to compute the air exiting temperature $h_{g,2}$, by the calculation of the generator heat exchanger thermal efficiency ε_g:

$$\varepsilon_g = 1 - \exp^{-}\left(\frac{S_g U}{\dot{m}_a \; c_{p,air}}\right) \tag{5.31}$$

The thermal power to be externally provided to the blade is:

$$\dot{Q}_{IPS,ave} = \frac{\dot{Q}_{ice}}{\eta_{IPS}} = \dot{m}_a \left(h_{g,in} - h_{g,1}\right)(1 - R) \tag{5.32}$$

The thermal power available by the electrical generator is the function of the electrical generator efficiency $\eta_{g,el}$:

$$P_{av} = P'_{el}\left(1 - \eta_{g,el}\right) S_g \varepsilon_g \tag{5.33}$$

Therefore, the steady-state energy conservation equation, neglecting kinetic terms and mechanical power of the fans promoting the air circulation:

$$P'_{el}\left(1 - \eta_{g,el}\right) S_g \varepsilon_g = \dot{m}_a \left(h_{g,in} - h_{g,1}\right)(1 - R) \tag{5.34}$$

If the data of Table 5.14 are input in the Eq. (5.34), the plots of Figs. 5.51 and 5.52 are obtained. Figure 5.51 shows the reheat factor, and the ratio of the externally supplied thermal power to the one required with no reheat (R = 0) as function of the blade hot air inlet temperature. The simulations are made for two different mass flows of air, 1 and 2 kg/s, respectively. The graph clearly shows that with 1 kg/s air mass flow, the regenerative effect approaches 100 % as a relatively low air temperature at the blade root. The external need of the electrical anti-icing power drops accordingly. As the mass flow increases, the benefit decreases, the heat extraction process through the electrical generator being less efficient.

Figure 5.52 shows the same result but the ratio of the power made available by the electrical generator to the anti-icing power required is evidenced.

Table 5.14 Input in the model of Eq. (5.34)

Parameter	Input
$T_{w,g}$ working temperature of the generator	100 °C
$T_{g,1}$ ambient temperature	−3 °C
P'_{el} specific power of the generator	125 kW/m^2
S_g inner surface of the electrical generator	10 m^2
$\eta_{g,el}$ efficiency of the generator	0.90
η_{IPS} thermal efficiency of open-circuit	0.30

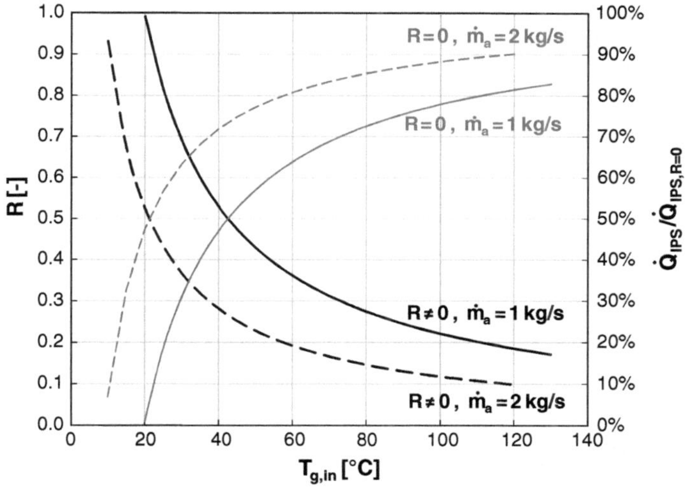

Fig. 5.51 Reheat factor and heat flux demand ratio for different air mass flows versus air inlet blade temperature

Fig. 5.52 Power made available by the electrical generator to the anti-icing power required versus inlet air temperature

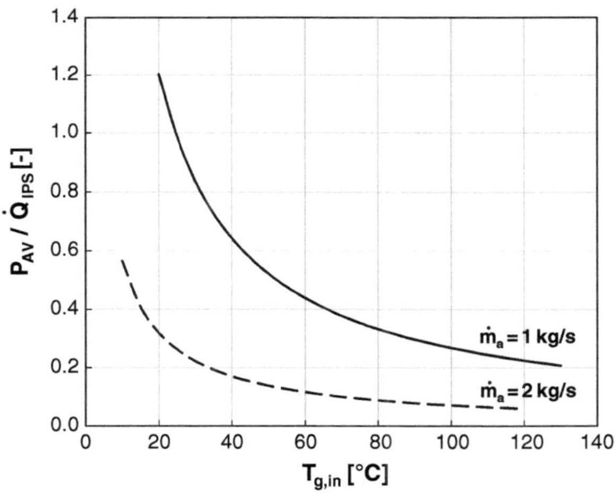

The lower the blade inlet air temperature required, higher the contribution of the thermal power available from the electrical generator. Intermittent heating realised without using intermittent electric heaters but through a continuous heating source (the electrical generator) would allow a decisive gain in availability.

Technically speaking, the warm-air stream could be automatically directed to the rotor by suitable cut-off windows, accordingly; for instance, to the arrangement shown in Fig. 5.53 [36, 37].

Another source of thermal power could be obtained from other ancillary electric devices as the converters are located into the tower or into the nacelle.

Fig. 5.53 Possible arrangement to automatically direct the hot air into the rotor [36]

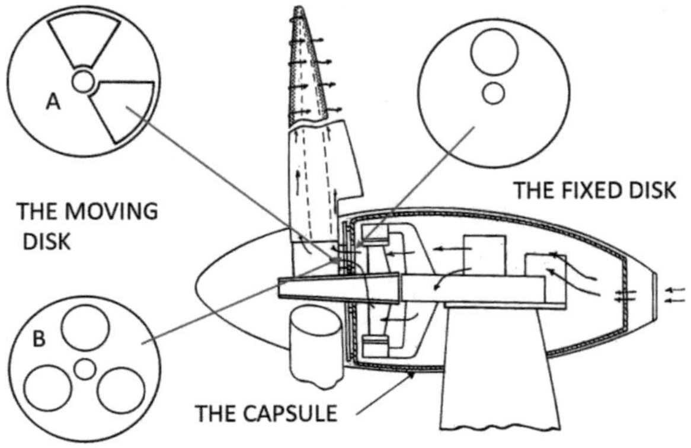

5.11.2.4 Film Heating

Film heating is a technology based on the known concept of the film cooling, already in use for gas turbine blades and combustion chamber cooling for more than 60 years. It is based on the generation of a spatially continuous film of air enveloping partially or totally the blade external surface, which reduces the heat fluxes toward the blade. Analogously, if one refers to ice prevention systems, a heating film process is required. The film is created by ejection of warm-air from the inside to the outside of the blades through slots or arrays of holes distributed into the wall, as schematically shown in Fig. 5.54 [36].

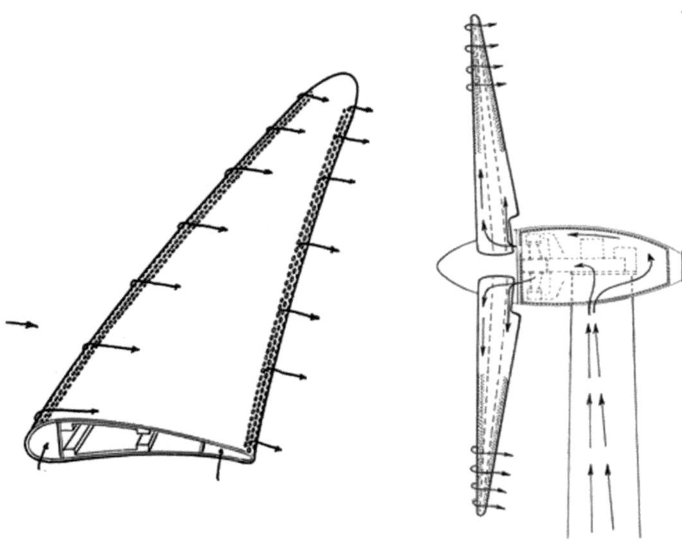

Fig. 5.54 Effusive air heating scheme and IPS general concept [36]

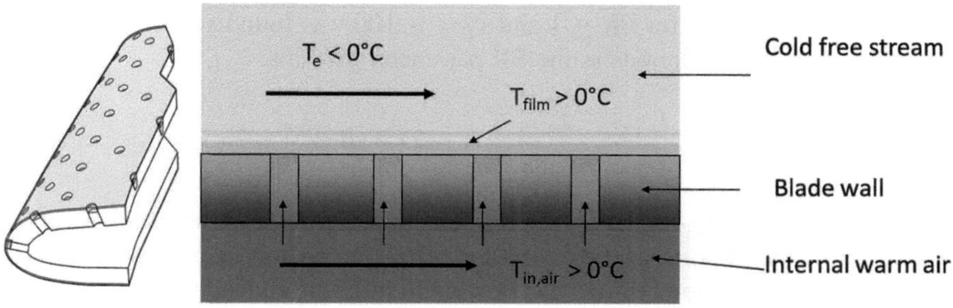

Fig. 5.55 Heat transfer mechanism for film heating

The warm-air is forced into the blade through the root and then directed by inner channel to the portions of the wall equipped with slots or holes. The ejection process can be either continuous or cyclic in time by setting a proper intermittent factor.

The heat transfer mechanism is shown in Fig. 5.55. The film heating determines a mix of inner warm and external cold air. The mixing process produces both an increase of the heat exchange coefficient by convection, dependent on the fluid's flow features, and an increase of the air temperature at the surface. The net effect leads to a decrease of the heat fluxes from surface to the cold stream, as the heat transfer mechanism allows bypassing the high thermal resistance of the glass fibre wall.

5.11.2.5 Model of Effusive Heating

In film heating, the warm-air is injected through the wall and diffused by turbulent mixing in the cold outside air, and its enthalpy is gradually destroyed on the downstream region after leaving the orifice (slot or hole). As downstream distance from the injecting orifice increases, the mixed air asymptotically approaches again the local outside temperature.

Weighardt [38], already in 1946 analysed in de-icing experiments hot air blowing through a slot opened in a flat plate. He used a temperature ratio parameter defined as:

$$\frac{T_{s,ext} - T_{a,in}}{T_{a,\infty} - T_{a,in}}$$

$T_{a,in}$ expresses here the static temperature of the injected air, and $T_{a,\infty}$ the static ambient temperature. This temperature ratio depends on the blowing ratio parameter, also called film to stream mass velocity ratio, and given by:

$$BR = \frac{s_s\, \rho_{a,in}\, v_{a,in}}{x\, \rho_{a,\infty}\, w_{a,\infty}}$$

being s_s the slot width and x the distance downstream the slot, and $s_s\, \rho_{a,in}\, v_{a,in}$ the heating mass ejected from the slot per unit time. The following experimentally determined relationship for $BR \leq 1$ and $x/s_s > 100$ was found valid. It shows that the wall temperature increases as the BR parameter growths:

$$\frac{T_{s,ext} - T_{a,in}}{T_{a,\infty} - T_{a,in}} = 1 - 21.8\,(BR)^{0.8} \tag{5.35}$$

To make this finding useful for the implementation of not only a single slot, but of a perforated wall portion of extension x, where an heated mass flow is injected through, the conservation of mass can be used to infer the equivalent mass velocity:

$$\rho_a\, v_a = \frac{s_s\, \rho_{a,in}\, v_{a,in}}{x}$$

Now one can introduce the mass velocity in Eq. (5.35), with the aid of the hypothesis that $T_a = T_{a,in}$, since the low thermal conduction of the blade wall (the flux through the wall can be can be considered adiabatic), thus getting:

$$\frac{T_{w,ext} - T_{a,in}}{T_{a,\infty} - T_{a,in}} = 1 - 21.8 \left(\frac{\rho_a\, v_a}{x\, \rho_{a,\infty}\, w_{a,\infty}} \right)^{0.8} \tag{5.36}$$

for

$$\frac{\rho_a\, v_a}{x\, \rho_{a,\infty} - w_{a,\infty}} < \frac{1}{100}$$

and

$$\frac{\rho_{a,in}\, v_{a,in}}{x\, \rho_{a,\infty}\, w_{a,\infty}} \leq 1$$

This equation deduced by manipulating Weighardt's equation (5.36) is valid only for modest temperature differences between heating gas and cooling air stream.

By putting:

$$\frac{\dot{m}_a}{A_a} = \rho_a\, v_a$$

Eq. (5.36), the surface external temperature $T_{s,ext}$ for given ambient conditions as function of the blowing ratio results:

$$T_{s,ext} = T_{a,in} + T_{a,\infty} - T_{a,in} \left[1 - 21.8 \left(\frac{\frac{\dot{m}_a}{A_a}}{\rho_{a,\infty} w_{a,\infty}} \right)^{0.8} \right] \tag{5.37}$$

One can define now the film temperature T_{film} as the mixing temperature of the injected film and the external free stream:

$$T_{film} = T_{a,\infty} - \eta_{film} \left(T_{a,\infty} - T_{a,in} \right)$$

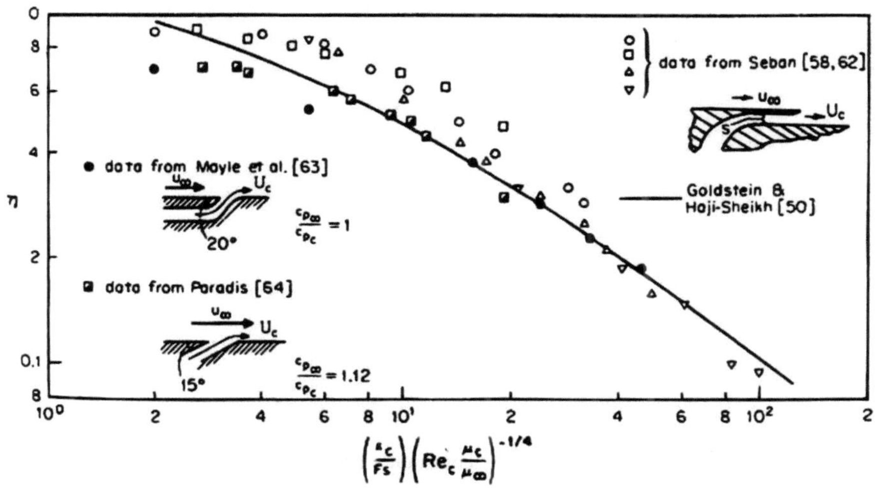

Fig. 5.56 A comparison of the predicted effectiveness with experimental results from Hartnett (1985) [39]

The film efficiency η_{film} can be derived from the well-known experimental relationship deduced for gas turbine film cooling [39], which is indicating a satisfactory prediction of slots and low injection rates as shown in Fig. 5.56.

$$\eta_{film} = \frac{1.9 Pr^{\frac{2}{3}}}{1 + 0.329 \frac{c_{p,air}}{c_{p,a,in}} \xi^{\frac{4}{5}}} \tag{5.38}$$

Due to the closeness of ambient and injection air temperature, $c_{p,air}/c_{p,a,in} = 1$. ξ is defined as:

$$\xi = \frac{\rho_{a,\infty} v_{a,\infty}}{\frac{\dot{m}_a}{A_a}} \left(\frac{\mu_a Re_s}{\mu_{a,\infty}} \right)^{-\frac{1}{4}} \tag{5.39}$$

being $Re_s = \rho_{a,\infty} v_{a,\infty} s_s / \mu_a$. This model assumes that the heating injection does not disturb the boundary layer and as a result entrainment takes place as for the turbulent boundary layer.

By the wall temperature expressed by Eq. (5.37) the parameter θ can be built as:

$$\theta = \frac{T_{a,\infty} - T_{a,in}}{T_{a,\infty} - T_{s,ext}}$$

This temperature ratio, together with the film efficiency is used to define the heat ratio:

$$\frac{Q_{film}}{Q_{nofilm}} = \frac{h_{c,film}}{h_{c,ext}} \left(1 - \theta \eta_{film}\right) \tag{5.40}$$

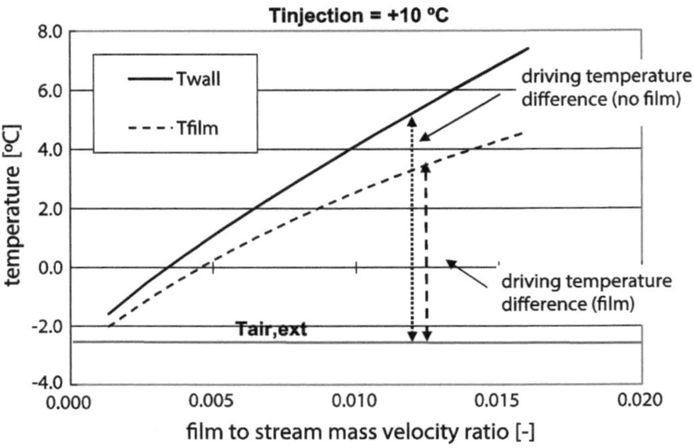

Fig. 5.57 Effect of blowing ratio on driving temperature difference

The effect of film heating on the convective heat loss is shown in Fig. 5.57, where the driving temperature gradient at the wall is shown as function of the blowing ratio BR. With a relatively low warm-air mass flow and low air heating temperatures the cooling driving temperature is considerably reduced. The heat flux at the wall is reduced compared to no film heating case (impermeable wall), as shown in Fig. 5.58, therefore the thermal power for anti-icing is reduced. This simple model shows that the anti-ice heat flux requirement at the blade surface is reduced up to 70–80 %, with a very reduced amount of heating air. Only warm-air is needed for anti-ice purpose and the regenerative schemes can take fully advantage from this solution.

In fact, since very low enthalpy is required for anti-icing purpose, compared to conventional systems, the air forming the film can by sucked through the nacelle rear part or the tower bottom, and after flushing the electric devices, as the trafo/inverter within the tower or the electric generator within the nacelle, is warmed up and directed in to the rotor.

Fig. 5.58 Effect of blowing ratio on cooling heat flows

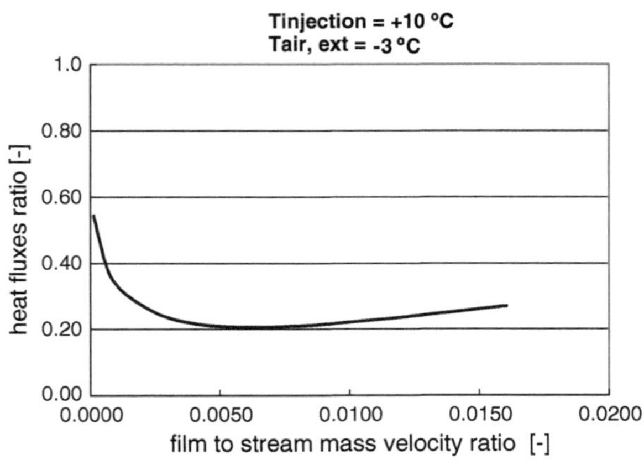

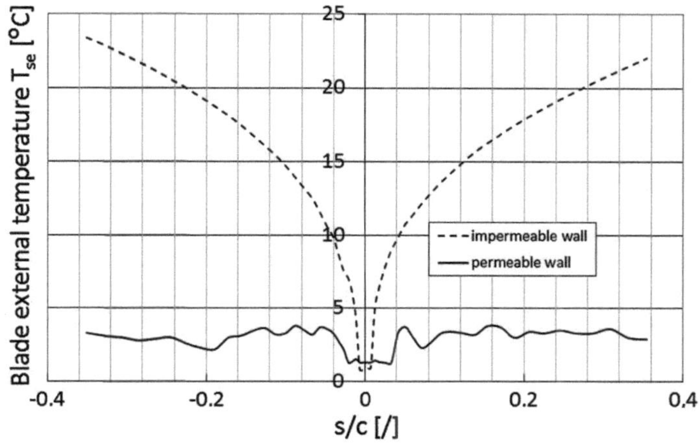

Fig. 5.59 Comparison of the external wall temperatures distribution at the NACA 4415 LE region among impermeable wall (current technology) and the permeable wall (AoA = 0°)

The local effect on the blade external wall temperature is finally shown in Fig. 5.59 where a comparison of the wall temperatures distribution at the LE region among impermeable wall (current technology) and the film heated wall for a typical tip section is presented. Film heating allows achieving very smooth and isothermal like blade skin temperature, compared to traditional hot gas systems. This computation derives from the model of thermal heating presented in this Chapter, modified in order to account for film heating and conjugate heat transfer of one blade section. The injection temperature of the air was set to +10 °C.

About the concerns on the disturbances of the effusing air on the boundary layer stability, it is well proven by decades of application in gas turbines that the employment of proper BR is preventing jet formation and fluid dynamic losses.

The air ejection reduces also the collection efficiency of the blade profile, thus decreasing the quantity of water impinging the surface. From the economic point of view, the operating costs can be advantageously reduced, since air heating can be obtained by regenerative effect. In this configuration, the system does not need icing detection system to establish icing onset, since as the wind turbine rotates the flow is automatically established with minor air pumping power demand.

5.11.3 Low Adhesion Coating Materials

Scarce adhesion of ice on the blade surface can be accomplished by means of:

- hydrophobic surfaces (reduce the surfaces wettability);
- ice-phobic surfaces (reduce ice adhesion).

There are two primary benefits of reducing water or ice on the surface:

- ice prevention systems requires less energy;
- the removal of ice is less partial.

Early studies in aeronautic field on static ice adhesion to surfaces ended up with the following conclusions:

- adhesive shear stress increases linearly with decreasing surface temperature;
- it is uncertain if surfaces with hydrophobic properties will have also ice-phobic capability. Hydrophobicity does not necessarily produce ice phobicity;
- a given coating will behave differently in different icing condition, rime or glaze that means in dry or wet ice growth;
- ice-phobic characteristics of a material can change with time and repeated removal.
- Silicon-based materials reduce the adhesion of the ice, but if provided as a coating layer, it is stripped away with each ice removal, thus not providing a permanent coating;
- Teflon coatings are prone to being electrically charged; therefore, humid dust or wet particles are attracted to the surface, creating the condition to promote ice growth.

In ice glaze accretion process, the water coalesces into beads. Small defects in the coating on the surface lead water beds to turn to ice locally, with the result to mask the coating. More ice will then grow over these local spots. This phenomenon is what I called "the frying pan like defects", as it can be observed even on quite new Teflon-coated frying pans. When wet, some droplets of water or even rivulets will remain on the coating without slipping away as the pan is hung. These wet areas indicate the presence of coating defects that hampers the surface to act hydrophobically.

The performance of the coatings in time (durability) depends on the degradation of the original surface bonding (i.e. on site conditions), thus even if very low adhesion can originally being achieved, durability must be demonstrated both for repeated ice removals and for rain impingement at blade speed.

5.11.3.1 Adhesion Strength

Let us consider the generic ice blade data listed in the first four columns of Table 5.15. Ice mass is subjected to centrifugal and aerodynamic forces. As the latter acting on the ice accretion are an order of magnitude less than the former, they will not be considered further. The mass of ice accreted at the tip can be computed as

$$m_{ice} = \rho_{ice} \, A_{ice} \, t_{ice}$$

The centrifugal force and the resulting superficial adhesion strength are:

$$F_c = \Omega^2 \, R \, m_{ice}$$

Table 5.15 Data of a generic iced blade

Rated power (kW)	Ω (rad/s)	R (m)	c (m)	Stand still		Operation	
				t_{ice}, CL_2 (m)	F_c/A_{ice} (kPa)	t_{ice}, $k_s/c = 20 \times 10^{-4}$ (m)	F_c/A_{ice} (kPa)
100	4.12	17	0.3	0.9	16	6×10^{-4}	0,16
1,000	2.77	27	0.4	0.12	15	8×10^{-4}	0,15
2,500	1.70	41	0.8	0.24	17	1.6×10^{-3}	0,17

$$\left.\frac{F_c}{A_{ice}}\right|_{tip} = \rho_{ice}\, t_{ice}\, \Omega^2 R$$

The value of t_{ice} depends on the wind turbine operation. When the turbine starts with iced blades it can be deduced by the contamination levels introduced in Chap. 4, while during operation it can be deduced by using the relative roughness parameter k_s/c. As the aerodynamic performances degrade considerably when a limit relative roughness parameter $k_s/c = 20 \times 10^{-4}$ is achieved, this figure is used to obtain the data of the last four columns of Table 5.15. Ice density was set at 900 kg/m^3.

For ice to shed spontaneously, the adhesion strength should be less than 17 kPa for all turbine sizes at starting with iced blades, while during operation the adhesion strength will be two order of magnitude less. Typical adhesion strength of ice to various materials at −10 °C are listed in Table 5.16.

By comparing these figures with the ones of Table 5.15 it is apparent that with current materials, centrifugal forces are generally insufficient to cause ice shedding from any surface, even at the tip of the blade.

As a consequence, the following conclusions can be drawn:

- icing prevention on wind turbine blades by coatings alone is not realistic;
- the adhesion strength of an ice accretion could, however, be reduced when blade are treated with the appropriate coating;
- Ice phobicity capability of a surface is dominated by local, molecular characteristics of the surface bonding. A macroscale approach is useless for practical purposes. The coating application technique and the final surface homogeneities are of paramount importance. This is a challenging task for the large surface of a wind blade;

Table 5.16 Typical adhesion strength of ice to various materials at −10 °C

Materials	Adhesion strength (kPa)
Steel	900
Epoxy paint	400
Polyurethane resin varnish (urethane)	411
Acrylic silicone resin (water repellant)	198
PVC	90
Teflon	40

- combinations of coatings with de-or anti-icing devices should be taken into consideration since a reduced adhesion might be helpful in improving the performance of ice protection systems.

5.12 Offshore Ice Prevention Systems

The turbine components need to be resistant to vibration under time-varying environmental or operational loads. Ideally, the wind turbine should incorporate sensors that monitor the environmental loads, including both those due to icing and to the sea ice, and the state of the structure. This concept needs more advanced diagnostic tools to be developed compared to current state-of-the-art ones. Based on this information, semi-active countermeasures are activated automatically to prevent excessive vibrations.

A method to mitigate the vibrations could be obtained by embedding damping and smart elements in the supporting structure of the wind turbine. These would be used to reduce the dynamic response and to increase the fatigue life of the structure. Structural damping is almost always an effective solution against excessive vibrations. The foundations of the wind turbines have also an effect on preventing ice-induced movements. In appropriate sites, the natural land-fast ice zone can be extended in the wind park such that the static and dynamic ice forces on individual foundations remain small.

References

1. Al-Khalil KM, Miller DR, Wright WB (2001) Validation of NASA thermal ice protection computer codes: part 3-the validation of antice. NASA Langley Research Center, NASA/TM-2001-210907
2. Makkonen L, Laakso T, Marjaniemi M, Finstad KJ (2001) Modeling and prevention of ice accretion on wind turbines. Wind Eng 25(1):3–21
3. Beaugendre H, Morency F, Habashi WG (2003) FENSAP-ICE's three-dimensional in-flight ice accretion module. J Aircr 40(2):239–247
4. Battisti L, Fedrizzi R, Rialti M, Dal Savio S (2005) A model for the design of hot-air based wind turbine ice prevention system. In: Conference WREC05, Aberdeen, Germany, 22–27 May 2005
5. Ronsten G (2004) Svenska erfarenheter av vindkraft i kallt klimat nedisning. Vindkast Ochavisning, Elforsk rapport 04:13–40
6. Peltola E, Marjaniemi M, Stiesdal H (1999) An ice prevention system for the wind turbine blades. In: Proceedings of European wind energy conference, Nice, France, 1–5 March 1999
7. Seifert H (2003) Technical requirements for rotor blades operating in cold climate. In: Proceedings of Boreas VI, DEWI Deutsches Windenergie-Institut GmbH, p 5
8. Mansson J (2004) Why de-icing of wind turbine blades? In: Proceedings of global windpower, Chicago, 18-21 March 2004
9. Pederson E (2008) Wind turbine ice protection system (WTIPS). Kelly aerospace thermal systems. Winterwind. Norrkping, 9–10 December 2008

10. Woigt H (1949) Deutsche Patentschrift n.842330
11. Enercon International GmbH (2003) Enercon E-66 20.70 Technical Description
12. Battisti L (2006) Ice prevention systems selection and design.DTU special course, Master of Science in Wind Energy, June 2006
13. Battisti L, Dal Savio S (2003) Sistema antighiaccio per pale di turbine eoliche parte 1: valutazione del fabbisogno energetico. In: 58th congresso ATI, Padova, Italy, 8–12 September 2003
14. Battisti L, Soraperra G (2003) Sistema antighiaccio per pale di turbine eoliche parte 2: sistemi a circolazione di aria. In: 58th congresso ATI. Padova, Italy, pp 8–12 September 2003
15. Thomas SK, Cassoni RP, MacArthur CD (1996) Aircraft anti-icing and deicing techniques and modeling. J Aircr 33(5):841–853
16. Albers A (2011) Summary of a technical validation of ENERCON's rotor blade de-Icing system. Deutsche wind guard consulting Gmbh, PP11035-V2
17. Enercon International GmbH (2010), ENERCON ice detection system power curve method. Technical description, D0154426-2
18. Krenn A, Winkelmeier H, Wlfler T, Tiefenbacher K (2011) Technical assessment of rotor blade heating system in the Austrian Alps. In: Winterwind 2011 conference. Umeå, Sweden, 9–10 February 2011
19. Jonsson C (2012) Further development of ENERCON's de-icing system. In: Winterwind 2012 Conference, Skelleftea, 7–8 February 2012
20. Kays WM, Crawford ME (1980) Convective heat transfer and mass transfer. McGraw-Hill, New York
21. Ruff GA, Berkowitz BM (1990) Users manual for the NASA Lewis ice accretion prediction code (Lewice), NASA Langley Research Center, Technical report, NASA CR 185129
22. Incropera F, DeWitt D (1996) Fundamentals of heat and mass transfer, 5th edn. Wiley, New York
23. Halpin JC (1992) Primer on composite materials analysis, 2 Revised edn. Technomic publication, Lancaster
24. Øye S (1988) Project K 30 m Glasfibervinge Teknik Beskrivelse. Afdelingen for Fluid Mekanik Den Politekniske Lreanstalt, Lyngby, Denmark
25. Botura G, Fisher K (2003) Development of ice protection system for wind turbine applications. In: Proceedings of the VI BOREAS conference, Pyhatunturi, Finland, 9–11 April 2003
26. Mayer C (2007) Systme lectrothermique de Dgivrage pour une Pale d'Eolienne, UQAR, Rimouski, Canada
27. Dalili N, Edrisy A, Carriveau R (2009) A review of surface engineering issues critical to wind turbine performance. Renew Sustain Energy Rev 13:428–438
28. Andersen E, Börjesson E, Vainionp P, Undem LS (2011) Wind power in cold climate. WSP Environmental 2011
29. Battisti L, Baggio P, Fedrizzi R (2006) Warm-air intermittent de-icing system for wind turbines. Wind Eng 30(5):361–374
30. Battisti L, Fedrizzi R (2007) 2D numerical simulation of a wind turbine de-icing system using cycled heating. Wind Eng 31(1):33–42
31. Yasilik AD, De Witt KJ, Keith TG (1992) Three-dimensional simulation of electrothermal deicing systems. J Aircr 29(6):1035–1042
32. Gray VH, Bowden DT, von Glahn U (1952) Preliminary results of cyclical de-icing of a gas-heated airfoil. NASA Langley Research Center, Technical report, NACA-RM-E51J29
33. Mingione G, Brandi V (1998) Ice accretion prediction on multielement airfoils. J Aircr 35(2):240–246
34. Özisik MN (1994) Finite difference methods in heat transfer. CRC Press, Boca Raton
35. Battisti L, Fedrizzi R, Dal Savio S, Giovannelli A (2005) Influence of the and size of wind turbines on anti-icing thermal power requirement. In: Proceedings of EUROMECH 2005 wind energy colloquium, Oldenburg, Germany, 4–7 October 2005
36. Battisti L (2002) Anti-icing system for wind turbines. Patents US7637715B2, EP1552143B1 et al., priority 2002

37. Battisti L (2006) Method for implementing wind energy converting systems, Patents US8398368 et al., priority 2006
38. Weighardt K (1946) Hot air discharge for de-icing. Air material command—AAF Trans, technical publication, F-TS-919 RE
39. Rohsenow WM, Hartnett JP, Ganic EN (1985) Mass transfer cooling. Handbook of heat transfer applications, 2nd edn. McGraw-Hill, New York

Suggested Readings

List of Books

Adamson AW (1990) Physical chemical of surfaces, 5th edn. Wiley, New York

Cebeci T, Smith A (1974) Analysis of turbulent boundary layers, 1st edn. Academic Press, New York

Mason BJ (1971) The physics of the clouds. Oxford University Press, Oxford

Schmid PJ, Henningson DS (2001) Stability and transition in shear flows. Springer, New York

Spalding DB (1963) Convective mass transfer, an introduction. McGraw-Hill, New York

Thwaites B (1960) Incompressible aerodynamics: an account of the theory and observation of the steady flow of incompressible fluid past aerofoils, wings, and other bodies, 1st edn. Dover Publications Inc., New York

White FM (2000) Viscous fluid flow, 2nd edn. McGraw-Hill, New York

List of Articles

Abid R (1993) Evaluation of two-equation turbulence models for predicting transitional flows. Int J Eng Sci 31(6):831–840

Abu-Ghannam B, Shaw R (1980) Natural transition of boundary layers—the effects of turbulence, pressure gradient and flow history. J Mech Eng Sci 22(5): 213–228

Achenbach E (1977) The effect of surface roughness on the heat transfer from a circular cylinder to the cross flow of air. Int J Heat Mass Transf 20:359–369

Addy H (2000) Ice accretions and icing effects for modern airfoils. NASA report, TP-2000–210031

Addy H et al (2003) A wind tunnel study of icing effects on a business jet airfoil. NASA report, TM-2003–212124

L. Battisti, *Wind Turbines in Cold Climates*, Green Energy and Technology,
DOI 10.1007/978-3-319-05191-8

Addy H, Chung J (2000) A wind tunnel study of icing on natural laminar flow airfoil. AIAA paper 2000–0095, 38th aerospace sciences meeting and exhibit, Reno (NY), 10–13 Jan 2000

Aihara T (1990) Augmentation of convective heat transfer by gas-liquid mist. In: International heat transfer conference, vol 1. Hemisphere Publishing Co., New York, pp 445–461

Al-Khalil KM (1991) Numerical simulation of a an aircraft anti-icing system incorporating a rivulet model for the runback water. Ph.D thesis—University of Toledo, Ohio, USA

Al-Khalil KM, Horvath C, Miller DR, Wright W (2001) Validation of NASA thermal ice protection computer codes. Part 3—Validation of ANTICE. Contractor Report, 2001–210907, National Aeronautics and Space Agency, Cleveland, OH, p 18

Al-Khalil KM, Keith TG, De Witt J (1994) Development of an improved model for runback water on aircraft surfaces. J Aircr 31(2):271–278

Amick JL (1950) Comparison of the experimental pressure distribution on an NACA 0012 profile at high speeds with that calculated by the relaxation method. National Advisory Committee for Aeronautics, Technical Note, TN-2174

Anderson DN (2004) Manual of scaling methods. Technical report, NASA/CR 2004-212875

Arnal D (1971) Description and prediction of transition in two-dimensional incompressible flow. Minimum thickness of a draining liquid film. Int J Heat Mass Transf 14:2143–2146

Bankoff SG (1971) Stability of liquid flow down a heated inclined plate. Int J Heat Mass Transf 14:377

Baxxter DC, Reynolds WC (1958) Fundamental solutions for heat transfer from nonisothermal plates. J Aeronaut Sci 25:403–404

Beaugendre H, Morency F, Habashi WG (2003) Fensap-ice three-dimensional in-flight ice accretion module: Ice3d. J. Aircr 40(2):239–247

Bentwich M, Glasser D, Kern J, Williams D (1976) Analysis of rectilinear rivulet flow. AIChE J 22(4):772–779

Bernardin JD, Mudawar I, Christopher F, Walsh B, Frensesi EI (1997) Contact angle temperature dependence for water droplets on practical aluminum surfaces. Int J Heat Mass Transf 40(5):1017–1033

Bragg MB, Cummings SL, Henze CM (1996) Boundary-layer and heat transfer measurements on an airfoil with simulated ice roughness. 34th Aerospace sciences and meeting, Reno, pp 1–16

Bragg MB, Heinrich D, Valarezo W (1994) Effect of underwing frost on a transport aircraft airfoil at flight Reynolds number. J Aircr 31(6):1372–1379

Bragg MB (1981) Rime ice accretion and its effect on airfoil performance. Ph.D Thesis, The Ohio State University

Bragg MB (1986) An experimental study of the aerodynamics of a NACA 0012 airfoil with a simulated glaze ice accretion. Urbana, Ohio

Bragg MB, Broeren AP, Blumenthal LA (2005) Iced-airfoil aerodynamics. Prog Aerosp Sci 41(July):323–362

Cebeci T, Hefazi H, Roknaldin F, Carr LW (1995) Predicting stall and post-stall behavior of airfoils at low mach numbers. AIAA J 33(4):595–602

Cebeci T, Kafyeke F (2003) Aircraft icing. Ann Rev Fluid Mech 35:11–21

Collyer M, Lock R (1979) Prediction of viscous effects in steady transonic flow past an airfoil. Aeronaultical Q 30:485–505

Crawford ME, Kays WM (1976) A program for numerical computation of two-dimensional internal and external boundary layer flows. National aeronautics and space administration, Washington, Contractor Report, 2742. p 140

Croce G, Beaugendre H, Habashi WC (2002) Fensap-ice conjugate heat transfer computations with droplet impingement and runback effects. In: Proceedings of 40th aerospace sciences meeting & exhibit, Reno, Nevada, America institute of aeronautics and astronautics paper 2002-0386:1–10

Dey J (2000) On the momentum balance in linear-combination for the transition zone. J. Turbomach 122:587–588

Dhawan S, Narasimha R (1958) Some properties of boundary layer flow during the transition from laminar to turbulent motion. J Fluid Mech 3:418–436

Diprey DF, Sabersky RH (1963) Heat and momentum transfer in smooth and rough tubes at various prandtl numbers. Int Heat Mass Transf 6:329–353

Doenhoff AEV, Horton EA (1956) Low-speed experimental investigation of the effect of sandpaper type roughness on boundary—layer transition report NACA TN 3858

Domingos RH, Pustelnik M, Trapp LG, Silva GAL, Campo W, Santos LCC (2007) Development of an engine anti-ice protection system using experimental and numerical approaches. Proceedings of SAE aircraft and engine icing international conference, Society of Automotive Engineers, Warrendale, SAE paper 2007-01-3355

Downs SJ, James EH (1988) Heat transfer characteristics of an aero-engine intake fitted with a hot air jet impingement anti-icing system. Proceedings of 25th national heat transfer conference, American society of mechanical engineers, New York 1:163–170

Drazin PG, Reid WH (2004) Hydrodynamic Stability, 2nd edn. Cambridge University Press, Cambridge

Drela M, Giles M (1987) Viscous-inviscid analysis of transonic and low Reynolds number airfoils. AIAA J 25(10):1347–1355

Dukhan N, Masiuániec KC, De Witt KJ (1999) Experimental heat transfer coefficients from ice-roughened surfaces for aircraft deicing design. J Aircr 36(6): 948–956

Eckert ERG (1955) Engineering relations for friction and heat transfer to surfaces in high velocity flow. J Aeronaut Sci 22:585–587

El-Genk MS, Saber HH (2001) Minimum thickness of a flowing down liquid film on a vertical surface. Int J Heat Mass Transf 44:2809–2825

El-Genk MS, Saber HH (2002) An investigation of the break-up of an evaporating liquid film, falling down a vertical, uniformly heated wall. Trans. ASME J Heat Transf 124:39–50

Emmons H (1951) The laminar-turbulent transition in a boundary layer—Part I. J Aeronaut Sci 234(2348):490–498

Flemming R, Lednicer R (1985) High speed ice accretion on rotorcraft airfoils. Sikorsky Aircraft Division, CR 3910

Frick CW, McCullough GB (1942) A method for determining the rate of heat transfer from a wing or streamline body. Moffet Field, National Advisory Committee for Aeronautics, NACA Report 830

Gelder TF, Lewis JP (1951) Comparison of heat transfer from airfoil in natural and simulated icing conditions. Ashington, National Advisory Committee for Aeronautics, Technical Note 2480)

Gent R, Trajice A (1990) Combined water droplet trajectory and ice accretion prediction program for aerofoils, Farnborough, Royal Aerospace Establishment, Technical report 90054

Gent RW, Dart NP, Cansdale J (2000) Aircraft icing. Phil Trans Roy Soc Lond A 358:2873–2911

Gile-Laflin BE, Papadakis M (2001) Experimental investigation of simulated ice accretions on a natural laminar flow airfoil. Proceedings of 39th American institute of aeronautics and astronautics meeting and exhibit, AIAA Paper 2001–0088, Reno, Nevada, 8–11 Jan 2001

Gray VH (1958) Correlations among ice measurements, impingement rates, icing conditions, and drag coefficients for unswept, NACA 65A004 Airfoil, United States. National advisory committee for aeronautics, technical note 4151

Gray VH (1964) Prediction of aerodynamic penalties caused by ice formations on various airfoils. NASA-TN-D-2166

Gray VH, von Glahn UH (1964) Aerodynamics effects caused by icing of an unswept NACA 65A004 airfoil. United States. National Advisory Committee for Aeronautics, Technical note NACA TN 4155

Hartley DE, Murgatroyd W (1964) Criteria for the break-up of thin liquid layers flowing isothermally over solid surfaces. Int J Heat Mass Transf 7:1003

Havugimana P, Lutz C, Saeed F, Paraschivoiu I, Kerevanian G, Sidorenko HCM, Bragg MB, Kim HS (1998) Freestream turbulence measurements in icing condition. Proceedings of 36th American institute of aeronautics and astronautics meeting and exhibit, AIAA Paper 1998–1996, Reno, 12–15 Jan 1998

Hess J, Smith A (1967) Calculation of potential flow about arbitrary bodies. Prog Aeronaut Sci 8:1–138

Jackson D, Bragg M (1999) Aerodynamic performance of an NLF airfoil with simulated ice. American institute of aeronautics and astronautics, Reno, Nevada, EUA, AIAA Paper 99–0373

Johnson MW, Fashifar A (1994) Statistical properties of turbulent bursts in transitional boundary layers. Int J Heat Fluid Flow 15(4):283–290

Kays WM, Crawford ME, Weigand B (2004) Convective heat and mass transfer, 4th edn. McGraw-Hill, New York

Kays WM, Moffat RJ (1975) The behaviour of transpired turbulent boundary layers. In: Lauder B (ed) Studies in convection theory, measurements and applications, vol 1. Academic Press, New York, pp 223–319

Kennedy J, Marsen D (1976) Potential flow velocity distributions on multi-component airfoils ections. Can Aeronaut Space J 22(5):243–256

Kerho MF, Bragg M (1997) Airfoil boundary-layer development and transition with large leading-edge roughness. AIAA J 35(1):75–84

Kim H, Bragg M (1999) Effects of leading-edge ice accretion geometry on airfoil aerodynamics. American institute of aeronautics and astronautics, Reno, Nevada, EUA, AIAA Paper 99–3150

Korkan Jr KD, Cross Jr EJ, Cornell CC (1985) Experimental aerodynamic characteristics of an NACA 0012 airfoil with simulated ice. J Aircr 2(22):130–134

Kuhns IE, MAson BJ (1968) The supercooling and freezing of small droplets falling in air and other gases. Proc Roy Soc A302:437–452

Langmuir I (1961) Supercooled water droplets in rising of cold saturated air. The collected works of iorving Langmuir, the atmospheric phenomena. Pergamon Press Reprints 10:199–334

Leary WM (2002) We freeze to please. A history of NASA's icing research tunnel and the quest for flight safety. National aeronautics and space administration, The NASA history series, NASA SP-2002-4226

Ludlam FH (1951) The heat economy of a rimed cylinder. Q J Roy Meteorol Soc 77(334):663–666

Macarthur C, Keller J, Luers J (1982) Mathematical modeling of airfoil ice accretion on airfoils. 20th Aerospace sciences meeting and exhibit, America institute of aeronautics and astronautics, Reno, AIAA paper 82-36042

Makkonen L (2000) Models for the growth of rime, glaze, icicles and wet snow on structures. Philos Trans Roy Soc 358(1776):2913–2939

Makkonen L (1985) Heat transfer and icing of a rough cylinder. Cold Reg Technol 10:105–116

Mateer GG, Monson DJ, Menter FR (1996) Skin-friction measurements and calculations on a lifting airfoil. AIAA J 34(2):231–236

Mayle R (1991) The role of laminar-turbulent transition in gas turbine engines. ASME J Turbomach 113(5):509–537

Messinger BL (1953) Equilibrium temperature of an unheated icing surface as a function of air speed. J Aeronaut Sci 20(1):29–42

Mikielewicz J, Moszynsky JR (1978) An improved analysis of breakdown of thin liquid films. Arch Mech 30:489–500

Mikielewicz J, Moszynsky JR (1982) Breakdown and evaporation of thin shear driven liquid films. Proceedings international centre of heat and mass transfer, Gdanski, 467–481

Morency F, Tezok F, Paraschivoiu I (1999) Anti-icing system simulation using Canice. J Aircr 36(6):999–1006

Morency F, Tezok F, Paraschivoiu I (1999) Heat and mass transfer in the case of an anti-icing system modelisation. 37th Aerospace sciences meeting and exhibit, America institute of aeronautics and astronautics, Reno, AIAA paper 99-0623

Moretti PM, Kays WM (1965) Heat transfer to a turbulent boundary layer with varying free-stream velocity and varying surface temperature - an experimental study. Int J Heat Mass Transf 8:1187–1202

Mortensen K (2008) CFD simulations of an airfoil with leading edge ice accretion. Ph.D. thesis, Technical University of Denmark

Narasimha R (1957) On the distribution of intermittency in the transition region of a boundary layer. J Aerosp Sci 24:711–712

Newton JE et al (1988) Measurement of local convective heat transfer coefficients from a smooth and roughened NACA 0012 airfoil: flight test data. National aeronautics and space agency, Cleveland, Technical Memorandum 1000284, 17

Neel JCB, Bergrun NR (1947) The calculation of the heat required for wing thermal ice prevention in specified icing conditions. National advisory committee for aeronautics, Washington, Technical Note 1472

Owen PR, Thomson WR (1963) Heat transfer accross rough surfaces. J Fluid Mech 15:321–334

Papadakis M, Alansatan S, Seltmann M (1999) Experimental study of simulated ice shapes on a NACA 0011. America institute of aeronautics and astronautics, Reno, Nevada, EUA, AIAA Paper 99-0096

Papadakis M, Alansatan S, Wonng S (2000) Aerodynamic characteristics of a symmetric NACA section with simulated ice shapes. America institute of aeronautics and astronautics, Reno, Nevada, EUA, AIAA Paper 2000-0098

Potapczuk MGA (1999) Review of NASA LEWIS' development plans for computational simulation of aircraft icing. AIAA-99-0243

Raw M, Schneider GA (1985) New implicit solution procedure for multidimensional finite-difference modeling of the Stefan problem. Numer Heat Transf 8:559–571

Rothmayer AP, Tsao JC (2000) Water film runback on an airfoil surface. 38th aerospace sciences meeting and exhibit, America Institute of Aeronautics and Astronautics, Reno, AIAA Paper 2000-0237

Rothmayer AP, Tsao JC (2001) On the incipient motion of air driven water beads. 39th aerospace sciences meeting and exhibit, America Institute of Aeronautics and Astronautics, Reno, AIAA 2001-0676

Ruff GA, Berkowitz BM (1990) Users manual for the NASA Lewis ice accretion prediction code (LEWICE). National aernautics and space administration, Cleveland

Saber HH, El-Genk MS (2004) On the break-up of a thin liquid film subject to interfacial shear. J Fluid Mech 500(113)

Schlichting H, Gersten K (2000) Boundary-layer theory, 8th edn. Springer, Berlin

Schlichting H (1937) Experimental investigation of the problem of surface roughness. Technical memorandum n.823, Washington, DC, April 1937

Schmuki P. and Laso M. (1990) On the stability of rivulet flow. J. Fluid Mechanics, 215:125–143

Schubauer G, Klebanoff P (1955) Contributions on the mechanics of boundary-layer transition. National advisory committee for aeronautics, Washington, Contractor Report 1289

Schubauer G, Skramstad H (1948) Laminar boundary layer oscillations and transition on a flat plate. National advisory committee for aeronautics, Washington, Technical Report 909

Sharma OP (1987) Momentum and thermal boundary layer development on turbine airfoil suction surfaces. 23th AIAA/SAE/ASME joint propusion conference, America Institute of Aeronautics and Astronautics, AIAA Paper 87-1918:1–11
Shin J, Bond TH (1994) Repeatability of ice shapes in the NASA LEWISis icing research tunnel. J Aircr 31(5):1057–1063
Shin J, Chen HH, Cebeci TA (1992) Turbulence model for iced airfoils and its validation. National Aeronautics and Space Administration, Washington, Technical Memorandum 105373
Silva GAL, Silvares OM, Zerbini EJGJ (2003) Airfoil anti-ice system modeling and simulation. 43rd aerospace sciences meeting and exhibit, America Institute of Aeronautics and Astronautics, Reno, AIAA Paper 2003-734
Silva GAL, Silvares OM, Zerbini EJGJ (2005) Simulation of an airfoil electro-thermal anti-ice system operating in running wet regime. 41st aerospace sciences meeting and exhibit, America Institute of Aeronautics and Astronautics, Reno, AIAA Paper 2005-1374
Silva GAL, Silvares OM, Zerbini EJGJ (2006) Water film breakdown and rivulets formation effects on thermal anti-ice operation simulation. 9th AIAA/ASME joint thermophysics and heat tranfer conference, San Francisco, AIAA Paper 2006-3785
Silva GAL, Silvares OM, Zerbini EJGJ (2007) Numerical simulation of airfoil thermal anti-ice operation. Part 1: mathematical modeling. J Aircr 44(2):627–633
Silva GAL, Silvares OM and Zerbini EJGJ (2006) Numerical simulation of airfoil thermal anti-ice operation. Part 2: implementation and results. J Aircr 44(2): 634–641
Silva GAL, Silvares OM, Zerbini EJGJ (2008) Aircraft wing electrothermal anti-icing: heat and mass transfer effects. 5th European thermalsciences conference, Eindhoven, European committee for the advancement of thermal sciences and heat transfer
Silva GAL, Silvares OM, Zerbini EJGJ (2008) Boundary-layers integral analysis - heated airfoils in ice conditions. 46th AIAA aerospace sciences meeting and exhibit, Reno, AIAA Paper 2008-0475
Smith AG, Spalding DG (1958) Heat transfer in a laminar boundary layer with constant fluid properties and constant wall temperature. J Roy Aeronaut Soc 62:60–64
Sogin HH (1954) A design manual for thermal anti-icing systems. Wright air development center technical, illinois, technical report, pp 54-13
Spalart PR (1996) Topics in industrial viscous flow calculations. Transitional boundary layer in aeronautics. Amsterdam, North-Holland, pp 269–282
Spaldig DB (1958) Heat transfer from surfaces of non-uniform temperature. J Fluid Mech 4:22–32
Steelant J, Dick E (1996) Modeling of bypass transition with conditioned average Navier Stokes equations coupled to an intermittency transport equation. Int J Numer Methods Fluids 23:193–220
Stefanini LM, Silva GAL, Silvares OM, Zerbini EJGJ (2007) Convective heat transfer effects in airfoil icing. 19th international congress of mechanical engineering.

Procedings of COBEM 2007, Rio de Janeiro: Brazilian Society of Mechanical Sciences and Engineering

Stefanini LM, Silva GAL, Silvares OM, Zerbini EJGJ (2008) Boundary-layers integral analysis—airfoil icing. 46th AIAA aerospace sciences meeting and exhibit, reno, AIAA Paper 2008-474

Thwaites B (1949) Approximate calculation of the laminar boundary layer. Aero Quart 10:245–279

Tobaldini NL, Pimenta MM, Silva GAL (2008) Laminar-turbulent transition modeling strategies for thermally protected airfoils. ASME fluids engineering conference, Proceedings of FEDSM2008

Towell GD, Rothfeld lB (1966) Hydrodynamics of rivulet flow. AICHE J 12(5): 972–980

Trela M, Mikielewicz J (1992) An analysis of rivulet formation during flow of an air/water mist across a heated cylinder. Int J Heat Mass Transf 35(10):2429–2434

Vargas M, Tsao JC (2007) Parametric study of ice accretion formation on a swept wing at sld conditions. SAE Aircraft and Engine Icing international Conference, SAE Paper 2007-01-3345

Vinod N, Govindarajan R (2007) The signature of laminar instabilities in the zone of transition to turbulence. J Turbul 8:2

Volino RJ, Simon TW (1995) NASA CR. measurements in transitional boundary layers under high free-stream turbulence and strong acceleration conditions, National aeronautics and space administration, Washington, Contractor Report 198413

Von Doenhoff AE, Horton EA (1956) A low-speed experimental investigation of the effect of a sandpaper type of roughness on boundary-layer transition. National advisory committee for aeronautics, Washington, Technical Note 3858

Wade SJ Modeling of the performance of a thermal anti-icing system for use on aero-engine

Wright W, Gent R, Guffond D (1997) DRA/NASA/ONERA collaboration on icing research part II—prediction of airfoil ice accretion, National aeronautics and space administration, Cleveland, Contractor Report 202349

Wright WB (1995) NASA CR. user manual for the improved NASA Lewis ice accretion code LEWICE 1.6, National aeronautics and space administration, Cleveland, Contractor Report 198355

Wright WB (1999) User manual for the NASA Glenn ice accretion code LEWICE version 2.0. National aeronautics and space administration, Cleveland, Contractor Report 209409

Glossary

A

a	Axial induction factor or annuity factor.
$\mathbf{a}_d$	Droplet acceleration.
A	Swept area by the rotor or general area.
$\mathbf{A}_c$	Section channel area.
$\mathbf{A}_{ext}$	External surface panel area.
$\mathbf{A}_{heat}$	Heated blade area.
$\mathbf{A}_{int}$	Internal surface panel area.
$\mathbf{A}_m$	Mean surface panel area, between the external and the internal surface area.
$\mathbf{A}_p$	Generic surface panel area.
$\mathbf{A}_w$	Generic panel area covered by water.
AoA	Angle of Attack.

B

BR	Blowing ratio parameter.

C

$\mathbf{c}_{\mathrm{el}}$	Electrical energy cost priced from the net.
$\mathbf{c'}_{\mathrm{el}}$	Cost of electric energy.
$\mathbf{c}_f$	Friction coefficient.
$\mathbf{c}_{\mathrm{p}}$	Specific heat coefficient at constant pressure.
$\mathbf{c}_{\mathrm{p,air}}$	Dry air specific heat coefficient at constant pressure.
$\mathbf{c}_{\mathrm{p,eq}}$	Equivalent specific heat coefficient at constant pressure.
$\mathbf{c}_{\mathrm{p,vap}}$	Vapour specific heat coefficient at constant pressure.
$\mathbf{c}_{\mathrm{opt}}$	Ideal chord lenght.
$\mathbf{c}_{\mathrm{opt,P}}$	Optimum chord lenght.
$\mathbf{c}_{\mathrm{u}}$	Linearized chord lenght.
$\mathbf{c}_{\mathrm{v}}$	Specific heat coefficient at constant volume transformation.
$\mathbf{c}_{\mathrm{w}}$	Water specific heat coefficient.

L. Battisti, *Wind Turbines in Cold Climates*, Green Energy and Technology,
DOI 10.1007/978-3-319-05191-8

C	Scale parameter of the Weibull function.
$\mathbf{C_D}$	Drag coefficient.
$\mathbf{C_L}$	Lift coefficient.
$\mathbf{C_{L,max}}$	Maximum lift coefficient.
$\mathbf{C_m}$	Aerodynamic moment coefficient.
$\mathbf{C_{P,aero}}$	Aerodynamic power coefficient.
$\mathbf{C_{P,elmax}}$	Maximum electrical power coefficient.
$\mathbf{C_{P,R}}$	Power coefficient at rated power.
$\mathbf{C_{P,max}}$	Power coefficient at maximum power.

D

d	Maximum throwing distance or droplet diameter.
$\mathbf{d_b}$	Diameter of surface water bead.
D	Drag force or rotor diameter or nose diameter.
$\mathbf{D_{ev}}$	Diffusivity of water vapour in air.
$\mathbf{D_h}$	Hydraulic diameter.
$\mathbf{D_s}$	Sphere drag force.
DoI	Degree of Irregularity.

E

$\mathbf{e_v}$	Air vapour pressure.
$\mathbf{e_{v,s}}$	Air vapour pressure at the surface.
$\mathbf{e_{v,s}^{sat}}$	Air saturated vapour pressure at the surface.
E	Total collection efficiency.
$\mathbf{E_{ice,ave}}$	Energy production during the IPS period of operation.
$\mathbf{E_I}$	Anti-icing energy.
E_{IPS}	Ice prevention system annual electric consumption.

F

f	Freezing fraction or friction factor.
$\mathbf{f}_g$	Freezing fraction for glaze ice.
$\mathbf{f}_r$	Freezing fraction for rime ice.
f()	Probability density function.
F(V)	Cumulative probability distribution function.
$\mathbf{F_c}$	Centrifugal force.
$\mathbf{F_u}$	Capacity factor.
Fo	Fourier number.

G

g	Gravitational acceleration.

H

$\mathbf{h}_{\mathrm{amb}}$ Overall heat exchange coefficient.
$\mathbf{h}_{\mathrm{b}}$ Height of surface water bead.
$\mathbf{h}_{\mathrm{c}}$ Convective heat exchange coefficient.
$\mathbf{h}_{\mathrm{c,ext}}$ External convective heat exchange coefficient.
$\mathbf{h}_{\mathrm{c,int}}$ Internal convective heat exchange coefficient.
$\mathbf{h}_{\mathrm{eq,ext}}$ Equivalent external heat exchange coefficient.
$\mathbf{h}_{\mathrm{in}}$ Incoming specific enthalpy.
$\mathbf{h}_{\mathrm{m}}$ Vapour mass exchange coefficient.
$\mathbf{h}_{\mathrm{out}}$ Outgoing specific enthalpy.
$\mathbf{h}_{\mathrm{w,imp}}$ Impinging water specific enthalpy.
$\mathbf{h}_{\mathrm{w,in}}$ Incoming water specific enthalpy.
$\mathbf{h}_{\mathrm{w,out}}$ Outgoing water specific enthalpy.
$\mathbf{h}_{\mathrm{w,sh}}$ Water shedding specific enthalpy.
$\mathbf{h}^0$ Total specific enthalpy.
$\mathbf{H}$ Turbine hub height or the Heaviside step function.
$\mathbf{H}_i$ Chordwise length of the heated panel.

I

$\mathbf{I}$ Turbulence intencity or current through a conductor.
$\mathbf{I}_{\mathrm{icing}}$ Instrumental icing event.
$\mathbf{I}_{\mathrm{IPS}}$ Investment cost for a de-ice/anti-ice device.
$\mathbf{I}_{\mathrm{WT}}$ Investment cost for a WT.
$\mathbf{I}_0$ Magnetic field intensity.
$\mathbf{I}_{15}$ Turbulence intensity at 15 m/s.

K

$\mathbf{k}$ Shape parameter of the Weibull function or surface roughness.
$\mathbf{k}_{\mathrm{air}}$ Air thermal conductivity.
$\mathbf{k}_{\mathrm{b}}$ Blade material thermal conductivity.
$\mathbf{k}_{\mathrm{s}}$ Equivalent sand grain roughness or safety coefficient.
$\mathbf{K}_0$ Modified inertia parameter.
$\mathbf{K}_{\mathrm{A}}$ Area ratio.
$\mathbf{K}_{\mathrm{I}}$ Kinetic energy of turbulence.
$\mathbf{K}_{\mathrm{L}}$ Modified Langmuir parameter.
$\mathbf{K}_{\mathrm{st}}$ Stokes number.

L

$\mathbf{L}$ Generical width or Lewis number.
$\mathbf{L}_i$ Spanwise length of the heated panel.

M

$\mathbf{m}$	Polytropic exponent in thermodynamic processes or mass.
$\mathbf{m}_a$	Mass of hot air flowing into duct.
$\mathbf{m}_{air}$	Mass of dry air.
$\dot{m}_{ev,subl}$	Water mass of evaporation or sublimation.
$\mathbf{m}_{ice}$	Mass of the blade ice layer.
$\mathbf{m}_{ice,avg}$	Average mass of the cast ice pieces.
$\mathbf{m}_{idy,avg}$	Average ice accretion mass per day on the rotor.
$\mathbf{m}_{w,vap}$	Mass of water vapour in the air.
$\mathbf{m}_w$	Mass of water in the air.
$\dot{m}$	mass flux.
$\dot{m}_{w,ev}$	Mass flux of runback water entering a given section.
$\dot{m}_{w,ev}^{pot}$	Potential water mass flow.
$\dot{m}_{in}$	Inlet mass flux.
$\dot{m}_{ice}$	Mass flux of water freezing in a given section.
$\dot{m}_{w,imp}$	Mass flux of impinging water.
$\dot{m}_{w,imp,t}$	Total mass flux of impinging water.
$\dot{m}_{out}$	Outlet mass flux.
$\dot{m}_{w,sh}$	Mass flux of water shedding.
$\dot{m}_{w,st}$	Mass flux of standing water due to surface tension.
$\dot{m}_{w,in}$	Incoming mass flux of runback water.
$\dot{m}_{w,out}$	Outgoing mass flux of runback water.
$\mathbf{mm}_w$	Water molecular mass.
$\mathbf{M}$	Bending moment or Mach number or total number of chordwise blade panels.
$\mathbf{M}_{icing}$	Meteorological icing event.
$\mathbf{M}_{IPS}$	Ice prevention system annual maintenance costs.
$\mathbf{Ma}$	Mach number.

N

$\mathbf{n}_{idy}$	Number of icing days per year.
$\mathbf{N}$	Number of ice strikes or total number of spanwise blade stations.
$\mathbf{Nu}$	Nusselt number.
$\mathbf{Nu}_c$	Chord Nusselt number.

O

$O\&M_{IPS}$	Ice prevention system annual operation and maintenance costs.

P

$\mathbf{p}$	Air pressure.
$\mathbf{p}_e$	Local pressure at the edge of the boundary layer.
$\mathbf{p}_{out}$	Outgoing pressure of hot air flowing into duct.

$\mathbf{p}^0$ Total air pressure.
$\mathbf{p}^0_\infty$ Total free stream pressure.
$\mathbf{p}_\infty$ Free stream pressure.
$\mathbf{P}$ Power.
$\mathbf{P}_{\mathrm{ab}}$ Power absorbed in liquid water or ice.
$\mathbf{P}_{\text{clean blade}}$ Average electrical power with ice free rotor.
$\mathbf{P}_{\mathrm{el,max}}$ Maximum electrical power.
$\mathbf{P}_{\text{i. blade,i}}$ Average electrical power with iced rotor.
$\mathbf{P}_{\mathrm{ice}}$ Average electrical power delivered during icing period.
$\overline{P}_m$ Average mechanical power to promote the hot air flow.
$\mathbf{P}_{\mathrm{R}}$ Rated power.
PI Performance index.
Pr Prandt number.

Q

q_{aer} Aerodynamic specific heat.
$q_{\mathrm{conv,ext}}$ Convective specific flux at the external side of the blade wall.
q_{ev} Evaporative specific heat.
q_{sens} Sensible specific heat.
$\dot{q}_{\mathrm{ab}}$ Heat flux absorbed in liquid water.
$\dot{q}_{\mathrm{cond}}$ Conduction heat flux.
$\dot{q}_{\mathrm{conv}}$ Convective heat flux.
$\dot{q}_{\mathrm{conv,int}}$ Convective heat flux at the inner side of the blade wall.
$\dot{q}_{\text{de-icing,ave}}$ Average de-icing specific thermal power.
$\dot{q}_{\mathrm{gen}}$ Generated heat power.
$\dot{q}_{\mathrm{ice}}$ Ideal specific anti-icing thermal power, supplied to the blade wall.
$\dot{q}_{\mathrm{IPS,ave}}$ Anti-icing specific thermal power.
$\dot{q}_{\mathrm{IPS,ave}}$ Average anti-icing specific thermal power.
$\dot{q}_{\mathrm{k}}$ k-th heat flux.
$\dot{q}_{\mathrm{kin}}$ Droplet kinetic energy heat.
$\dot{q}_{\mathrm{lr}}$ Longwave radiation flux.
$\dot{q}_{\mathrm{out}}$ Specific heat exchange at the outer side of the blade wall.
$\dot{q}_{\mathrm{sr}}$ Shortwave radiation flux.
$\dot{Q}_{\mathrm{ice}}$ Ideal anti-icing thermal power.
$\dot{Q}_{\mathrm{IPS}}$ Anti-icing thermal power.
$\dot{Q}_{\mathrm{IPS,ave}}$ Average anti-icing thermal power.

R

$\mathbf{r}$ Recovering factor or local radius.
$\mathbf{R}$ Ideal gas constant or maximum radius or wall thermal resistance or reheat factor.
$\mathbf{R}_{\mathrm{eq}}$ Equivalent fatigue load or rotor radius.
$\mathbf{R}_0$ Non-profiled rotor radius.

$\mathbf{R}_G$	Ice mass barycentre radius.
$\bar{\mathrm{R}}_{\mathrm{air}}$	Air gas constant.
Rh	Air relative humidity.
$\bar{\mathrm{R}}_{\mathrm{vap}}$	Vapour gas constant.
$R^{i,j}$	(i, j) panel thermal resistance.
Re	Reynolds number.
$\mathbf{Re}_{\mathrm{c}}$	Chord Reynolds number.
$\mathbf{Re}_{\mathrm{c\ ott}}$	Reynolds number corrisponding to ideal chord lenght.
$\mathbf{Re}_{\mathrm{c\ ott,P}}$	Reynolds number corrisponding to optimum chord lenght.
$\mathbf{Re}_{\mathrm{c\ u}}$	Reynolds number corrisponding to linearized chord lenght.
$\mathbf{Re}_d$	Water particle Reynolds number.
$\mathbf{Re}_{D_h}$	Reynolds number evaluated with the hydraulic diameter.
$\mathbf{Re}_k$	Critical roughness based Reynolds number.
$\mathbf{Re}_s$	Reynolds number evaluated at the coordinate s.
$\mathbf{Re}_\infty$	Free stream Reynolds number.
$\mathbf{Re}_\theta$	Reynolds number evaluated at the coordinate θ.
$\mathbf{R\ R}_{\mathrm{eq}}$	Equivalent fatigue loads ratio.

S

$\mathbf{S}_s$	Incoming global diffuse radiation flux.
SRO	Specific Rated Output.

T

t	Blade thickness or time.
$\mathbf{t}_b$	Blade shell thickness.
$\mathbf{t}_{\mathrm{heat\text{-}on}}$	Period of time when the anti-icing systems is switch-on.
$\mathbf{t}_{\mathrm{heat\text{-}off}}$	Period of time when the anti-icing systems is switch-off.
$\mathbf{t}_i$	Total icing duration event.
$\mathbf{t}_{\mathrm{i.\ blade,i}}$	Icing duration event during i-period.
$\mathbf{t}_{\mathrm{ice}}$	Ice layer thickness.
$\mathbf{t}^*_{\mathrm{ice}}$	Critical rime ice layer thickness.
$\mathbf{t}^{**}_{\mathrm{ice}}$	Critical glaze ice layer thickness.
$\mathbf{t}^*_{\mathrm{w}}$	Critical water film thickness.
$\mathbf{t}_{\mathrm{w}}$	Critical water film thickness.
$\mathbf{t}_0$	Initial time.
T	Air temperature or aerodynamic thrust.
$\mathbf{T}_a$	Temperature of hot air flowing into duct.
$\mathbf{T}_{a,in}$	Temperature of hot air injected into duct.
$\mathbf{T}_e$	Local air temperature at the edge of the boundary layer.
$\mathbf{T}^i$	i-th panel surface temperature.
$\mathbf{T}_{\mathrm{ice}}$	Ice temperature.
$\mathbf{T}_{\mathrm{heat\text{-}on}}$	Blade surface temperature to switch-on the anti-icing systems.

$\mathbf{T}_{\text{heat-off}}$ Blade surface temperature to switch-off the anti-icing systems.
$\mathbf{T}_{\text{rec}}$ Recovery temperature.
$\mathbf{T}_{\text{max}}$ Maximum allowable material working temperature for IPS continuous operations.
$\mathbf{T}_{\text{min}}$ Minimum external surface temperature preventing ice formation.
$\mathbf{T}_{\text{out}}$ Outgoing temperature of hot air flowing into duct.
$\mathbf{T}_s$ Surface temperature.
$\mathbf{T}_{\text{s,ext}}$ External surface temperature.
$\mathbf{T}_{\text{s,int}}$ Internal surface temperature.
$\mathbf{T}_w$ Water temperature.
$\mathbf{T}_{\text{w,imp}}$ Impinging water temperature.
$\mathbf{T}_{\text{w,in}}$ Incoming water temperature.
$\mathbf{T}_{\text{w,out}}$ Outgoing water temperature.
$\mathbf{T}^0$ Total temperature.
$\mathbf{T}_\infty$ Total free stream temperature.
$\mathbf{T}_\infty$ Free stream temperature.
$\bar{T}_{a,m}$ Mean temperature of hot air flowing into duct.
$T^{i,j}$ (i, j) panel temperature.
TSR Tip speed ratio.

U

$\mathbf{u}_{ice}$ Ice internal energy.
$\mathbf{u}_w$ Water internal energy.
U Overall heat transfer coefficient.

V

$\mathbf{v}$ Air volume.
$\mathbf{v}_d$ Droplet velocity.
$\mathbf{v}_e$ Local flow velocity at the edge of the boundary layer.
$\mathbf{v}_k$ Surface local flow velocity.
$\mathbf{V}$ Wind velocity.
$\mathbf{V}_0$ Far upstream wind velocity.
$\mathbf{V}_{\text{ave}}$ Average site wind velocity.
$\mathbf{V}_{\text{cut,in}}$ Cut-in wind velocity.
$\mathbf{V}_{\text{cut,out}}$ Cut-out wind velocity.
$\mathbf{V}_{\text{design}}$ Rotor design wind velocity.
$\mathbf{V}_{\text{e1}}$ 1-year extreme wind velocity.
$\mathbf{V}_{\text{e50}}$ 50-year extreme wind velocity.
$\mathbf{V}_{\text{hub}}$ Wind velocity at hub height.
$\mathbf{V}_{\text{R}}$ Rated wind velocity.
$\mathbf{V}_{\text{ref}}$ Reference wind velocity.
$\mathbf{V}_\infty$ Free stream velocity.

W

$\mathbf{W}_0$	Droplet terminal velocity.
$\mathbf{W}$	Relative wind velocity.
$\mathbf{W}_{\mathrm{ave}}$	Average profile relative wind velocity.
$\mathbf{W}_{\mathrm{e}}$	Relative air velocity at a given location at the edge of the boundary layer.
$\mathbf{W}_{\mathrm{in}}$	Inlet relative flow velocity.
$\mathbf{W}_{\mathrm{out}}$	Outlet relative flow velocity.
We	Weber number.
$\mathbf{We}_{\mathrm{cr}}$	Critical Weber number.

Z

$\mathbf{z}$	Height above ground.
$\mathbf{z}_{\mathrm{h}}$	Hub height above ground.
$\mathbf{Z}$	Number of rotor blades.

Coordinate systems

s	Airfoil surface coordinate.
$x,\ y,\ z$	Generical coordinates system.
$x_0,\ y_0,\ z_0$	Generical initial coordinates values.
$r,\ \theta,\ z$	Cylindrical polar coordinates system.
u, v, w	Coordinate system used for the wind field description.

Greek symbols

α	Absorption coefficient per length.
β	Local collision efficiency.
$\beta_{\mathrm{C\ ott}}$	Blade twist angle corresponding to ideal chord lenght.
$\beta_{\mathrm{C\ ott,P}}$	Blade twist angle corresponding to optimum chord lenght.
$\beta_{\mathrm{C\ u}}$	Blade twist angle corresponding to linearized chord lenght.
β_{twist}	Blade twist angle.
β_0	Stagnation collision efficiency.
β_{2D}	Local 2D collection efficiency.
β_{3D}	Local 3D collection efficiency.
γ	Isoentropic exponent in thermodynamic processes or angle.
$\Gamma(V)$	Gamma function of V.
δ	Power law exponent or boundary layer height.
Δh_{ev}	Vaporisation specific heat.
Δh_f	Fusion latent specific heat.
Δh_{subl}	Sublimation specific heat.
Δh_f	Fusion latent heat.
$\Delta \dot{Q}_{ice}$	Anti-icing thermal power in the segment of blade.
Δt	Time interval.
Δt_0	Initial time interval.

ΔW_d	Friction losses.
Δx	x-position interval.
Δy	y-position interval.
ε	Isoentropic coefficient or body emissivity.
ε_g	Generator overall thermal efficiency.
ε_{IPS}	IPS heating efficiency.
ε_τ	Energy consumption ratio.
η_{aux}	Auxiliary devices efficiency.
η_{IPS}	Ice prevention system overall efficiency.
η_{el}	Electrical generator efficiency.
$\eta_{el,R}$	Electrical generator efficiency at rated power.
$\eta_{el,g}$	Electrical generator thermal efficiency.
η_{film}	Film efficiency.
η_m	Mechanical efficiency.
$\eta_{m,R}$	Mechanical efficiency at rated power.
θ	Azimuthal or radial position.
λ_d	Drop range in absence of gravity.
λ_D	Drag range.
$\lambda_{d,ST}$	Drop range in absence of gravity within Stokes law validity.
λ_m	Thermal conductive of blade material.
μ	Flow dynamic viscosity.
μ_{air}	Air dynamic viscosity.
μ_{ICE}	Ice mass distribution per meter of blade.
ξ	Fraction of extra power required for de-icing.
Φ	Langmuir parameter.
Ω	Rotational velocity.
Ω_{ave}	Average rotational velocity.
$\Omega_{\max}$	Maximum rotational velocity.
$\Omega_{\min}$	Minimum rotational velocity.
ρ	Flow density.
ρ_0	Air density in ISO standard atmosphere (1.225 kg/m^3).
ρ_{air}	Air density.
ρ_{e}	Air density at a given location at the edge of the boundary layer.
ρ_{ice}	Ice density.
$\rho_{ice,g}$	Glaze ice density.
$\rho_{ice,r}$	Rime ice density.
ρ_{mat}	Material density.
ρ_v	Water vapour density.
$\rho_{v,e}$	Water vapour density at the boundary layer edge.
$\rho_{v,s}$	Water vapour density at the surface.
ρ_w	Water density.
ρ_∞	Free stream density.
σ	Surface tension between water and air or Stefan–Boltzmann constant.
σ_k	Standard deviation of k-parameter.
τ	Anti-icing systems intermittent factor.

MIX
Papier aus verantwortungsvollen Quellen
Paper from responsible sources
FSC® C105338

If you have any concerns about our products,
you can contact us on
ProductSafety@springernature.com

In case Publisher is established outside the EU,
the EU authorized representative is:
Springer Nature Customer Service Center GmbH
Europaplatz 3, 69115 Heidelberg, Germany

Printed by Libri Plureos GmbH
in Hamburg, Germany